EMBODIED ENVIRONMENTAL RISK IN TECHNICAL COMMUNICATION

This collection calls for improved technical communication for the public through an embodied, situated understanding of environmental risk that promotes social justice.

In addition to providing a series of chapters about recent issues on risk communication, this volume offers a diverse look at methodological practices for students, researchers, and practitioners looking to address embodied aspects of crisis and risk that incorporate UX, storytelling, and dynamic text. It includes chapters that bring embodiment to the forefront of risk communication, highlighting the cycle of content creation, dissemination, public response and decision making, continuing iterations of educational efforts, and recovery, toward increasing adaptive capacity as a whole. In addition, this work directs necessary attention to overcoming perceptual difficulties, memory lapses, definitional differences, access issues, and pedagogical problems in the communication of risks to diverse publics.

This collection is essential reading for scholars and can be used as a supplemental text or casebook for courses in technical communication, environmental communication, risk and crisis communication, science communication, and public health.

Samuel Stinson

Samuel Stinson is assistant professor of English with Minot State University where he also serves as the director of the Northern Plains Writing Project and coordinator of the English concentration in the MEd program. He also serves as a list manager for the WritingStudies-L listserv and currently co-coordinates the Writing about Writing special-interest group with the Conference on College Composition and Communication. His research interests include professional writing, multimodality, game studies, and pedagogy. His current research focuses on writing transfer and online platforms.

Mary Le Rouge

Mary Le Rouge is director of writing at the Cleveland Institute of Music. She is an active member of the Conference on College Composition & Communication and its Environmental Special Interest Group, among other organizations. Her research lies at the intersection of the humanities and the sciences, looking for ways to improve communication between experts, policymakers, and the public.

ATTW Series in Technical and Professional Communication

Tharon Howard, Series Editor

For additional information on this series please visit www.routledge.com/ATTW-Series-in-Technical-and-Professional-Communication/book-series/ATTW, and for information on other Routledge titles visit www.routledge.com.

EMBODIED ENVIRONMENTAL RISK IN TECHNICAL COMMUNICATION

Problems and Solutions Toward Social Sustainability

Edited by
Samuel Stinson and Mary Le Rouge

Routledge
Taylor & Francis Group

NEW YORK AND LONDON

Cover Image: © Getty Images

Published 2022
by Routledge
605 Third Avenue, New York, NY 10158

and by Routledge
4 Park Square, Milton Park, Abingdon, Oxon, OX14 4RN

Routledge is an imprint of the Taylor & Francis Group, an informa business

© 2022 Taylor & Francis

First edition published by Routledge 2017

Library of Congress Cataloging-in-Publication Data
Names: Stinson, Samuel, 1980- editor. | LeRouge, Mary, 1976- editor.
Title: Embodied environmental risk in technical communication : problems and solutions toward social sustainability / Edited by Samuel Stinson and Mary LeRouge.
Description: New York : Routledge, 2022. | Series: ATTW series in technical and professional communication | Includes bibliographical references and index. | Identifiers: LCCN 2021044575 | ISBN 9781032210582 (hardback) | ISBN 9781032155494 (paperback) | ISBN 9781003266549 (ebook)
Subjects: LCSH: Communication in the environmental sciences. | Sustainability—Social aspects. | Risk communication.
Classification: LCC GE25 .E43 2022 | DDC 333.7201/4—dc23/eng/20211117
LC record available at https://lccn.loc.gov/2021044575

ISBN: 9781032210582 (hbk)
ISBN: 9781032155494 (pbk)
ISBN: 9781003266549 (ebk)

DOI: 10.4324/9781003266549

Typeset in Bembo
by codeMantra

This collection is dedicated to the memory of our colleague Bolanle Olaniran whose work continues to inspire.

CONTENTS

FIGURES

TABLES

CONTRIBUTORS

Elizabeth Baddour, PhD is a freelance writer and member of the English Department adjunct faculty at the University of Memphis. Her research interests and publication areas range from technical writing and medical rhetoric to analyses of the historical, social, and political dimensions of pedagogical trends in writing instruction and modern manifestations of the 1974 Students' Right to Their Own Language Resolution. She is a frequent presenter at national rhetoric and communication academic conferences, and in addition to writing and research, she enjoys competing in triathlons and earning a spot on the podium.

Courtney Cox is a doctoral candidate studying rhetoric, composition, and technical communication at Illinois State University. She is the coeditor of *Digitally Mediated Composing and You: A Beginner's Guide to Understanding Rhetoric and Writing in an Interconnected World*, as well as assistant editor of *Rhetoric Review* and associate managing editor of *Xchanges*. Courtney's research considers required university genres—including crisis communication—as they relate to student literacy, institutional equity initiatives, and social justice teaching practices. Her scholarship can be found in *Enculturation*, as well as several other edited collections.

Cooper Day is a PhD candidate in rhetoric and composition at the University of Louisville, Department of English. His research interests include environmental rhetoric, science and technical communication, writing across the curriculum (WAC), writing in the disciplines (WID), transfer, and professional writing. He has taught Composition I & II, Science and Technical Communication, and Business Communication. Currently, he is working on a larger research project with Bernheim Arboretum and Research Forest located just outside of Louisville, Kentucky.

Huiling Ding teaches technical communication at North Carolina State University. Her research focuses on intercultural professional communication, technical communication, risk communication, and epidemic communication. Her recent projects have been exploring the connections between artificial intelligence, communication technologies, job screening, risk communication, and social justice.

Zachary Garrett (John A. Logan College, Carterville, Illinois) is a rhetoric and composition scholar who teaches graduate-level courses in English pedagogy and technology and first-year English composition. He holds a doctorate in English pedagogy from Murray State University (KY). Additionally, he has a business administration background in the professional services sector. In addition to English education, his scholarship focuses on the interaction of climate change and rhetoric, especially among hard-to-reach groups. Garrett lives in downstate Illinois with his wife and son.

Barbara George is an assistant professor in English at the Kent Salem campus. Her research areas include linguistic explorations of the intersections of environmental risk and communication, issues of intersectionality within environmental deliberation, and environmental justice. Barbara also explores the use of narrative in finding new ways to "reimagine" the environment, particularly in rust belt communities.

Uma S. Krishnan is a professor, program director of the Writing Internship Program, and associate writing program coordinator in the Department of English at Kent State University. She teaches professional writing, editing and publishing, argumentative prose, and other writing courses. She also teaches South-Asian literature and non-Western rhetorics courses to graduate and undergraduate students, both in the English department and Honors College. Her research interests include cross-cultural ethnographic case studies, literacy studies in indigenous and ethnic population, language acquisition pedagogy, theories of globalization, and research methodologies. Her recent research has been focused on the importance of self-actualization and its impact on the role of "I" in terms of the individual's cognitive perceptions of oneself ("I"dentity) and of the community around them.

Sarah Lavallee is a PhD candidate in the School of Environmental Studies at Queen's University in Kingston, Ontario, with a background in Biological Sciences (BScH). Her current research position takes place at Public Health Ontario Laboratories Kingston, and her project focuses on groundwater quality, private well-water stewardship, and human health. More specifically, her research investigates cognitive factors (i.e., awareness, attitudes, risk perceptions, beliefs) and health behaviors among private well-water users that can potentially contribute to drinking water contamination and consequent waterborne infection. Sarah's research interests include groundwater contamination, sustainable water management, and environmental public health.

Mary Le Rouge is director of writing at the Cleveland Institute of Music. She also volunteers as associate publisher for monographs, collections, and conference proceedings at the WAC Clearinghouse, an open-source publisher for the field of writing studies. She is an active member of the Conference on College Composition & Communication and its Environmental Special Interest Group, among other organizations. Her research lies at the intersection of the humanities and the sciences, looking for ways to improve communication between experts, policymakers, and the public.

Anna Majury is a public health microbiologist with Public Health Ontario, in Ontario, Canada, and an assistant professor in the Departments of Biology and Molecular Sciences, Public Health and Environmental Studies at Queen's University, in Kingston, Ontario. Her research interests include the study of private well water in the context of public health, health risk, ecohealth, and one health in support of rural communities challenged with access to reliable and sufficient potable water sources.

Heather Manzo is the executive director of the Allegheny County Conservation District (ACCD), which is a regulatory entity that works at the nexus of soil and water to protect our natural resources. Prior to ACCD, she worked in higher education as a regional food systems and community economic development consultant. She is also cofounder and emeritus board member of the Pittsburgh Food Policy Council and serves on the board of Hilltop Urban Farm.

Diane Martinez is an associate professor of English at Western Carolina University in Cullowhee, North Carolina, where she teaches technical and professional communication graduate and undergraduate courses. She previously served as the director of the Professional Writing Program, and her research interests include environmental and nature writing, social justice in technical and professional communication, internationalizing curriculum, intercultural communication, the Bologna Process, and online writing instruction.

Simon Mooney is a final year PhD candidate based in Technological University Dublin. The objective of his research is to develop a modular, evidence-based water communication framework to promote private well stewardship and sustainable socio-hydrogeological interactions among nonexpert Irish householders vulnerable to groundwater contamination risk. The term *socio-hydrogeology* refers to the dynamic feedback and interactions between social and hydrogeological systems. Through a combination of multimodal research methodologies (literature review and meta-analysis, expert interviews, and surveys), he hopes to integrate risk-communication approaches with groundwater science and environmental policy and, ultimately, reduce private groundwater contamination.

Jean O'Dwyer is deputy head of environmental science at University College Cork (UCC) and head of environmental geoscience at the Irish Centre for Research in Applied Geoscience (iCRAG). Jean's research focuses on the links between the natural environment and human health. She currently leads the Environment and Health Lab at the Environmental Research Institute in UCC where she is the principal investigator on projects ranging from groundwater contamination to the impact of climate change on human health and well-being.

Bolanle A. Olaniran passed away on June 17, 2021. He was professor in the Communication Department at Texas Tech University (TTU) for more than 20 years and was an internationally recognized scholar in organizational communication, intercultural communication, small group communication, conflict management, and electronic-mediated communication. Among his many publications, he coedited *Pre-Crisis Planning, Communication, and Management: Preparing for the Inevitable* (2012). He was honored with the American Communication Association's Outstanding Scholar in the Communication Field Award, TTU's 2006 Office of the President's Diversity Award, and TTU's 2007 President's Excellence in Teaching Award.

Beverly A. Sauer (PhD Brandeis) investigates risk communication in difficult cross-cultural and technical environments. She was professor of management in the Carey School of Business, Johns Hopkins University, and associate professor of English at Carnegie Mellon. Sauer wrote the prize-winning book *The Rhetoric of Risk: Technical Documentation in Hazardous Environments* (Routledge, 2003). Sauer is president of Beverly Sauer Consultants, specializing in strategic risk communication in aerospace, financial, and transportation systems. She is currently completing a book on the role of gesture in transnational communication in South African coal mines supported by grants from the National Science Foundation.

Christopher Scheidler earned his PhD in Rhetoric and Composition from the University of Louisville. He is currently visiting assistant professor of professional and technical writing at Weber State University. His past research has dealt with digital media, modality, and the people who put those concepts to work.

Lately, he has been thinking about digital interfaces, the environment, and the constraints and consequences of the infrastructures that put both together. When he is not working, he is hiking, climbing, snowboarding, or paddle boarding.

Erika M. Sparby (she/they) is an assistant professor of digital rhetorics and technical writing at Illinois State University, where she researches and teaches on digital aggression, memes, ethics, and risk/crisis communication. She has received national acclaim for her research—including the 2016 Gloria Anzaldúa Rhetorician Award, the 2017 Hugh Burns Dissertation Award, and the 2019 Distinguished Book Award (with Jessica Reyman for the coedited collection *Digital Ethics: Rhetoric and Responsibility in Online Aggression*), and she cofounded the Digital Aggression Working Group in 2019. Erika's work has appeared in *Computers and Composition, Enculturation,* and *Methods and Methodologies for Research in Digital Writing and Rhetoric* (edited by Crystal VanKooten and Victor Del Hierro).

Samuel Stinson is assistant professor of English with Minot State University where he also serves as the director of the Northern Plains Writing Project and coordinator of the English concentration in the MEd program. He also serves as a list manager for the WritingStudies-L listserv and currently cocoordinates the Writing about Writing special-interest group with the Conference on College Composition and Communication. His research interests include professional writing, multimodality, game studies, and pedagogy. His current research focuses on writing transfer and online platforms.

Scott Weedon is assistant professor of English at Texas Tech University. He researches science and health communication, engineering communication and education, and the intersections between design and rhetoric.

Joseph Williams currently teaches Technical Communication + Rhetoric (TCR), Business Communication, Cultural Rhetorics, Ethics in TCR, Research Methods, Risk Communication, and Technical Presentations at Louisiana Tech University. Dr. Williams previously worked at Texas A&M University at Qatar, where he initially developed his research in Petroleum Risk Communication. While he worked on his masters, he served as a graduate exchange student to Bogazici University in Istanbul for two full semesters, where he cultivated his passion for intercultural communication. His academic interests include publications and presentations in the fields of intercultural technical communication, risk communication, applied linguistics, ethics, and visual rhetoric.

FOREWORD

Huiling Ding

Risk communication has never been so relevant and so widely examined before. We are living in a world increasingly affected by manufactured risks (Beck, 1992; 1999), be it climate crisis, automation and technological unemployment, or emerging pandemics such as COVID-19 partially caused by the deteriorating environment and the increasing proximity of humans, animals, and microbes.

When Drs. Samuel Stinson and Mary Le Rouge invited me to write a foreword, I was fascinated by the book's timely and much needed focus on embodied risk communication. My book explored transcultural risk communication about SARS among China, the United States, and the World Health Organization in 2003 (Ding, 2014). The communication technologies used back then were mostly pre-social media: print fliers, public notifications, official documents circulated in governmental and public health divisions, traditional media, such as newspapers, radio, and TV stations, as well as alternative media such as individual blogs and websites, text messages, and discussion forums. It was challenging for concerned citizens, professionals, and communities to convey and communicate embodied health risks due to the limited channels that supported such risk messages. Tracing such embodied messages was difficult too, which required patience, triangulation of perspectives, and corroborating evidence that functioned as different and scattered pieces of the puzzle.

The world of risk communication has been radically transformed in the past 20 years with the widespread use of technologies such as social media, interactive maps, data visualization, artificial intelligence, AI-powered chatbots, podcasts, vlogs, and user-generated, crowdsourced content, such as videos and tutorials. AI tools also accelerate the creation and dissemination of misinformation and biased content. Through algorithmic amplification (Beer, 2009; Bucher, 2012; Wood, 2017), ideologically driven content and misinformation then get widely disseminated and circulated in social media platforms and helps sustain the filter bubbles (Pariser, 2011) or echo chambers (Del Vicario et al., 2016; Quattrociocchi et al., 2016) in which individuals are insulated by personalized algorithms against ideas they disagree with. Such algorithmically mediated exposure helps reinforce aggregation of selective views, opinion manipulation, and thus group polarization.

This proliferation of communication technologies has the potential to further blur the expert-public boundaries discussed in existing risk-communication studies (Grabill & Simmons, 1998; Sauer, 2003; Frost, 2013; Ding, 2014). It can also make the three spheres of argumentation identified by Goodnight (2012), namely, public, technical, and personal spheres, less distinguishable when individuals and organizations playing multiple roles as witnesses, activists, advocates, spokespersons, or whistle-blowers with competing interests, competing knowledge claims, and thus different levels of authority.

The foundational issues underlying risk-communication processes remain little changed, however. The emerging communication technologies do little to break down the expert-non-expert divide, the one-directional, top-down, and technocratic information flow from authorities to the public. These technologies also have limited impacts in generating the so-called "rational responses" from the public to reduce potential harm done by "irrational" responses, i.e., fear, anger, anxiety, distrust, despair, and resistance, from concerned, and sometimes panicked citizens, which are "sometimes a more appropriate and reasonable response than logic" (Katz & Miller, 1996, p. 131; see also Price-Smith, 2009; Leiss & Powell, 2004). Here it helps to revisit Beck's (1992, 1995, 1999) theory of world risk society, hierarchy of risk knowledge, and risk perceptions.

For Beck (1992), risk is "a systematic way of dealing with hazards and insecurities induced and introduced by modernization itself" (p. 21). He later distinguished between "the immateriality of mediated and contested definitions of risk and the materiality of risk as manufactured by experts and industry world-wide" (Beck, 1999, p.5). Living in the age of unintended consequences, uncertainty and ambivalence, we face the ultimate question about knowledge: who knows them and on what basis, which denotes a conflict of knowledge and rationality (p.119). Beck (1999) further explains:

> [In the hierarchy of social credibility,] the claims of different expert groups [not only] collide with one another, [but also] with the claims of ordinary knowledge and of the knowledge of social movements. The knowledge of side-effects thus opens up a *battleground of pluralistic rationality claims [...] and thus conflict of* rationalities with an enlarged (possibly

difficult-to-delimit) horizon of competing agents, producers and interested parties for knowledge.

(p. 120)

Beck identifies two perspectives in framing risks: "natural objectivism" and "cultural relativism" (1995a, p. 162). Based upon scientific knowledge and economic calculation, the natural objectivist approach has dominated institutional risk-assessment practices with the power of expert rationality (Beck, 1999, p. 99). This approach corresponds to Grabill and Simmons' (1997) technocratic paradigm that prioritizes expert knowledge while delegitimizing non-experts' claims. Cultural relativism, in contrast, considers risks as a social reality shaped by institutional discourses, shared cultural beliefs and values, and individual perceptions (Mythen, 2004). Risk perceptions, thus, are culturally situated (Beck, 1996; 1999), discursively mediated, and structured by social contexts (Mythen, 2004).

In the realm of risk communication, only one type of immaterial labor is revered: the intellectual labor of "symbolic analysis and problem solving" that generates data, scientific results, and public health recommendations (Hardt & Negri, 2000, p. 17). As a part of the abstract labor that "manipulates symbols and information," communication labor is rendered secondary and often takes place at the end of intellectual labor (p. 293). In risk communication, affective labor, featured by "human contact and interaction," is often done by risk communicators, health educators, and concerned citizens to help create "a feeling of ease, well-being, satisfaction, excitement, or passion" (p. 293). Affective labor often remains unacknowledged and invisible, if not institutionally excluded (see Hardt & Negri, 2000; Greene, 2004; Ding, 2019; Kong, 2021).

Also falling into the category of unofficial communicative labor and affective labor, public perceptions of risk are often shaped by political outlook and personally held values. Both qualitative and quantitative studies have demonstrated that multiple factors help shape individual perception of and responses to risks, including class (Graham and Clemente, 1996; Douglas 1985), gender (Gustafson, 1998; Flynn et al. 1994; Weaver et al., 2000), age (Hinchcliffe, 2000: 127; Mooney et al., 2000; Jackson & Scott, 1999), and ethnicity (Caplan, 2000; Mackey, 1999; Finucane et al., 2000).

Embodied Environmental Risk in Technical Communication provides a timely addition to the existing scholarship on risk communication by highlighting the embodied human experiences in environmental and health risk communication and by acknowledging the legitimacy of our local/petite narratives, to borrow de Certeau's language, in responding to and resisting master narratives about the risks faced by our shared communities today. It makes an important contribution to culturally relative approaches in studying risk perceptions and the interaction, contradiction, and competition between expert knowledge and ordinary, embodied knowledge. Embodied risk communication offers a productive way to articulate what can be dismissed by the responses that are lumped together and labeled as "irrational." Such "irrational" responses can be personal, situated, localized, or emotional, which are shaped by biological, psychological, socio-economic, and politico-cultural contexts we are embedded in and thus deeply influential at communal and individual levels. Such responses often invoke immaterial labor in communities and individuals, whether it is intellectual labor of trying to understand risks around us, communication labor in terms of explaining our risks perception to others and sharing them with others, as well as emotional labor, i.e., living with uncertainties knowing hazards may be lurking around us without knowing what action can be taken to effectively reduce such risks.

A wide range of topics are covered in the edited collection, with the first two sections focusing on representation of human bodies in health risk communication, i.e., COVID-19, and representation of the earth's body in environmental risk communication, i.e., river, ground water, water rights and water quality, and climate change. Exploring the representations of human and earth, the third section argues that, in public deliberation about environmental issues, both communication practices and listening practices should be incorporated, with the latter paying attention to embodied language, rhetoric, place, cognition, viewpoints, responsibilities, knowledge, and practices. Diverse research methods, both qualitative and quantitative, are represented in the edited collection, which include critical rhetorical analysis, case studies, interviews and surveys, visualization critique, public policy analysis, discourse analysis, auto-ethnography, public science communication, and teacher research. Thanks to the breadth of the topics, the collection features a broad range of stakeholder groups with their embodied experiences, communication strategies, and knowledge: college students, governmental agencies, private owners, nonprofit and activist groups, epidemiologists, patients, end-users of technologies and policies, and actors involved in the right to repair movement.

It covers many key themes in risk communication: power, culture, institution, pedagogy, public engagement, uncertainty management, argumentation practices, embodied experiences, policy deliberation, and knowledge-making and knowledge dissemination, which demonstrate the unique contributions that humanities and social science researchers can bring to the study of environmental studies and environmental and health risk communication.

It is enjoyable, enlightening, and inspiring to read the wide variety of chapters in *Embodied Environmental Risk in Technical Communication*. For me, the edited collection opens up new space for examining, articulating, and sharing how citizens and communities perceive, communicate, respond to, and resist environmental and health risks with immaterial labor and petite narratives that require intellectual, communicative, and affective work. By acknowledging, describing, archiving, and commending such embodied risk-communication efforts, the editors and authors reveal possible ways for communities and citizens to introduce productive resistance and to open up space for positive alternative action in our shared world risk society, which, sadly, has been witnessing accelerating emergences of new unintended consequences. In interrogating and reframing risks as embodied experiences, *Embodied Environmental Risk in Technical Communication* makes meaningful efforts in formulating new questions about the global and local risks we are facing today and, more importantly, the ones we will confront in the future.

References

Beer, D. (2009). Power through the algorithm? Participatory web cultures and the technological unconscious. *New Media & Society, 11*(6), 985–1002.

Bucher, T. (2012). Want to be on the top? Algorithmic power and the threat of invisibility on Facebook. *New Media & Society, 14*(7), 1164–1180.

Caplan, P. (2000). *Risk revisited London*. Pluto Press.

de Certeau, M. (1984). *The practice of everyday life*. University of California Press.

Del Vicario, M., Vivaldo, G., Bessi, A, Zollo, F., Scala, A., Caldarelli, G. & Quattrociocchi, W.(2016). Echo chambers: Emotional contagion and group polarization on Facebook. *Scientific Reports, 6*, 37825 https://doi.org/10.1038/srep37825.

Ding, H. (2019). The materialist rhetoric about SARS sequelae in China: Networked risk communication, social justice, and immaterial labor. In L. Walsh and D. Gruber (Eds.), *Routledge handbook of language & science* (pp. 262–277). Routledge.

Douglas, M. (1985). *Risk acceptability according to the social sciences*. RussellSage.

Finucane, M. L., Slovic, P., Mertz, C., Flynn, J., & Satterfield, T. (2000). Gender, race and perceived risk: The white male effect. *Health, Risk and Society, 2*(2), 159–172.

Flaxman, S., Goel, S., & Rao, J. M. (2016). Filter bubbles, echo chambers, and online news consumption. *Public Opinion Quarterly, 80*(1), 298–320.

Flynn, J., Slovic, P., & Mertz, C. (1994). Gender, race and perception of environmental health risks. *Risk Analysis, 14*(6), 101–108.

Frost, E. A. (2013). Transcultural risk communication on Dauphin Island: An analysis of ironically located responses to the Deepwater Horizon disaster. *Technical Communication Quarterly, 22*(1), 50–66.

Goodnight, G. T. (2012). The personal, technical, and public spheres of argument: A speculative inquiry into the art of public deliberation. *Argumentation and Advocacy, 18*, 214–227.

Graham, J., & Clemente, K. (1996). Hazards in the news: Who believes what? *Risk in Perspective, 4*(4), 1–4.

Greene, R. W. (2004). Rhetoric and capitalism: Rhetorical agency as communicative labor. *Philosophy and Rhetoric, 37*(3), 201,

Gustafson, P. E. (1998). Gender differences in risk perception: Theoretical and methodological perspectives. *Risk Analysis, 18*(6), 805–811.

Hardt, M., & Negri, A. (2000). *Empire*. Harvard University Press.

Hinchcliffe, S. (2000). Living with risk: The unnatural geography of environmental crises In S. Hinchcliffe and K. Woodward (Eds.), *The natural and the social: Uncertainty, risk, change* (pp. 117–154). Routledge.

Jackson, S., & Scott, S. (1999). Risk anxiety and the social construction of childhood. In D. Lupton (Ed.), *Risk and sociocultural theory: New directions and perspectives* (pp. 86–107). Cambridge University Press.

Leiss, W., & Powell, D. (2004). *Mad cows and mother's milk: The perils of poor risk communication* (2nd ed.). McGill-Queen's University Press.

Lyotard, J. (1979). *The postmodern condition: A report on knowledge.* Manchester University Press.

Mackey, E. (1999). Constructing an endangered nation: Risk, race and rationality in Australia's native title debate. In D. Lupton (Ed.), *Risk and sociocultural theory: New directions and perspectives* (pp. 108–130). Cambridge University Press.

Mooney, G., Kelly, B., Goldblatt, D., & Hughes, G. (2000). *Tales of fear and fascination: The crime problem in the contemporary UK.* Routledge.

Mythen, G. (2004). *Ulrich beck: A critical introduction to the risk society.* Pluto P Press.

Pariser, E. (2011). *The filter bubble: What the internet is hiding from you.* Penguin Books.

Price-Smith, A. T. (2009). *Contagion and chaos: Disease, ecology, and national security in the era of globalization.* MIT Press.

Quattrociocchi, W., Conte, R., & Lodi, E. (2011). Opinions manipulation: Media, power and gossip. *Advances in Complex Systems, 14,* 567–586.

Quattrociocchi, W., Scala, A., & Sunstein, C. R. (2016, June 13). Echo chambers on Facebook. http://dx.doi.org/10.2139/ssrn.2795110.

Weaver, C. K., Carter, C., & Stanko, E. (2000). The female body at risk: Media, sexual violence and the gendering of public environments. In S. Allan, B. Adam, & C. Carter (Eds.), *Environmental risks and the media* (pp. 171–183). Routledge.

Wood, M. A. (2017). Antisocial media and algorithmic deviancy amplification: Analysing the id of Facebook's technological unconscious. *Theoretical Criminology, 21*(2), 168–185.

1

INTRODUCTION

Mary Le Rouge and Samuel Stinson

This edited collection emerged from a confluence of interest developed prior to and during the COVID-19 pandemic for the purpose of promoting a greater understanding of risk communication for the public good. Risk communication about mask wearing, social distancing, and "safe" activities during the 2020–2021 COVID-19 pandemic often failed to reach the public and change their behavior toward self-preservation. We were consistently reminded by events during the pandemic that effective risk communication is important and saves lives. While some of the exigence for collecting these chapters on risk communication comes from the COVID-19 pandemic, its true source is the problem of climate change and figuring out how to clearly communicate the risk that global warming poses to the public. Some would even argue that the COVID-19 pandemic was the product of increased human–animal interactions and global interconnectedness through international travel that have made the mutation and spread of the virus more likely. The reduction of natural habitat and increased collocation of human settlement near wild forests made possible by industrialization and necessitated by human population growth has not only adversely impacted animal populations but has also increased incidents of zoonotic disease outbreaks. International travel made possible by technological advances and necessary to conduct transnational business has become an opportunity for COVID-19 to spread widely around the globe. However, we have yet to find a practical way to communicate environmental risk to the public in local contexts, let alone on an international scale, which is a major problem.

What we will argue through this compilation of chapters is that although common problems with stasis, definition, access, and aggregation trouble the public's reception of technical communication about environmental issues, it is

DOI: 10.4324/9781003266549-1

also the lack of focus on human *embodiment* in the content of technical documents that reduces their functionality. Criticisms so often fall on the format of documents to the exclusion of the "messy" work of untangling the metaphors, symbolism, and basic ideologies that support the knowledge contained within. Technical communicators often work from the assumption that the public is (or should be) on the same page, prizing scientific objectivity over embodied knowledge, but the reverse is often true. People understand the environment primarily through their physical bodies and through metaphors and symbols that reflect this embodiedness.

Embodiment is a concept that was first developed by the 18th-century German aesthetic philosopher Alexander Gottlieb Baumgarten through *aesthetics* in the evaluation of art (using the 5 bodily senses, or "sensible cognition," to account for taste). His work was taken up by Emmanuel Kant, who used Baumgarten's book *Metaphysics* (1739) as a textbook during his lectures. Kant focused on methods for effective reasoning and judgment informed by human experience. The idea of embodiment was later more fully developed by *phenomenologist* Maurice Merleau-Ponty in his book the *Phenomenology of Perception* (1945), who studied how perception affects our understanding of the world. His work was then taken up by George Lakoff and Mark Johnson during the 1980s in a linguistic application of theories of embodied knowledge through metaphors that mirror the body's parts and movements in *Metaphors We Live By* (1980) and later, in *Philosophy in the Flesh: The Embodied Mind and Its Challenge to Western Thought* (1999). Francisco J. Varela, Evan Thompson, and Eleanor Rosch recommended application of the concept of embodied cognition to many different fields of knowledge, including primarily psychology, in the 1990s with their book *The Embodied Mind* ([1991] 2016). Since then, embodiment has been studied in rhetoric and technical communication more because of its absence than its presence, especially in the context of risk communication.

For the past 20 years, scholars in technical communications have been studying the difficulties of enacting effective risk and crisis communication policies to address local and global environmental problems such as pandemics, natural and manmade disasters, medical emergencies, and workplace and community dangers (Angeli, 2019; Ding, 2014; Frost, 2013; Potts, 2013; Sauer, 2003; Walker, 2016). Differences exist between embodied experience and abstract representation, separating those at risk from institutional policymakers, which can make effective risk communication more difficult (Sauer, 2003). This is especially true in circumstances in which embodied knowledge is dismissed as unimportant in official documentation of risky environments (Sauer, 2003, pp. 5–6). In addition, different bodies experience different levels of risk, depending on artificial social hierarchies based on race, body shape, gender markers, and the physical marks of socioeconomic differences. By focusing on embodiment (instead of ignoring physical differences out of convenience), it is possible to chart a more realistic view of the determinants of risk for diverse individuals for technical communicators. As Iris Ruiz describes in *Race, Rhetoric, & Research Methods* (2021):

> An embodied rhetorical practice is a praxis that acknowledges this reality for minoritized populations. It is one step closer to performing an antiracist method that relies on decolonial options, such as epistemic delinking, to achieve the goal of antiracist scholarship.
>
> *(2021, p. 63)*

The objective of this collection is to improve technical communication for the public through an embodied, situated understanding of risk that promotes social justice. We support the continually evolving social justice turn highlighted in the ATTW Series in Technical and Professional Communication's previous publication by Rebecca Walton, Kristen Moore, and Natasha Jones: *Technical Communication After the Social Justice Turn: Building Coalitions for Action* (2019). In addition to providing a series of chapters about recent issues on risk communication, this volume offers a diverse look at methodological practices for students, researchers, and practitioners looking to address embodied aspects of crisis and risk both locally and globally that incorporate UX, storytelling, and dynamic text with visuals. We include chapters that bring embodiment to the forefront of risk communication, throughout the cycle of content creation, dissemination, public response and decision-making, continuing iterations of educational efforts, and recovery, toward increasing adaptive capacity as a whole. In addition, we focus on topics such as overcoming perceptual difficulties, memory lapses, definitional differences, access issues, and pedagogical problems in the communication of risks to the public. Here, we review some recent work in technical communication and describe how chapters included in this edited collection work toward bridging this gap between technical communicators and the public.

Chapter Organization

The book is broadly organized into three sections: (I) Representations of the Human Body, (II) Representations of the Earth's Body, and (III) Representations of Human and Earth Together. The first section engages with representations of the human body as a site of public regulation through technical communication in response to health risks and crises. Chapters in this section include studies of health communication, especially responses to the COVID-19 pandemic. The second section deals with representations of the earth's body and its parts/functions through technical communication. Chapters in this section include research on water management and climate science communication. The earth is seen to hold a state of health and natural abundance or illness, decay, imbalance, or pollution. Systems theory views the planet as a working body of complex, intertwined parts, each affecting the other, as in the Gaia theory. Work in this section entertains notions of geography and spatial/mapping issues, renditions of natural earth cycles and processes, cosmology, and Earth's place in the universe. The third and final section brings together representations of natural and manmade disasters through technical communication, where both humans and the earth together are experiencing physical hardship and a symbiosis between the two is sought. Chapters in this section include concepts of balance and solutions that work toward synthesis of human/nonhuman ontologies. Relationships between humans and animals and nonhuman entities are investigated as key toward creating synergies. The topics in this edited collection's sections bring up questions such as: In what ways do representations of the earth's body, specifically its parts and functions, mirror embodied human lived experience? How can we better understand and communicate about global crises in the context of the human body?

Representations of the Human Body

We have seen in the current COVID-19 pandemic that official government communication about risk filters down to the public and is altered in many ways by the media and through unofficial interpersonal communication on social media. Huiling Ding's use of the term *transcultural* to denote a difference in culture between the ruling class and the public does offer some explanation for the incommensurability in communication about the epidemic in China. And this theoretical construct, when applied to communication failures about the risk of COVID-19 in the United States, works to highlight moments when conflicting reports about the efficacy of mask wearing, hand washing, and social distancing caused mass-spreader events. But we argue that by naming the cultural difference between official and public communication about the disease as *embodied* and attending to the physical experience of the public adds a deeper understanding of where that communication failed and why.

Huiling Ding uses the theoretical framework of transcultural professional communication to study the rhetoric surrounding the SARS epidemic in China (Ding, 2014, p. 21). She takes up Arjun Appadurai's five areas of global textuality: ethnoscape, mediascape, technoscape, financescape, and ideascape to study transcultural communication about the global epidemic (Ding, 2014, p. 22). She describes her research as using a "critical contextualized methodology" that provides "a theoretically informed framework both to examine transcultural rhetorics and to explore ways for participation and social intervention in global events" (Ding, 2014, p. 29). Ding seeks out and analyzes transformational moments in international events, tipping points that will reflect power relations in China and other countries that were threatened by the epidemic.

Her study is unique in that it focuses on events in a non-Western country and compares that with Western ways of dealing with SARS. It also stands out as the only book-length study in the field of technical communication that describes an epidemic.

Ding has also studied transcultural communication about HIV/AIDS and H1N1 comparatively between the United States and China (Ding, 2013, 2014). There are other studies of yellow fever and H1N1 and swine flu that focus on disease mapping and metaphors found in electronic media, respectively (Angeli, 2012; Welhausen, 2015). What these studies have brought to the table is a corpus of information regarding communication about epidemics that shows how social context in communication tactics and epistemology affects audience understandings of risk and their ability to participate in decision-making and self-protection strategies. According to Angeli, "The way in which pandemic flu is framed affects how millions of people understand and act on health concerns, thus affecting the public's medical decisions…" (Angeli, 2012, p. 219). Her disambiguation of the role of metaphor in constructing the public's understanding of disease is akin to how Ding describes the role of folklore in the circulation of information through the public when official communication sought to deny the reality of the epidemic.

Ding's book is partly focused on the environmental determinants of risk for becoming infected: geographical factors of country/city and the location/construction of medical facilities. It also speaks to the power relations between government rhetoric that initially sought to deny the problem and on-the-ground public deliberation of those who saw friends and family becoming sick. Therefore, although a critical assessment of Chinese medical rhetoric about epidemiology and disease risk factors is included in her study, Ding shows that it is altogether affected by the strong arm of the Chinese government and myths/gossip perpetuated by the general populace. She states, "The official rhetoric operated to calm down the public with partial disclosure about SARS situations … [leading to] the official claim of 'SARS is under control'" when it was not (Ding, 2014, p. 79). In contrast, the public conversation about SARS was characterized by the spread of gossip mongering, so that "An asymmetry existed between the absence of credible official information and the overwhelming number of rumors" (Ding, 2014, p. 116). This conclusion is often found in studies of top-down, official government or corporate communication to the public or employees about environmental risk and the vernacular response to that communication (Frost, 2013; Grabill, 2007; Potts, 2013; Sauer, 2003; Simmons, 2007).

We show through the chapters included in this section that an embodied understanding of medical risk is also necessary toward effective communication about health issues that affect the public. Chapter 2 investigates the disembodied nature of pandemic risk communication at Illinois State University during fall 2020, and how it erased cultural and ethnic differences and normalized power inequities between the student body and the administration. Chapter 3 considers the benefit of using the Anticipatory Model of Crisis Management with attention to embodiment toward more effective risk and crisis communication about the pandemic in a multi-factor environment. Chapter 4 investigates the rhetorical basis for development of deliberation and judgment practices in situations of health risk during the pandemic. Chapter 5 addresses the pedagogical factors that help to encourage personal responsibility in students toward their self-development and response to risk.

Representations of the Earth's Body

The recent focus on experience architecture in technical communication has been spearheaded by Liza Potts in her book *Social Media in Disaster Response*, and with Michael Salvo in their edited collection *Rhetoric and Experience Architecture*. Experience architecture offers a new way to tackle "wicked problems" that are defined as "social problems which are ill-formulated, where the information is confusing, where there are many clients and decision makers with conflicting values, and where the ramifications in the whole system are thoroughly confusing" (Salvo & Potts, 2017, p. 87). Whereas user experience in software design originally meant making sure that users were able to navigate a prebuilt program successfully, attending to found problems with post-production work and user reference "help" guides after the fact, user experience in contemporary rhetorical literature has come to mean an iterative process that takes place throughout the design process, from start to finish. This includes a focus on the user's persona, emotions, identity, and social boundaries that is meant to give users more control over the product and "provides users with an interpretive framework that allows them to successfully experience the interface" (Salvo & Potts, 2017, Chap. 7). Attention to users' mental modes can result in greater coherence, definitional agreement, and importance given to critical events in the community.

Potts takes up the idea of experience architecture in *Social Media in Disaster Response* to show "how everyday people deploy technologies to communicate with each other" (Potts, 2014, p. 4). She has found that "in times of disaster and crisis, people tend to gravitate toward the systems and networks that are most relevant and familiar to them" (Potts, 2014, p. 11). However, "sites for news agencies, government agencies, and NGOs are typically still closed systems in that they don't support the exchange of information beyond their boundaries" (Potts, 2014, p. 47). Potts ends up examining participants' "literate appropriations of tools and information" through the social web (Potts, 2014, pp. 61–66). Using actor network theory (ANT) and citing Bruno Latour's framework, she outlines the process that users engage in to respond to a disaster through (1) problematization, (2) interessement, (3) enrollment, and (5) mobilization online. This approach to the study of social media activity allows a fuller picture of the many actors and their tasks involved in an ecosystem of response to disaster. She claims that researchers need to participate as experience architects to both understand and support the flow of information across the web in a move toward social justice and civic participation. However, she cautions that designing web systems for participation can lead to reduced stability and the proliferation of potentially misleading information (Potts, 2014, p. 44). If sites that have traditionally been run in a top-down style of information flow, similar to the *New York Times,* open up to participatory news-making and dissemination, how is the validity of that information assured? Potts suggests validation through cross-referencing and triangulation and increased "systems literacy" from users, but this seems to ask a lot of users who are faced with the emotions and impending consequences of a disaster (Potts, 2014, pp. 59–60, 64–65). When complex scientific information is also included in such communication, the bar is set even higher for public participation in deliberations online.

When the "natural world" is extended to encompass not only biological and geographical earth systems but also the technological beings and surroundings constructed by humans, agency is given to nonhuman entities in an ontological shift that has primarily been discussed through activity theory and ANT by Peter Smagorinsky, Liza Potts, and Bruno Latour (2011, 2018, and 2017, respectively). Gee notes that "humans across the globe face serious risks, dangers, and disasters from interacting complex systems that are on the verge of going out of control. These are systems like massive inequality, environmental degradation, global warming, vast migration flows…" that call for greater "collective intelligence" (Gee, 2017, p. 4). The systems view of the writer does not see the author as an individual cut off from the outside world, but rather always interacting with the "natural world" and other authors, organizations, and issues that complicate the communication process.

Experience architecture is a positive step toward prioritizing the audience (or public) in technical communication through the construction of personas and acknowledging that the public does not exist in a vacuum. But experience architecture has its basis in the virtual world of software computer programs and disembodied experience. It is not just a matter of plugging in the socioeconomic determinants of a persona and the current sociohistorical context and coming up with a framework or architecture in which the relationships and their outcomes are predetermined. Computer games have only so many programmed pathways to completion but the human mind is not a computer, even though we have been operating under this assumption for quite a while. The human mind is *embodied*, and yes it exists in a complex world system, but going further than personality, is the foundational relationship that the mind's constructs have with the human body. This is most evident when we are communicating about the environment that exists outside of our bodies.

Grabill quotes John Dewey as writing that "the public has a problem: we can and should solve that problem by helping the public behave better, to be better informed, to be more deliberative" (Grabill, 2014, p. 255). This pragmatic take on education and civic duty influences Grabill to build or organize the public through practical activities and infrastructural characteristics that make public participation in deliberations about local water quality more possible. His work with Hart-Davidson on "Understanding and Supporting Knowledge Work in Schools" sets up a two-step approach to creating online public repositories. This includes building curated repositories where experts organize and provide access to primary materials, and secondary repositories that are user-driven and add value to primary content through reviews, comments, and derivatives. Hart-Davidson and Grabill state that in reflecting relationships in a given context, "life is textualized in digital environments," although not always in ways that aid public participation (Hart-Davidson & Grabill, 2012, p. 175). In his article "Citizens Doing Science in Public Spaces," Grabill writes with Stuart Blythe about the semiotic remediation practices used by a community environmental organization, where signs (and texts) are recontextualized in new situations through inventional activities (Grabill & Blythe, 2010, p. 185). They show a traceable relationship between the many texts used in public deliberation, including their remediation, and the conversations that take place in organizational meetings to promote group formation and social action.

Another way that citizen participation in environmental contexts has been studied is through the public use of social media such as Twitter, and the public construction of websites for information exchange. In an article that she published before her book on social media use during disasters was released, Liza Potts writes with Dave Jones about using ANT to analyze social web usage (Potts & Jones, 2011). They use the term *fire spaces* to describe moments where groups work together to solve problems, such as Twitter topics that become "hot" and generate high volumes of information movement, where "the more movement, the more visible it becomes..." (Potts & Jones, 2011, p. 344). They also show how on social media miscommunication happens easily because "extra layers of decisions between a participants' intention and that participant's desired outcome can negatively affect any understanding of the community and its context" (Potts & Jones, 2011, p. 349). They suggest that to promote inclusive, user-friendly fire spaces on social media, designers should take locatability, navigability, discoverability, and retrievability into account (Potts & Jones, 2011, p. 353). Their prescriptions mirror scholars in technical communication who suggest deployment of usability studies and user experience modifications to make the virtual technological space more just.

Technical communication in relation to the environment has distinct ramifications for populations who live in the *anthropocene* (an age when human activity has become the dominant environmental force on the planet) and must make decisions about how to live sustainably in the face of risks caused by climate change and global warming. Emphasizing transcultural communication that crosses boundaries between experts and nonexperts, governments and citizens, scientists and laypeople, brings difficulties in communication styles to the foreground. In addition, looking at environmental rhetoric as not only a form of technical communication, but as belonging to the subset of risk communication because of the exigencies of accelerated climate change, highlights the two-way flow of information that is critical in situations involving personal and community risk. Extending the field of experience architecture to encompass the rhetorical engineering that is necessary to communicate in an intercultural inquiry is the logical solution to introducing a more ethical, just way to reach those who are subjected to environmental risks due to global warming.

As Sidney Dobrin's *Ecology, Writing Theory, and New Media* shows, it is the establishment of stasis, "what is," and definition that makes public debate about an issue even possible (Dobrin, 2012). If climate change does not exist, then the whole argument against the use of fossil fuels and development of renewable energies falters, and the debate stays centered on matters of existence instead of formulating ways to move forward. If the study of risk communication necessitates the promotion of two-way communication, as described by Regina Lundgren and Andrea McMakin in their handbook, then the job of technical communicators in this subset of inquiry is to investigate how citizens can untangle the sources and purposes of environmental rhetoric and respond in alternative ways that will be recognized by others as authentic (Lundgren & McMakin, 2019). Ulmer describes this ability as "electracy," the digital form of literacy, and posits that this should be tempered by "ecosophy," in which wisdom of the ecological factors at play in communicative practices is brought to bear on contemporary problems such as climate change (Dobrin, 2012).

Although community participants might not have preexisting knowledge about climate change, they will have preexisting ideas about how they, and others should, interact with the environment and what should and should not be supported or allowed by the government. As Addison notes in Katrina Powell and Pamela Takayoshi's collection *Practicing Research in Writing Studies*, "cultural norms often motivate the preconstruction of scientific narratives" (Powell & Takayoshi, 2012, p. 380). Additionally, documents created by scientists and policymakers will have very specific purposes that are not necessarily objective and are very well meant to persuade their readers toward a certain path of action. As a mode toward promoting increased civic participation in local debates, participatory research projects in rhetoric and risk communication are critical to create an opening for dialogue between experts and citizens that might not otherwise exist.

We show through the chapters included in this section that engaging the public in deliberation about environmental issues represented by technical communication is more effective when the *embodiment* of the earth and its inhabitants is prioritized. Chapter 6 researches discourses surrounding framing of the Ohio River as a natural asset and body of water that requires protection. Chapter 7 investigates difficulties in private groundwater regulation and risk communication through *socio-hydrology* that includes public perceptions and belief models in scientific determinations of water risk. Chapter 8 describes how embodied cognition and metaphor affect public communication about a proposed wind farm on Lake Erie. Finally, Chapter 9 critiques the *Fourth National Climate Assessment Report* as disjointed and ineffective at communicating climate change as a holistic event that has been exacerbated by human activity.

Representations of Human and Earth Together

It is this disjuncture in communication styles and purposes between those in power and the public that Beverly A. Sauer investigates in *The Rhetoric of Risk*, which compares the ways that miners experience risk in the pit versus how official corporate understanding of risk from above operates to silence the miner's more embodied modes of communication. It is, as she claims, a difference between embodied experience and abstract representation that separates the two and makes effective risk communication difficult. Sauer uses the theoretical framework of Aristotle's Rhetoric as an "inventional art" combined with "psychological and linguistic studies of gesture" to represent embodied knowledge that is often dismissed as unimportant in official documentation of risky environments (Sauer, 2003, pp. 5–6). Sauer's work can be described as a grounded investigation into the mining industry and its risk communication practices. Her methodology is both pragmatic in looking for solutions to the problem of reducing risk for miners and feminist in the types of solutions that she provides, including attention to tacit understanding, embodied experience, and gestural modes of communication. She claims that "agencies create standards because experience is a poor teacher … But standards do not always define how individuals should respond to those crises" (Sauer, 2003, p. 37). When input and participation from stakeholders is not sought by organizations in a position of power, important information is missed and "the activity of reporting becomes a formulaic act that has no rhetorical force beyond the agency's requirements for documentation" (Sauer, 2003, p. 121).

The valuing of civic participation in public deliberation, especially in regard to situations of environmental risk, has been discussed in other studies as a counterpoint to risk communication that comes from positions of power and often overlooks everyday language practices (Grabill, 2007; Potts, 2014; Salvo & Potts, 2017; Simmons, 2007). In the field of technical communication, solutions to creating a context where civic participation can become more prevalent and accepted have included work on experience architecture in public forums and digital environments. In W. Michele Simmons's book *Participation and Power*, she finds that in the case of public deliberation on the disposal of VX nerve gas in one community is missing because "historically, the lack of publicity of public meetings has prohibited many citizens from even being made aware of environmental health issues and decisions" (Simmons, 2007, p. 16). If citizens do not even know about an issue or opportunities for participation, it is not possible to take part in deliberations. In addition, Simmons describes meetings that did take place as superficial and intimidating to civilians, where their input was not really taken into account, and their language was devalued as not being technical enough (Simmons, 2007, pp. 69–70). She calls this type of interaction "pseudoparticipation," where the audience is given the appearance of having a say in the deliberation process but is only silenced.

Similarly, Jeffrey Grabill notes in *Writing Community Change* that public meetings on a harbor-dredging project that threatened to pollute the local water supply were characterized by expert engineers talking to citizens, without leaving time for audience comments, questions, or concerns (Grabill, 2007, p. 89). Grabill recommends a wholesale restructuring of community meetings to encourage public participation and two-way invention work between the EPA and the local community. He also takes on an advisory and interventionist role by helping to set up a grassroots website and resources to inform the community about the dredging project. He argues that "Citizenship is knowledge work … supported by a complex information structure [that is] both fragile and poorly designed to support knowledge work in communities" (Grabill, 2007, p. 59). Therefore, the public mapping tool that he sets up for a local organization works to allow citizen input into its design, content, and function, ensuring that the website supports public knowledge-building and advocacy efforts.

Simmons and Grabill call for greater public participation in dialogues about environmental issues in local communities, and even go so far as to intervene and support local initiatives toward civic participation. But again, the reason for the incommensurability that they found at their sites of research is framed either as a problem of overly technical or expertise language that the public was not trained to understand, or as stemming from poorly organized informational resources. By focusing on format, the content of the environmental documentation in Simmons and Grabill's research was not evaluated for its attention to embodied understanding; however, Sauer describes doing just this in her research on miners' gestural communication. For this reason, we are working from Sauer's theoretical framework of embodied risk communication in the evaluation of public reception of technical communication about environmental issues.

The success of technical communication is predicated on the literacy of a community, which is supported or discouraged by sponsors. Literacy sponsorship is defined by Deborah Brandt as: "agents, local or distant, concrete or abstract, who enable, support, teach, model, as well as recruit, regulate, suppress, or withhold literacy – and gain advantage by it in some way" (Brandt, 2001). Here, we will take a broad definition of agents who modify access to literacy to include nonhuman actors and the context and sociopolitical systems that exist in an ecology of reading and writing (Dobrin, 2012; Syverson, 1999).

With this methodological approach, it is not just the capacity of the individual to understand complex information about climate change and the risk of natural disasters that needs to be developed, rather the knowledge system that surrounds the individual determines ways that the individual is able to respond to issues of risk (Latour, 2018; Rai & Druschke, 2018).

Situating disaster preparedness for the effects of climate change in the literature on development of public literacy moves the focus from study of literate responses to common events (Barton & Hamilton, 1998; Brandt, 2001; Flower, 2008) to evaluation of literate practices in preparation for extraordinary events (Blackburn, 2007; Gee, 2017; Selfe, 2007). Research done in terms of education's effects on resiliency is slim, but several studies reflect the importance of education to forwarding sustainability efforts and promoting useful responses to situations of risk (Friedman et al., 2010; Goto et al., 2018; Kanbara et al., 2016; Muttarak & Lutz, 2014). The lack of a forum for public discourse about climate change is evident, as we overlook that which is deemed too big, complex, or "scientific" for the public to comprehend or address. Just as racial issues are commonly subsumed as politically incorrect to bring up in polite society, as in Jenny Rice's ecological analysis of Austin's public housing debates and African American segregation practices during the 1930s, environmental decline is invisible to many who cannot see the direct effects or are protected in their carefully constructed upper-class environments. Rice states that "One of the major points of deliberation concerned whether slums even existed in Austin," because the streets that the well-to-do population commonly used did not cross into the "industrial" area where the slums existed (Dobrin, pp. 180–181). Such "invisibility" allows disparities in racial and socioeconomic situations to persist and worsen over time and perpetuates environmental injustices through simple questions of existence.

To move from debates about stasis, there are also arguments regarding definition that trouble public action on environmental issues. The way scientists and governments define a problem and the terms used in the process directly affects the kinds of responses that are possible. The EPA censored usage of the term "climate change" in official documentation because the Trump administration refuted the existence of the problem. The focus of the texts on its website was on protection of the U.S. population from the health effects of water, air, and ground pollution, attending to the direct effects of industrial production on human health, instead of taking a broader overview of the effects of fossil fuel consumption on earth systems (EPA, 2019). This defined the problem narrowly for the public and mirrors the effects of the existence debate. Through the Freedom of Information Act and its 1996 amendments, the public is supposed to have access to most unclassified government documents online, but processing times are prohibitively long. There is also a recent movement in technical documentation of governmental information for the public to become more readable. This is done through accessibility modifications, caps on reading levels of written materials, and use of "plain language," especially on official websites (Jones et al., 2012). How this accommodation of scientific and legal information "dumbs down" the meaning of the documents and serves to hide important facts is the reverse of Orwell's "doublespeak" that hides meaning through overly wordy language (Orwell, 1946).

However, it is not just access or the level of language found in official documents that can serve to limit the ability of the public to understand scientific statistics, laws, and regulations. Jeffrey Grabill and Michele Simmons note that in their studies of community organization against local environmental risks, the ability of community members to successfully use information provided by governmental bodies on their websites was extremely difficult as the data was not aggregated in user-friendly ways (Grabill, 2007; Simmons, 2008). Not only this, but the data could not be easily manipulated by the public to configure it in ways that would give them answers to their questions and provide evidence. So, the statistics to make a case against policies that threatened their communities were made available but were effectively useless to their causes. For this reason, Grabill established a website tailored to the needs of the local organization fighting a dredging project in their harbor that provided the functionality that they required to exchange and configure data. Similar public website projects and apps have been developed by other scholars of social action and citizens, including providing customizable mapping technologies to share information such as local safe routes and checking in after a disaster, communication sites such as blogs and instant messaging hubs or "fire spaces," and other databases and repositories for public documents. These bottom-up contexts for collaborative knowledge making act very differently from official documents in that they are constantly changing, flexible, and responsive to their environment, and are therefore adaptive and strong, as Liza Potts argues in *Social Media in Disaster Response* (Potts, 2013).

We show through chapters included in this section that public deliberation in environmental issues requires not only a "space" for communication, but also listening practices that are inclusive of nontraditional, *embodied* forms of communication, such as gesture, visual representation, and personal responsibility and practice through action. Chapter 10 describes how gesture and cognition in communication of mining risks in South Africa can overcome incommensurability of language difference. Chapter 11 takes the example of a global art installation of sculpted forest giants to show how the idea of physically reanimated risk helps to communicate theories of climate change. Chapter 12 calls on ANT to describe how students writing iFixit technical documentation leads to the repair and reuse of electronics and supports personal agency. Chapter 13 develops a theoretical basis for how notions of place affect perceptions of climate change communication and mitigation efforts. Chapter 14 questions the current use of disaster rhetoric in the classroom and calls for a reconceptualization of this method to openly discuss the problem of climate change.

We hope that the wide variety of approaches to investigating embodied environmental risk in technical communication that appear in this edited collection will provide fertile ground for future research on this topic. Therefore, we have added abstracts, definitions, and discussion questions/assignments to each chapter so that this text can be more easily adopted for use in the classroom to teach future scholars of technical communication and environmental rhetoric about the ways that embodiment affects understanding. We are grateful to series editor Tharon Howard and ATTW/Routledge for their support, Beverly A. Sauer for her contributions to the risk communication field in general and the development of this collection in particular, Huiling Ding for setting the tone with her forward to this book, and all the other authors who have contributed their voices to this collection. We would like this collection to act more as a gateway to the production of new research than as an end in itself, and as such we welcome feedback and commentary about the topics taken up in these chapters.

References

Angeli, E. L. (2012). Metaphors in the rhetoric of pandemic flu: Electronic media coverage of H1N1 and swine flu. *Journal of Technical Writing and Communication, 42*(3), 203–222.

Angeli, E. L. (2019). *Rhetorical work in emergency medical services: Communicating in the unpredictable workplace.* Routledge.

Barton, D., & Hamilton, M. (1998). *Local literacies: Reading and writing in one community.* Routledge.

Baumgarten, A. (2014). *Metaphysics: A critical translation with Kant's elucidations, selected notes, and related materials.* [1739] C. D. Fugate & J. Hymers (Trans.). Bloomsbury Academic.

Blackburn, M., & Clark, C. (2007). *Literacy research for political action and social change.* Peter Lang Publishing.

Brandt, D. (2001). *Literacy in American lives.* Cambridge University Press.

Ding, H. (2013). Transcultural risk communication and viral discourses: Grassroots movements to manage global risks of H1N1 flu pandemic. *Technical Communication Quarterly, 22*(2), 126–149.

Ding, H. (2014). *Rhetoric of a global epidemic: Transcultural communication about SARS.* Southern Illinois University Press.

Dobrin, S. (Ed.). (2012). *Ecology, writing theory, and new media.* Routledge.

Flower, L. (2008). *Community literacy and the rhetoric of public engagement.* Southern Illinois University Press.

Friedman, D. B., & et al. (2010). Disaster preparedness information needs of individuals attending an adult literacy center: An exploratory study. *Community Literacy Journal,* 55–73.

Frost, E. A. (2013). Transcultural risk communication on Dauphin Island: An analysis of ironically located responses to the Deepwater Horizon disaster. *Technical Communication Quarterly,* 50–66.

Gee, J. (2017). *Teaching, learning, literacy in our high-risk high-tech world: A framework for becoming human*. Teachers College Press.

Goto, A., Lai, A., Kumagai, A., Koizumi, S., Yoshida, K., Yamawaki, K., & Rudd, R. (2018). Collaborative processes of developing a health literacy toolkit: A case from Fukushima after the nuclear accident. *Journal of Health Communication*, 200–206.

Grabill, J. (2007). *Writing community change: Designing technologies for citizen action*. Hampton Press.

Grabill, J. (2014). The work of rhetoric in the common places: An essay on rhetorical methodology. *JAC: A Journal of Composition Theory, 34*(1–2), 247–267.

Grabill, J., & Blythe, S. (2010). Citizens doing science in public spaces: Rhetorical invention, semiotic remediation, and simple little texts. In P. A. Prior, & J. A. Hengst (Eds.), *Exploring semiotic remediation as discourse practice* (pp. 184–205). Palgrave Macmillan.

Hart-Davidson, W., & Grabill, J. T. (2012). Understanding and supporting knowledge work in schools, workplaces, and public life. In D. Starke-Meyerring, A. Pare, N. Artemeva, M. Horne, & L. Yousoubova (Eds.), *Writing in knowledge societies* (pp. 161–176). WAC Clearinghouse; Parlor Press.

Jones, N., McDavid, J., Derthick, K., Dowell, R., & Spyridakis, J. (2012). Plain language in environmental policy documents: An assessment of reader comprehension and perceptions. *Journal of Technical Writing and Communication, 42*(4), 331–371.

Kanbara, S., Ozawa, W., Ishimine, Y., Ngatu, N. R., Nakayama, Y., & Nojima, S. (2016). Operational definition of disaster risk-reduction literacy. *Health Emergency and Disaster Nursing*, 1–8.

Lakoff, G., & Johnson, M. (1980). *Metaphors we live by*. University of Chicago Press.

Lakoff, G., & Johnson, M. (1999). *Philosophy in the flesh: The embodied mind and its challenge to western thought*. Basic Books.

Lundgren, R. E., & McMakin, A. H. (2018). *Risk communication: A handbook for communicating environmental, safety, and health risks*. IEEE.

Merleau-Ponty, M. (1945). *Phénoménologie de la perception*. Gallimard.

Muttarak, R., & Lutz, W. (2014). Is education a key to reducing vulnerability to natural disaster and hence unavoidable climate change? *Ecology and Society, 42*.

Orwell, G. (1949). *1984*. Secker & Warburg.

Potts, L. (2013). *Social media in disaster response: How experience architects can build for participation*. Routledge.

Potts, L., & Jones, D. (2011). Contextualizing experiences: Tracing the relationships between people and technologies in the social web. *Journal of Business and Technical Communication, 25*(3), 338–358.

Powell, K., & Takayoshi, P. (Eds.). (2012). *Practicing research in writing studies: Reflexive and ethically responsible research*. Hampton Press.

Rai, C., & Druschke, C. (Eds.). (2018). *Field rhetoric: Ethnography, ecology, and engagement in the places of persuasion*. University of Alabama Press.

Ruiz, I. (2021). Critiquing the critical: The politics of race and coloniality in rhetoric, composition, and writing research traditions. In A. L. Lockett, I. D. Ruiz, J. C. Sanchez, & C. Carter (Eds.), *Race, rhetoric, and research methods* (pp. 39–79). WAC Clearinghouse; University Press of Colorado.

Salvo, M. J., & Potts, L. (Eds.). (2017). *Rhetoric and experience architecture*. Parlor Press.

Sauer, B. A. (2003). *The rhetoric of risk: Technical documentation in hazardous environments*. Lawrence Erlbaum.

Simmons, M. (2007). *Participation and power: Civic discourse in environmental policy decisions*. SUNY Press.

Smagorinsky, P. (2011). *Vygotsky and literacy research: A methodological framework*. Brill.

Syverson, M. A. (1999). *The wealth of reality: An ecology of composition*. Southern Illinois University Press.

Varela, F. J., Thompson, E., & Rosch, E. (2016). *The embodied mind: Cognitive science and human experience* (Rev. ed.). MIT Press.

Walker, K. C. (2016). Mapping the contours of translation: Visualized un/certainties in the ozone hole controversy. *Technical Communication Quarterly, 25*(2), 104–120.

Welhausen, C. A. (2015). Power and authority in disease maps: Visualizing medical cartography through yellow fever mapping. *Journal of Business and Technical Communication, 29*(3), 257–283.

I

Representations of the Human Body

2

TOWARD AN AUDIENCE-CENTERED APPROACH

Rhetorical Analysis of University Crisis Communication Emails

Courtney Cox and Erika M. Sparby

Following the quick-paced closures of college campuses in response to the initial surge of the COVID-19 pandemic in spring 2020, campus communicators paved the way by outlining evolving policies, protocols, and expectations for students. In general, communications were multi-pronged, but the bulk of these messages were sent via email. Illinois State University (ISU) was no exception, and lengthy crisis communication messages were sent to students, faculty, and staff in the months leading up to the hybrid return to campus in fall 2020 and following the return to campus in the ensuing months. Although crisis communication messages are an ever-present reality on college campuses, they became necessary for all campus community members during the COVID-19 pandemic, a lifeline for updates and policies in a rapidly changing moment of public risk.

Crisis communication is a type of technical communication under the umbrella of risk communication. While risk communication can deal with longer-term communication about potential emergencies, crisis communication addresses immediate and ongoing emergencies. Crisis messaging occurs across a range of different genres and media—including short-form text message alerts, announcements over loudspeakers, and posts on social media pages. Although crisis communicators may have a range of different media for sharing information, the choice of medium and how the message is shared will affect the type and depth of details they can include. A partial message may provide limited or inaccurate information, whereas a lengthy message may overwhelm readers with information overload. For mundane and nonthreatening events, such as a pre-scheduled fire alarm or an isolated power outage, a text message alert may be sufficient. However, when it comes to complex and unprecedented crises, such as evolving information about the COVID-19 pandemic, longer-form messages like email were more appropriate for regularly sharing updated policies and linking to additional resources.

DOI: 10.4324/9781003266549-3

Crisis communication on college campuses is particularly tricky because communicators address audiences with distinct needs and expectations. When attempting to reach a varied and complex audience—one with different needs, expectations, and spread across wide distances—crisis communication becomes especially difficult. This is because audiences will be looking for different details as they read a message. For instance, students who were displaced from their dormitories might be looking for information about housing or meal places, whereas faculty have other distinct concerns, such as how they will teach their courses and what affordances they need to make to accommodate students.

In their study of campus crisis communication, Kate Pantelides and colleagues write that they receive an average of 20 email messages a year, messages that "are so well-codified that, at first, they seem algorithmically produced, as though their presence in our inboxes indicates a simple truth about campus safety and the assailants who could do us harm" (2016, p. 1). It is difficult to remember that these messages are created by human beings—affecting all of us, and resulting in different types of risk and response. We do not see the communicators who send these messages, and in many cases, they do not interact with us either. As a result, we may not be critical of these messages, aware of the inequities that they can promote, and cognizant of how they cement a narrative of events on campus.

Problem: Although crisis communication messages may fall to the background when we think about dangerous events, they are integral for the actions that we take and how we understand the event and our relationship to it. Institutional crisis communication often does not account for the diverse embodied experiences of their full audience and instead treats thousands of people as if they approach crisis information from the same perspective and positionality.

Solution: Below, we examine a case-example email from our institution, performing a section-by-section rhetorical analysis while also providing concrete takeaways for how university crisis communicators can adapt their communications to account for the lived realities of every student on campus, not only those who are privileged and visible.

Institutional Context

Due to the uncertainty leading up to the Fall semester, ISU pivoted and adjusted their return plans several times, updating the campus community on these changes via email. Tensions rose to a high when, during the first few weeks of the semester, cases spiked to over 1,000 student positives in a one-week period, as COVID-19 rates increased on campus and among community members ("COVID-19 Campus Case Tracker," 2021). During this peak, the metropolitan area not only gained attention from *The New York Times* as one of the areas with the highest per capita cases ("Coronavirus in the U.S.," 2021), but also because of a visit from infamous YouTubers called the NELK Boys on September 8. The three YouTubers film lifestyle videos, and as they describe on their channel, they "prank people sometimes" (NELK, 2010). With their base of over five million followers, the NELK Boys arrived near the ISU campus, resulting in a party of several hundred maskless students that was ultimately broken up by police (Chen, 2020). Footage of the party went viral, stoking community anger and also capturing some of the student partygoers on film.

This gaffe was dangerous, potentially exposing hundreds of ISU students to COVID infection, as well as bringing negative press to the university. The story went viral, with reporting on such sites as *Buzzfeed*, *New York Daily News*, and other state and local publications (Chen, 2020; Henry, 2020; Schladebeck, 2020; Swiech et al., 2020). Following this exposure, university administrators addressed this situation in public press conferences, as well as in weekly coronavirus emails sent to students, faculty, and staff. Before the pop-up parties began, the town of Normal issued orders limiting gatherings near ISU's campus to 10 individuals or less, with violators facing fines up to $750 (Swiech et al., 2020). The university responded with policies for holding students accountable and deemed the party a violation of university Code of Conduct warranting university suspension. YouTube characterized the party as a public health risk, temporarily banning the NELK Boys from their platform.

The NELK Boys' visit escalated and shaped some of the messaging and rhetorical strategies coming from university technical and crisis communicators. The message we analyze below—as well as a range of other emails leading up to and following this event—makes assumptions about students, their knowledge and understanding of crisis protocols, and the types of risks that they bring to campus. In this way, crisis communicators reveal an implicit construction of the student body as monolithic, or as all the same, which results in messages that are tailored to keeping certain types of students safe and not others. This relates to the idea of *embodiment*—the features of our bodies and identities that impact how we understand texts and systems. In reality, though, each student's experience of the coronavirus pandemic is distinct and reflective of their embodiment, and to better understand the needs of audiences, university communicators should spend time interrogating this and adapting their messages accordingly.

Because crisis messages are technical documents, some communicators (and even recipients) may not consider the impact that these assumptions about embodiment can make on how the intended message is or is not received by audiences. Without broader optimization, crisis communicators normalize their audiences, meaning that they maintain a limited view of recipients that shapes their messages. The majority of ISU's campus community is domestic (from Illinois), white, middle-class, able-bodied, cissexual, and heterosexual. These identity features tend to imbue students with the standards of academic currency and prestige, in part because they often have parents, grandparents, and other family members who have attended college before them and can help them navigate institutional structures (For more on this, see Fearn-Banks, 2017; Frost, 2018; Grabill & Simmons, 1998; Heath & Miller, 2004; Sauer, 2002).

If university crisis communicators only write with this normative group in mind, they will make assumptions about access to institutional knowledge, context, and policy that many may not have. Further, they will make assumptions about students' lived realities and how it will impact their ability to comprehend and act upon crisis communications; for instance, a student with a chronic health condition experiences the unique risks of a pandemic much differently than one without, and emails that assume all students are in perfect health will exclude these students who may be living with higher levels of anxiety and/or may be taking different steps to protect their health and safety. When institutional communication—including but not limited to crisis communication—fails to accommodate the full range of its students, they sustain a dangerous power inequity that does not uphold their mission statements to accommodate diversity and keep students safe (Powell, 2004). College and university campuses are increasingly attempting to diversify, but also often fall short of supporting these students when they arrive because the systemic structures do not have space for them (Figure 2.1).

FIGURE 2.1 Email from Illinois State University: Coronavirus Weekly Update.

This is a social justice issue: As technical communicators, and particularly as crisis communicators, a main goal should be identifying and changing institutional contexts that perpetuate inequalities (Colton & Holmes, 2018). In particular, with a social justice approach, communicators are equipped to not only identify and change these inequities, but also to rectify them, via the four R's: recognize, reveal, reject, and replace injustices (Walton et al., 2019). Kristin Moore also highlights how communicators are the agents of this change because they can "become citizen advocates, seeking to create more equitable and just decisions" (2017, p. 239). As such, we recognize that crisis communicators have the power to shape narratives and actions on university campuses, and we offer several suggestions below for how they can do so equitably for all students and their diverse bodies and experiences.

Personal Context

As of this writing, Courtney is a graduate student at ISU and Erika is a pre-tenure faculty member. Both received the weekly messages from distinct positionalities and perspectives that are shaped, in part, by their embodied identities. Like many other students and faculty, both felt a lot of uncertainty and anxiety about evolving COVID policies. Navigating the linked resources, keeping up with the changing expectations, and making sense of the often-convoluted email updates were difficult work complicated by the precarity and unprecedented nature of a pandemic.

Because she lives off campus, Courtney was not directly exposed to the same conditions of many other students, but she knew that some necessary activities such as visiting her campus mailbox and picking up books from the university library invited a certain level of risk. Yet, because the pandemic restrictions began at the end of her third year as a doctoral student, Courtney also had a fair amount of institutional knowledge: she can locate the buildings on campus necessary for testing, is fairly certain of who she can contact if she has questions, and also has trusted mentors who she can ask for help. Beyond this, Courtney occupied additional intersections of privilege as a straight, white, cis-gender woman: she has a steady stipend from the university, social support, and is able-bodied despite some elevated health risks if she were to contract COVID. However, like all students on campus, Courtney is balancing more than just her university identity, and her embodiment extends beyond campus, as well. For instance, Courtney works part-time at a local public library alongside her teaching assistantship at ISU. This work put her at higher risk for contracting COVID, as well as increased anxieties that come with a public-facing job during the pandemic. With elderly and high-risk family members, as well as concerns for older coworkers, like other students, staff, and faculty on campus, Courtney's experience of the pandemic is both embodied and embedded, reflective of who she is and varying factors that individualize risk.

Similarly, Erika's experience of the pandemic is both embodied and embedded. As a queer nonbinary white professor with a tenure-track job, she experiences some level of privilege in certain spaces; while she is able-bodied, she experiences anxiety and depression and also has asthma, which would put her at risk of a more severe case of COVID-19 should she contract it. As a faculty member, Erika not only lives off campus, but in an entirely different town from ISU, so her exposure to COVID conditions differed greatly from others who lived in the Bloomington-Normal area. As such, her embodied experience of risk also differed from many of the recipients of the weekly emails in that she was straddling two communities: the campus and the town where she lived. As a campus community member, Erika engaged in regular Zoom meetings during the latter half of spring 2020 with her fellow faculty, both in the department and across the university, and sought to provide input on how ISU should provide instruction in the upcoming fall 2020 semester; these suggestions were largely ignored, much to the frustration of many across campus who did not feel the university had their health and safety in mind. As a community member of her local town, she largely isolated herself from friends and family to protect both herself and others, although she set up a rigorous schedule of virtual hangouts. Finally, as someone with three degrees and nearly three years as a university professor under her belt, she was well-prepared with the institutional knowledge to understand how to approach university crisis communications and often helped her students parse them out and understand what they should do or what was expected of them.

The authors of this chapter are only two people out of thousands on ISU's campus, each approaching the pandemic from unique embodied experiences. Like all people on campus, our risks, knowledge, and experiences are always centered in how we understand the information the university communicates through crisis emails. Campus communities are not monoliths, and due to unique concerns and risks, each recipient will bring different questions and distinct needs when reading each university email. Yet, by sending crisis messages to the entirety of the campus community, communicators must make assumptions about these questions and needs, in effect approaching the campus audience like a monolith. This information is important: a matter of not only institutional function, but also widespread health and safety. For this reason, interrogating the assumptions and considering what perspectives are marginalized in these messages are not only imperative to promote campus equity, but they also offer important considerations for improving technical and crisis communication processes in other contexts.

In the remainder of this chapter, we will examine one of these email messages and introduce possible solutions for how these messages could better account for the diverse realities of students, including their unique experiences and bodies. In the case example that follows, university crisis communications often construe their student bodies as disembodied monoliths. We argue that the people who strategize and write these communications need to recognize that their student body has a wide range of people with different races, ethnicities, ages, genders, sexualities, physical and mental abilities, religions, and income levels, and that each of these identity characteristics may require different approaches to responding to crises. In particular, through considering a large-scale crisis event such as the COVID 19 pandemic, which placed extreme pressures on university communication, technical communication students, scholars, and practitioners can learn strategies to better mitigate day-to-day communications.

Analyzing a Case Example

Although weekly coronavirus messages were sent through the 2020–2021 academic year, following the NELK Boys' visit to campus, campus communicators addressed a particularly tense and complicated rhetorical moment for ISU. After the viral event with the NELK Boys, campus communicators went on damage control, ensuring that events like this would not continue while also enacting further protocols to keep the campus community safe. As a result, this message served not only to update students on the ever-evolving policies on campus, but also to reprimand the actions and address the damage that occurred following the documented off-campus parties that took place on September 8. Despite the fact that only a small fraction of students were involved in the NELK event and, presumably, many of the campus members practiced social distancing and acted in good faith, our below case example expresses wide-ranging blame, as well as presenting some of the important crisis-related messages in a convoluted and poorly optimized manner. In this way, university communicators seem to address recipients as a monolithic group, failing to acknowledge many of the complicated embodied experiences of the thousands of students, faculty, and staff who received this email.

Below, we will complete a section-by-section rhetorical analysis of the "Weekly Update" from September 10. A *rhetorical analysis* looks closely at a text to determine the rhetorical choices authors made to address their audience, often determining how successful they were (or not) at meeting their needs. It is a way of better understanding the actions and implications of crisis texts in the world, as well as a method for ensuring that future texts can respond to and represent audiences more completely in future crisis events. Because we are also both critical feminists who are dedicated to examining the ways in which rhetorical power is exercised in communication, our analysis also pays particular attention to the ways in which the email failed to account for the audience's diverse embodiments and identities. We also suggest some strategies and considerations for future crisis communications.

Weekly update, September 10: Disregard of Public Health Guidance, Town of Normal Ordinance Extension, Registration and

Withdraw Dates, Testing, Test Results, Student Absence Process, On-Campus Quarantine and Isolation, Housing and Meal Plan Contracts

This week's update includes information about consequences of not following public health guidance, the Town of Normal ordinance extension, testing, test results, the student absence process, clarification regarding on-campus quarantine and isolation requirements, and housing and meal plan contract cancellation.

Analysis

As we mentioned earlier, university messages like this one were sent weekly to Illinois State students. By the time this message was sent on September 10, the presence of emails like this one was commonplace in campus inboxes, and yet, many conventions dissuade student reading and preclude content clarity. From the start, the lengthy title fails to account for audience cognitive load and may overwhelm readers who are embedded in an already stressful situation of extreme information overload.

Suggestion 1: Revise Title

Although it is important that readers have a sense of the information that follows, redundancy may encourage readers to skim the email, or to assume that there is no new content in this message. A more optimized title, perhaps only including "Coronavirus Weekly Update: September 10" may encourage more thorough reading, especially since the subheadings of this message provide the thematic context in a clear manner in the email body.

Suggestion 2: Add Internal Hyperlinks to Subsections

Should communicators choose to frontload the email content like this, they could use hyperlinks to specific subsections to help enhance readability and usability; readability refers to the ease with which readers can understand and process information, and usability refers to the ways in which they can learn and use the information for future actions. For instance, if a student needs information about absence policies or quarantine dorms following a recent positive test, they could follow a hyperlink directly to those sections in the email and get access quickly to the information they need.

Disregard of COVID-19 Public Health Guidance

Student participation earlier this week in large off-campus gatherings instigated by YouTube pranksters has the potential to significantly compromise the University's efforts to mitigate the impacts of COVID-19 on the Illinois State University campus and within the community. President Larry Dietz said, "I am extremely disappointed in the actions of some of our students who gathered on Tuesday evening at various off-campus locations."

The University is working with the Town of Normal, the Normal Police Department, and the Illinois State University Police Department to investigate these events. Students will be charged with violations of the Student Code of Conduct and are subject to Town of Normal ordinance violations for their participation. Violations of the Student Code of Conduct can result in suspension from the University. "Most students are following health and safety guidance, and for that I am proud. Unfortunately, the actions of some will undoubtedly contribute to a significant rise in coronavirus cases in our community," Dietz said.

"Although you may not experience serious symptoms of this disease, your behavior puts others at risk," said Dietz. Anyone who participates in a large gathering is more likely to come into contact with someone who is COVID-19 positive. It is everyone's responsibility to protect those in our community who are most vulnerable to this disease.

Analysis

Given the publicity following the NELK Boys' visit to ISU's campus, it is not surprising that the response to this event is the first section. However, rather than foregrounding resources that students need, this email leads with blame and shame, which may instead serve to discourage and disenfranchise students. There is an explicit threat here, an acknowledgment of ongoing investigation and extreme disappointment. From the start, the phrase "student participation" serves to generalize the events, and although university President Dietz acknowledges that most students are doing the right thing, this does not come until the second paragraph. There is no acknowledgment that attending parties like these is typical college student behavior and that ISU bears some of the blame because they brought students back to campus for the Fall semester instead of remaining fully online.

The direct quotations set an editorialized tone that creates narrative distance. This section reads like a newspaper article, which creates further distance between communicators and recipients. Also, using direct quotes from President Dietz serves to localize his sentiments rather than making them official stances of ISU, which has the potential to impact their rhetorical power overall. While we do not necessarily agree with the stance that Dietz has taken here, it seems that when the entire university stands behind the campus's health and safety, recipients may be more likely to listen.

Finally, this email was sent to all students, faculty, and staff. While many students were involved in this party, faculty and staff were not, and so this section of the email feels misplaced considering the majority of the college campus is not its intended audience. This could discourage faculty and staff from reading further, even though the rest of the email contains important information to campus and course policies.

Suggestion 1: Decenter Blame

Crisis communicators could have crafted a more embodied response here through acknowledging the severity of the situation without centering blame as a foundational takeaway for students. Perhaps empirical evidence links to other university statements on this subject, and reminders of the health and safety guidelines would make this a proactive rather than reactive crisis message. In addition, ISU should shoulder some of the blame for deciding to invite thousands of 20-somethings back to campus. By focusing on a few bad actors, they absolve themselves of responsibility in ways that many recipients found to be frustrating and unproductive.

Suggestion 2: Recenter Those with Heightened Risk

For students with heightened risk, especially students with disabilities or chronic illness, these off-campus gatherings made attending ISU even riskier. Simultaneously, messaging like this centers the irresponsible students more than the at-risk students, enacting wholesale blame for all recipients. By placing higher focus on the students who are obeying the guidelines and doing what they should be, ISU's crisis messaging could lead with positive examples rather than alienate those who were not involved and are now at a higher risk of contracting COVID-19 because of the actions of a few.

Suggestion 3: Appeal to a Sense of Community

The tone in this opening is sterile and distances the communicators and audience from each other. Most campus community members are not positioned as active agents in the campus fight against COVID-19. Because email is already an overly formal medium, more direct appeals to a sense of community and student experience may also be needed to bridge the disconnect.

Suggestion 4: Send This Section in a Separate Email

Typically, the weekly coronavirus updates were reserved for information on things like testing, quarantining, and other relevant information for all campus community members. Some information might be specialized to particular subsets of the campus community, but for the most part, it was useful to everyone. As such, it seems like communicators should have sent two emails separate from the rest of the information: one for students, but with the suggestions we made incorporated, and the other for faculty and staff to inform them of what happened and how the university is responding.

Town of Normal Ordinance Extension

The Town of Normal extended two emergency orders limiting gatherings in and around Illinois State University to 10 people or fewer and requires customers of establishments with liquor licenses to be seated to be served. The orders are extended through December 31. Fines for those found in violation can be up to $750.

If the University is made aware of a student violating a Town of Normal ordinance, the student can also face repercussions for violating the Student Code of Conduct.

The rules state that any member of the Normal Police, Fire, or Inspections Department, any member of the Illinois State University Police Department, or any agent of the McLean County Health Department can enforce these emergency orders.

Analysis

Following the response to the campus partying, the blame and shame extend into the next subsection, albeit more subtly through explaining the extension of town ordinances. While it does not explicitly say that these extensions are the result of the NELK parties, the implication is that the town of Normal has decided to keep restrictions in place to discourage students from gathering. In addition, there is not much clarification on some of the repercussions.

Suggestion 1: Clarify the Link between Events

Crisis communicators should clarify if the town ordinance extensions are connected to the NELK parties. If not, they might consider putting this information in a separate email along with the sections below. Although overwhelming students with messages and information is a concern, the NELK Boys' response and ongoing public safety conventions are related but separate issues. Although the weekly emails were instituted to limit information overload, length and divergent emails provide similar risks for students' comprehension.

Suggestion 2: Add External Hyperlinks

To provide more context here, communicators could provide external hyperlinks to relevant websites for the town of Normal, periodicals (such as the *Pantagraph*, the local newspaper), and/or ISU for more contextualizing information.

For instance, this section construes students as both students and community members, a positionality that requires them to think about citizenship beyond their campus, so links to the resources could help them do that.

Testing

As stated by the CDC, individuals who have tested positive and re-covered may continue to have low levels of the virus in their bodies for up to 3 months after diagnosis. Therefore, if a person who has recov-ered from COVID-19 is retested within 3 months of initial infection, they may continue to have a positive test result, *even though they are not spreading COVID-19.* However, anyone experiencing COVID-19 symptoms should contact Student Health Services or their primary healthcare provider. Research is ongoing regarding the possibility of re-infection.

Anyone testing positive for coronavirus cannot "test out of isolation." The 10-day isolation period must be completed even for individuals who are asymptomatic. In addition, individuals cannot "test out of quaran-tine." The 14-day quarantine period must be completed even if a nega-tive test result is received.

Asymptomatic testing for students continues at two on-campus test-ing sites, currently located in the Brown Ballroom in the Bone Student Center and the former fire station at [address redacted], from 8 a.m.–4:30 p.m., Monday through Friday.

Students experiencing symptoms can call Student Health Services at [phone number redacted] for an initial screening and to make an ap-pointment to be tested. Students wishing for an antibody test can also call Student Health Services and make an appointment.

Testing is also available for students, faculty, and staff at the Interstate Center testing site. Students testing at this site should share positive test results with Student Health Services by uploading them to the *Secure Patient Health Portal.*

Analysis

Following the relatively short section on university and town responses, the in-formation on testing is much longer: consisting of five paragraphs and the first hyperlinks of the email. Testing procedures on campus are quite nuanced, and combined with the information on CDC policies, make for potentially over-whelming context. The paragraphs themselves are not organized in a manner that foregrounds the most important information first. Recipients might read this section a couple of different ways. Some might read it to stay informed on policy updates, but others will want to find information about how to get tested, and especially if they are actively experiencing COVID-19 symptoms and/or the anxiety and fear of uncertainty after potential exposure, they may not have optimal capacity for reading and processing information. Testing information is located throughout the final three paragraphs, but without any textual indicators (bold or italicized words, for instance) to find it quickly. In addition, this section contains a lot of contact information, but few hyperlinks to aid recipients in following up. Although some of this content is redundant from prior weekly emails, it should not be assumed that this information is intuitive for students, or that the panic and uncertainty surrounding the pandemic makes this situation less complicated.

Suggestion 1: Foreground How to Get Tested

This section should be able to cater to all audiences, well and unwell. Crisis communicators should foreground *how* the campus community can procure a test if they need one. Information about what the testing means can come afterward.

Suggestion 2: Add Textual Indicators

To optimize information delivery, crisis communicators could make information about how to get tested stand out more. There are many options for this. They could bold or italicize words and phrases like "symptomatic testing" or "off-campus testing" to help recipients skim the text for key information. Or they could add subheadings to indicate information about the different types and locations of testing, breaking up this information and making it easier to locate. This way, if a recipient were experiencing symptoms, they could easily identify the subheading and take the proper recourse, rather than skim an unorganized list of paragraphs.

Suggestion 3: Add External Hyperlinks

Currently, various points of contact and links are strewn across paragraphs, making it difficult for readers to find and use this information quickly and easily. Many recipients could be checking their email on smartphones where they could click a hyperlinked phone number and immediately call or click on an address to be directed to another app for directions. This way, for instance, if students need information about absence policies or quarantine dorms following a recent positive test, they would be able to locate this on their phone or computer first, rather than wading through the rest of the email.

Test Results

The University is currently reporting test results, not cases. It is worth noting that some populations are required to test regularly such as students participating in clinical experiences and student-athletes. If an individual has tested multiple times, all results are included in the total results number.

The University has engaged Reditus Labs to conduct all on-campus testing.

Analysis

Although test results are referenced in the previous subsection, this individual "Test Results" section stands alone, referring to the data reports the university aggregates by providing some more context on the statistics. This section is short, and, given the heading, which could be seen to refer to individual test results, misleading. In addition, the final sentence, which acknowledges the on-campus lab, seems disconnected from the rest of the section.

Suggestion 1: Rename the Section

The heading could be renamed something like "Test Result Reporting" since the information presented here is about how the results are reported. This heading would be more accurate for readers skimming for information quickly.

Suggestion 2: Clarify this Information

While some who paid close attention to test results and case reporting on and beyond campus would likely understand the difference between "tests" and "cases," most casual readers may not. In this instance, a test refers to a single test result. The example of the student-athletes begins to clarify what this means, but does not quite get there: basically, if a student takes two tests and receives a positive result both times, ISU counts that as two positive COVID tests, not as a single COVID case. This has a big impact on test result interpretation.

Suggestion 3: Move this Information to Other Communication Locations

It is not clear why this final sentence of information is included here. Perhaps some people on campus and in the local community were incorrectly interpreting the test results. But it seems that this information would be best placed on the university COVID dashboard, which is what this subsection is referring to without naming it. In an already long and detailed email, this information might be better suited elsewhere.

Student Absence Process

Students who are tested on campus at Student Health Services or either on-campus testing site will have their test results shared with Student

Health Services. Students tested at any other location should share positive test results with Student Health Services by uploading them to the *Secure Patient Health Portal.*

For those who need to quarantine or self-isolate, Student Health Services provides documentation, including an "Isolation/Quarantine Letter," to students through the Secure Patient Health Portal. This letter should be shared with the Dean of Students Office at [*email redacted*] in order to have instructors notified of the duration of the excused absence. Students are copied on the notification to instructors, which is sent by email. Similarly, upon release from quarantine or isolation, students may download a "Release Letter" from their Secure Patient Health Portal and email it to the Dean of Students Office. This prompts a return to class notice to instructors.

Questions about the process of notifying instructors about a student's need to quarantine or isolate should be directed to the Dean on Duty in the Dean of Students Office at [phone number redacted].

Analysis

By this point in the email, readers have waded through so much information—sometimes connected, sometimes not—on COVID-19 policy and procedure at ISU, after being scolded for an event that they may or may not have been part of. The average reader by this point is tired and will begin losing focus. As a result, this section is too wordy, especially for someone who may be ill and returning to the email for information about next steps after a positive test.

Suggestion 1: Cut Back on Wordiness

Crisis communicators should keep information like this concise so readers can find relevant information quickly.

Suggestion 2: Add Numbered Steps

This section of the email is entirely action oriented. A numbered step-by-step process would help readers navigate it.

On-Campus Quarantine and Isolation

University Housing Services is notified by Student Health Services when a student who is living in on-campus housing must isolate or quarantine. Once notified, Housing staff will call the student to assess their needs and to discuss their housing options.

"University Housing Services provides individualized attention to each student and many times their family members as they discuss their housing options for quarantine or isolation," said Director of University

Housing Services Stacey Mwilambwe. "Housing staff listen closely for concerns regarding safety, lack of resources, or other extenuating circumstances." Questions regarding medical concerns are referred to Student Health Services.

When possible, students are asked to return home to quarantine or isolate in order to accommodate students who cannot. The CDC has provided information about *safely quarantining at home with family members*. There are a variety of factors considered in determining why a student may not be able to return home such as food or housing insecurity. Students are provided with two hot meals and continental breakfast each day. Upon arrival to their assigned room, students will be provided a treat bag with a variety of refreshments and hearty snacks along with contact information for necessary dietary accommodations. "Our team, comprised of staff at all levels of the University, work long days and nights, including weekends," said Mwilambwe. "They work tirelessly to provide support and resources."

It is extremely important for the health and safety of our residential environments and the campus community that quarantine and isolation procedures are followed on- and off-campus. Students in on-campus housing who violate the terms of the isolation or quarantine agreement are at risk of housing contract cancellation and disciplinary sanctions.

Analysis

By this point in the analysis, you should be noticing a pattern in this email: wordy sections with editorializing and quotations from university officials (this time from the Director of University Housing Services Stacey Mwilambwe). Overall, this stylistic choice has made this section dense and difficult to unpack.

Suggestion 1: Cut Down Wordiness and Remove Sterile Editorializing

As with other sections, crisis communicators could streamline information here by removing editorialized quotations, which would cut back on wordiness. Stick to the relevant and necessary information at hand, keeping in mind the multiple audiences and their needs when reading and/or revisiting this section.

Housing and Meal Plan Contracts

Students can cancel their housing and meal plan contracts at any time during the fall term without financial penalty by visiting the *Housing and Dining portal*.

Questions regarding contract cancellation, can be directed to Housing at [phone number redacted] or Dining at [phone number redacted] from 8 a.m.–4:30 p.m., Monday–Friday.

Analysis

After a long email with several lengthy sections and information that was at times difficult to find and/or follow, this may be the most usable section in the email. Information is brief but comprehensive and provides an external link for readers to click and follow up as needed.

Practical Solutions

At no point does this email acknowledge the differential experiences of students, faculty, and staff, some of whom are likely concerned and scared now that they know students have gathered in large groups and have greatly increased the risk of infection on campus. This email makes three assumptions:

1   That everyone has the same goals each time they read this email. That is, this email is formulated to be an informational reference point with the assumption that the person reading it does not yet need this information. It does not account for people who might need specific information and so would want to skim to find what they are looking for, such as how to report their test results or how to begin their quarantine after testing positive.

2   That everyone is healthy and able to understand and process this information in the same way. An email like this could put someone who is already under additional stress and anxiety from a pandemic under more. It is not a comforting email and does not even acknowledge that some recipients are likely in need of mental health aid. In addition, recipients who are physically ill—with COVID-19 or even a head cold that they may for a few days believe to be COVID-19—may have more difficulty parsing this information. Fever, nausea, and headaches, among the general uncertainty and disorientation that comes with experiencing a novel virus, can impair a person's capacity for reading and comprehension.

3   That everyone has equal institutional knowledge to navigate not only the email but also the university website. This email assumes that all students, including those who have only been attending ISU for less than a month, will be able to understand and navigate all of the institutional offices and locations mentioned in the email. For instance, if a student is unfamiliar with how to contact the Dean of Students with a positive test result that they procured off campus, then they are unable to procure an excused absence and/or begin their quarantine process. We mention this specific example because Erika experienced this exact situation in her fall 2020 course, so the student continued to do coursework while ill and stayed in their dorm with their roommates, inadvertently spreading the infection to them as well.

Every recipient on campus will approach this information from a differently embodied standpoint or positionality that will impact how they are able to process this information, and this email does not account for that. It assumes everyone is privileged enough to feel safe, be healthy, and understand institutional contexts.

As such, in addition to the specific localized suggestions we made during the analysis above, we have a few general suggestions for this email as a whole, as well as any future crisis communications.

1   Use less text overall with more signposting to help recipients navigate the information. In many cases above, this could mean breaking information down into numbered steps, or creating subheadings within the main sections to enable skimming.
2   Provide more hyperlinks to relevant web pages that have more follow-up information. The fact that each university body mentioned in this email does not have their own regularly updated COVID-19 page is a major concern. Emails should be for providing the most crucial information, and the webpages should be for providing the details.

While these suggestions are helpful, they are likely not the only ways to optimize university crisis communications. As such, we also suggest that communicators perform usability testing with a group of target recipients so they can provide feedback on how to optimize the email. Usability testing is a way to evaluate how well a document performs its intended goals by testing it on a subset of the actual audience. In the case of emails like this, crisis communicators should show drafts of their emails to demographic representatives of their audiences so they can ensure that they are providing information in a way that is usable and useful. The communicators at ISU did not do this, and it shows in emails like these that assume one monolithic campus community.

The bottom line is that if crisis communicators are not accounting for the diverse campus community to whom they are writing, then they are not communicating effectively. Some will misunderstand or misinterpret information for various reasons connected to their positionalities, embodiments, and institutional backgrounds, some will choose to ignore it instead of wading through impenetrable walls of text. Some will understand it without issue, but they are not the only people on campus and their health and safety is not more important than everyone else's.

Discussion Questions

1   Consider your own embodied positionality as a campus community member. What are the identity factors that you would list for yourself? What are your needs and expectations for communication during a crisis? Do you feel like those needs have

been met in prior experiences? How could crisis communications meet your needs? Or, if they have met their needs, how do you think your positionality and embodiment have impacted your ability to parse communications?

2 Spend some time researching risk communication policies and resources on your campus website. Is this information easy or difficult to find? What are the implications of this? What kind of assumptions are made about students, their bodies, and safety needs?

3 In small groups, discuss what a task force between students and campus communicators that considers optimizing crisis communication strategies would look like. How should campus communicators solicit a group like this, and what kinds of topics should they discuss to make communications more responsive to student needs?

Assignments

1 Undergraduate Assignment

Building on what you learned about some of the complicating factors related to crisis communication from this chapter, try your own hand at writing as a campus crisis communicator. In this assignment, you'll be asked to revise a crisis communication message that reflects what you've learned about the implications of embodied rhetorical analysis on responding to crisis events.

Begin by studying a campus crisis communication email message—either provided for you by your instructor or another message you've received during your time as a student. As you analyze this message, consider the specifics of crisis communication email messaging from your own institution. Keep in mind the embodied experience of other students and the specific rhetorical situation of your campus community.

Part I: Complete a full revision of a crisis communication email from your own academic institution. As you revise, consider factors such as word choice, audience needs, document design, digital interactivity, and usability. Remember that an effective crisis message must clearly advise on future action, provide resources for further information, and acknowledge the complex and diverse lives of the recipients.

Part II: After you've finished your revision, discuss your revisions in a short response memo. Address your specific changes that you've made and detail *why* this is a more effective rhetorical response than in the original message. Provide specific support addressing the embodied realities of you and your campus community members.

Part III: In the final component of this assignment, develop some specific suggestions that you think campus communicators should implement in messages sent at your campus. Include at least five considerations specific to your campus that you would suggest for campus communicators. This can be a bulleted list but should address specific genre conversations and strategies related to messages like you've analyzed for this project.

2 Graduate Project

Prior to class discussions on crisis communication, audience, and embodiment, graduate students could read relevant texts to ground them. We suggest a few here, but others would certainly also be appropriate per student, professor, and/or course needs:

- Huiling Ding, "Rhetorics of alternative media in an emerging epidemic: SARS, censorship, and extra-institutional risk communication"
- Erin A. Frost, "Apparent feminism and risk communication: Hazard, outrage, environment, and embodiment"
- Jefferey Grabill and Michelle Simmons, "Toward a critical rhetoric of risk communication: Producing citizens and the role of technical communicators"
- Kristen R. Moore, "The technical communicator as participant, facilitator, and designer in public engagement projects"
- Beverly A. Sauer, *The rhetoric of risk: Technical documentation in hazardous environments.*

Unit Project: In groups of three, develop a generalized university crisis communication strategy for multiple media. The audience for this strategy is your university communications unit (the people who send campus-wide emails, not the department). In a 20-page report, advise them on how to communicate with students, providing some guidelines that are applicable to multiple crisis situations but that could be easily adapted for more specific contexts.

This project will be research-heavy. You will need to understand who is part of your campus community and research the best communication strategies for reaching everyone. For instance, what considerations must be made for students with physical disabilities during a crisis? Or what resources need to be made available for first-generation students during a crisis? You will need to consult with focus groups and perform usability testing on sample communications to ensure maximum optimization. You will need to research various media and platforms to determine what information should be shared where and when.

Your final report should include your research and outline a concrete plan for how to communicate a crisis to your entire campus community.

References

Chen, T. (2020, September 11). *Youtube is demonetizing popular frat channel NELK boys for encouraging people to ignore COVID safety.* Buzzfeed News. Retrieved 5 July 2021 from https://www.buzzfeednews.com/article/tanyachen/illinois-state-university-investigating-nelk-boys-parties.

Colton, J. S., & Holmes, S. (2018). A social justice theory of active equality for technical communication. *Journal of Technical Writing and Communication, 48*(1), 4–30.

Coronavirus in the U.S.: Latest map and case count (2021). *New York Times*. Retrieved 11 February 2021 from https://www.nytimes.com/interactive/2020/us/coronavirus-us-cases.html.

COVID-19 campus case tracker (2021). *Illinois State University*. Retrieved 11 February 2021 from https://coronavirus.illinoisstate.edu/dashboard.

Ding, H. (2009). Rhetorics of alternative media in an emerging epidemic: SARS, censorship, and extra-institutional risk communication. *Technical Communication Quarterly, 18*(4), 327–350.

Fearn-Banks, K. (2017). *Crisis communications: A casebook approach* (5th ed.). Routledge.

Frost, E. A. (2016). Apparent feminism as a methodology for technical communication and rhetoric. *Journal of Business and Technical Communication, 30*(1), 3–28.

Frost, E. A. (2018). Apparent feminism and risk communication: Hazard, outrage, environment, and embodiment. In A. M. Haas & M. F. Eble (Eds.), *Key theoretical frameworks: Teaching technical communication in the twenty-first century* (pp. 23–45). Utah State University Press.

Grabill, J. T., & Simmons, M. W. (1998). Toward a critical rhetoric of risk communication: Producing citizens and the role of technical communicators. *Technical Communication Quarterly, 7*(4), 415–441.

Haynie, J. M. (2020). Last night's selfish and reckless behavior. *Syracuse University News.* Retrieved 5 July 2021 from https://news.syr.edu/blog/2020/08/20/last-nights-selfish-and-reckless-behavior.

Heath, R. L., & Miller, D. P. (2004). *Responding to crisis: A rhetorical approach to crisis communication.* Lawrence Erlbaum Associates Publishers.

Henry, S. (2020, October 9). Nearly 100 Illinois State University students face discipline in NELK Boys visit; police issue no fines. *Pantagraph.* Retrieved 5 July 2021 from https://www.pantagraph.com/news/local/education/nearly-100-illinois-state-university-students-face-discipline-in-nelk-boys-visit-police-issue-no/article_9b2f6fbc-8b22-575f-8833-1e16dc3fcd51.html.

Moore, K. R. (2017). The technical communicator as participant, facilitator, and designer in public engagement projects. *Journal for the Society of Technical Communication, 64*(3), 237–253.

NELK (2010). NELK Filmz. *YouTube.* Retrieved 17 June 2021 from https://www.youtube.com/user/NelkFilmz/about.

Pantelides, K., Muller, D. N., & Green, G. (2016). Eight years a "wooden opponent": Genre change (and its lack) in campus timely warnings. *Present Tense, 5*(3). Retrieved from https://www.presenttensejournal.org/volume-5/eight-years-a-wooden-opponent-genre-change-and-its-lack-in-campus-timely-warnings/.

Planning, Research, and Policy Analysis (2019). *Fall 2019 enrollment report.* Illinois State University. Retrieved 11 February 2021 from https://prpa.illinoisstate.edu/downloads/student/enrollment/2019/Fall2019-Enrollment_Report.pdf.

Sauer, B. (2002). *The rhetoric of risk: Technical documentation in hazardous environments.* Erlbaum.

Schladebeck, J. (2020, September 10). Illinois State University students could face legal consequences for attending massive party thrown by YouTube stars The NELK Boys. *New York Daily Mail.* Retrieved 5 July 2021 from https://www.nydailynews.com/coronavirus/ny-coronavirus-illinois-state-university-nelk-boys-party-youtube-20200910-yfsuunqlmfhe7ai2x4tqvhyyyy-story.html.

Swiech, P., Petty, A., & Henry, S. (2020, September 10). Watch now: NELK boys parties could bring consequences for Illinois State University students. *Pantagraph.* https://www.pantagraph.com/news/local/watch-now-nelk-boys-parties-could-bring-consequences-for-illinois-state-university-students/article_f48ac98d-43b9-56f7-b9c0-5556e8e1a5c6.html.

Walton, R., Moore, K. R., & Jones, N. N. (2019). *Technical communication after the social justice turn: Building coalitions for action.* Routledge.

3

EMBODIED RISK COMMUNICATION IN THE COVID-19 PANDEMIC ENVIRONMENT

Bolanle Olaniran and Joseph Williams

Problem: Risk and crisis communication and management often go hand in hand; however, underlying any crisis or risk communication is the contextual factor or circumstances surrounding it. For instance, a study looked at how perceived inequality shifts the focus of COVID-19 from just a health crisis to a crisis for feminism (Yarrow & Pagan, 2020). How much control do stakeholders have? Have relationship complexities been aptly addressed? An embodied framework, in tandem with the Anticipatory Model of Crisis Management (AMCM), amplifies the multi-factor variables that have direct and indirect implications for how stakeholders relate and perceive risk (Olaniran, 2007). While case study analysis in crisis appears to address the contextual factor, it can fail to consider the complex nature of crisis or ensuing risk communication that undermines effective management.

Solution: To address this issue, we argue for the incorporation of embodied discourse as a conceptual framework for crisis and risk communication; additionally, we argue that this framework works effectively with AMCM.

During crisis, an effective communication mechanism is paramount in providing accurate and timely information to the public. However, how the public and affected stakeholders interact with this information is seldom known; ultimately, readers interpret this information to support their own viewpoint, accurate or inaccurate. Moreover, as people seek and access information through their technological devices especially during the current COVID-19 pandemic, there is an increased pressure, politically and socially, on government agencies to accurately communicate risk messages and survival techniques to audiences. This situation brings to the forefront of crisis management the need to understand users' interaction with provided information or data while understanding factors that inhibit the effectiveness of these messages. With new and social

DOI: 10.4324/9781003266549-4

media, individuals are connected to different sources for information. Discerning which information is appropriate and legitimate often takes a backstage to real evidence. Dr. Anthony Fauci, Director of Infectious Disease at the National Institute of Health, stresses this discrepancy when he claims that even a certain traditional media outlet (Fox News) offers outlandish comments that are not supported by facts, the effects of which are devastating in a pandemic such as COVID-19.

This chapter will explore contextual factors that foster or hinder effective risk and crisis communication while bolstering greater compliance using the AMCM alongside an embodied framework. Furthermore, we plan to examine how communication technologies, in particular, artificial intelligence (AI) can be deployed to help in crisis by generating embodied conversation (Goh et al., 2006) and in considering vulnerable populations. Finally, we will address implications for technical communication and crisis practitioners.

Overview of Anticipatory Model of Crisis Management

The AMCM (Olaniran & Williams, 2001; Scholl et al., 2005) introduces strategies in pre-crisis, crisis, and post-crisis to facilitate stakeholder control and buy-in; ultimately, this control empowers those affected by the crisis to take action as needed. Additionally, the AMCM aims to minimize as much harm and damage as humanly possible. At the core of the *embodiment* is the idea that crisis stakeholders feel connected to crisis and risk communication in a safe way that helps them express their needs, desires, fears, and wants as they see fit. A component of this model is *enactment*, decisions or actions that represent a process where a given action is brought about (Smircich & Stubbart, 1985). Enactment is crucial due to the fact that failure to install a crisis plan may invariably hinder management of the crisis (Olaniran, 2007). Alongside enactment is the idea of stakeholder expectations. The control component intersects with expectation and enactment to the extent that expectations influence enactments, and actions exert some degree of control over the crisis (Adler & Bartholomew, 1992; Konsynski & McFarlan, 1990; Olaniran & Williams, 2001).

The AMCM is a pragmatic tool for handling the task of crisis readiness or management, terrorism, and disaster prevention and health risk or communication and management (Olaniran & Scholl, 2020; Olaniran & Williams, 2001, 2004). Before exploring the model, we need to understand what a crisis is. In keeping with the key principles of the AMCM, a *crisis* has the following properties: (a) results in high-stakes consequences; (b) has a low probability of occurring; (c) consists of high risk and uncertainty; (d) operates under a time pressure; (e) interferes with normal business; and (f) potentially affects organizational reputations (Gregory, 2005; Pearson & Clair, 1998). The anticipatory model draws from the "enactment perspective" (e.g., Weick, 1988), which focuses on the prevention of possible crises, including disasters, to reduce their magnitude when they happen. The anticipatory model focuses on planning and prevention as primary components in the process of crisis, health risks, and other disaster management.

The notion of vigilance is key to planning and prevention (e.g., technologies). Therefore, AMCM suggests that crisis teams and practitioners pay attention to technologies that are considered failsafe (Olaniran & Scholl, 2014). When applying the model, it is important to understand the concept of first- and second-order effects of technology. The first-order effect addresses the notion of technology being used for the reason that it is originally designed (e.g., using email to send a message to people in a short amount of time). On the other hand, the second-order effect of technology addresses the adaptive and often negative use of technology (e.g., using email to infect computer networks with malicious software, hacking). Crisis emanates from either of the two effects (Olaniran & Scholl, 2016). Lerbinger (2012) stresses this problem when he points to the complexity of technologies, which makes their effect in crisis management difficult to predict.

Furthermore, technologies by nature are fallible because they are designed by humans and are subject to both human error and deliberate attempts to cause disaster. Evidence indicates that a majority of industrial disasters tend to be linked to a combination of human and technological errors (Jaques, 2006; Lukaszewski, 2013; Olaniran & Scholl, 2014; Shrivastava, 1987; Shrivastava & Mitroff, 1987; Shrivastava et al., 1988; Weick, 1988). In essence, technologies and other issues considered as failsafe may not be failsafe as originally thought. Those technologies would benefit from guidance that would provide the best recommendations for using and monitoring them. Technological failures lead to larger crisis events. For example, companies can provide employees with periodic training and reminders about cybersecurity and ways to safeguard against cyberattacks (e.g., phishing) and downloading malicious software. More attention should be devoted to understanding technologies and anticipating other factors that hold crisis potential or have negative consequences in an attempt to prevent crisis occurrence and escalation.

The Role of Technologies in Crisis and Health Risk Communication

The AMCM's framework can help practitioners and decision makers identify areas of vulnerability (e.g., susceptibility to natural disasters, organizational misdeeds) along with practices to manage them. For example, a company in California that handles hazardous waste cleanups could determine if it needs to prepare for earthquakes, which could damage waste containers and cause hazardous spills. That same organization might discover dangerous safety shortcuts that cause environmental threats to the surrounding community. Both of these crisis types (natural disaster and organizational misdeed) are plausible and must be anticipated in terms of both management actions and crisis communication strategies. However, the same preparations and communication strategies would not be effective for both crisis types. Rather, crisis management is an ongoing process of continuous monitoring, assessing, and reassessing crisis triggers or risks. A practitioner then needs to look into the community, organizational, industrial, and cultural history to best prepare for crises. This "detective work" negates a one-size-fits-all strategy that assumes that a single crisis management strategy will work in any community. On the contrary, a community-specific approach like a Crisis Communication Center (CCC) considers what organizational, environmental, and cultural factors are unique to a community that would prompt crises.

The AMCM was initially developed to aid public relations practitioners and researchers who must deal specifically with the threat of technological crises (Olaniran & Williams, 2001). To that end, the CCC is based on the AMCM principle that effective crisis management is prevention-oriented. The AMCM consists of five major components: understanding, environment, enactment, expectation, and control.

Understanding is critical to the CCC because it focuses on analyzing community-specific conditions that could result in crisis. At the same time, actions enabling people to understand potential failures in technology or human error can also help to de-escalate crises (Jaques, 2006; Olaniran & Williams, 2001; Weick, 1988). Consequently, understanding the *environment* in which CCCs serve is essential.

Understanding as a process encompasses the interrelated concepts of enactment, expectation, and control. *Enactment* involves taking some action (or inaction) that brings about one or more consequences (see Smircich & Stubbart, 1985). An enactment might involve a large retail store hiring security guards to patrol the store's parking lot in response to increased crime in the area. Olaniran and Williams (2001) contend that anticipation of the crisis in itself is action because it constrains an organization's choices in response to a particular piece of information. The appropriateness or effectiveness of an action (i.e., enactment) depends greatly on scanning the environment, which is another component of the AMCM. Sethi (1987) argues that one should anticipate nearly all the unthinkable before these factors become realities (see also Lindell & Perry, 2004).

Weick (1988) extends enactment to include the consequences certain actions impose. Enactment constitutes a retrospective sense-making process such that anticipating a crisis (or failing to do so) in and of itself is an action, given that it determines the choices people make based on available information. Communities and organizations alike must be adept in their decision-making processes; they must anticipate opportunities, threats, strengths, and weaknesses in their environment and take appropriate measures to safeguard their interests. Furthermore, individuals' actions (or lack of actions) result in different crisis outcomes. When people act one way versus another, they face different types of problems, opportunities, and constraints (Jaques, 2006; Kilduff, 1996; Lukaszewski, 2013; Weick & Sutcliffe, 2007).

Expectation includes the assumptions people make about certain events or objects. Assumptions about a given situation, environment, technology, or combinations of all can influence whether or not tragedies occur as well as the magnitude of their occurrences. In essence, decision makers' assumptions about technologies represent a critical element and a starting point for preventing disasters and crises in general. Olaniran and Williams (2001) note that assumptions at times bring about self-fulfilling prophecies. For example, the assumption that a particular technology is failsafe could cause a relaxation in safety mechanisms and lead a decision maker to avoid the necessary actions to prevent a crisis altogether. Furthermore, decision makers set in motion actions that are consistent with their assumptions (Olaniran & Williams, 2001, 2004; Weick, 1988). Thus, the ensuing potential crises from such oversight increase when people only take actions that are consistent with their assumptions (see Perrow, 1984; Weick, 1988). While early warning signs and environmental scanning are important in helping to prevent a crisis, they may have no value when management ignores, denies, or tries to suppress them (Jaques, 2006; Olaniran & Scholl, 2016). Expectation and enactment are not mutually exclusive concepts in the AMCM and consequently the CCC. This is because expectation influences enactment and vice versa. In other words, enactment could set into motion a condition where expectation leads to actions an organization takes.

Control represents the degree of influence that CCC members have at their disposal to manage crises proactively. Control is multifaceted and is presented in the model to address how well institutions and decision makers are in command of the crisis. Control, however, is measured in relative terms, based on the constraints faced by stakeholders and relevant environments (e.g., the public's perception of an organization). Moreover, control is linked with enactment and expectation. Control over certain aspects of a particular crisis event can affect decision makers' expectations of, for example, how much one can counter the risk factors that feed the potential crisis. Similarly, the level of perceived control can inform the enactments decision makers generate about a crisis situation.

Control, therefore, is elusive because it has to do with perception, especially when the influence is indirect in nature, which often is the case. Control influences crises in the sense that it reflects how individuals, leaders, and decision makers at large respond to crisis situations. For example, when individuals see themselves having the ability and power to do something about a crisis, it is more likely that they will take appropriate action. Thus, ability and authority go hand in hand and establish empowerment, which could extend to a vigilant response in crisis management. Crisis planning is not an end in effective crisis management; however, it provides a good starting point to have control over how one prepares for and handles the crisis. This, in turn, brought forth the issue of embodiment in health risk communication and environment.

Embodied Risk Communication and Crisis Management

When exploring the idea of embodiment in social media, emphasis should be on how these technologies make the "self" salient. Previous research on embodied behavior in virtual reality (VR) has explored the impact of social media platforms. Similarly researchers in immersive media need to identify opportunities and risks that may arise as embodied experiences become social (Kruzan & Won, 2019). Ultimately, our self-representation affects the conclusion or perspectives people conclude about COVID-19 health risks. The following indicates how VR, AI, and social media impact embodied interactions in health risk or crisis communication.

VR is used in the treatment of body image distortions and eating disorders (Ferrer-Garcia et al., 2013). Research indicates that an individual's point of view is manipulated. There are changes in both perceptual self at the level of body schema and the conceptual self, at the level of the body image (Riva, 2011, Serino et al., 2016). In other words, individuals with eating disorders become locked into viewing their bodies from an observer's perspective. In a study by Serino et al. (2016), individuals with eating disorders were embodied in avatars and viewed as having a thin stomach in place of their actual stomach. Seeing the thin stomach caused extreme distortion in the stored memory of the body (Serino et al., 2016). Therefore, this inconsistency opened opportunity for a therapeutic protocol known as body rescripting (Kruzan & Won, 2019; Riva, 2008).

AI affects how people seek and access information. With pandemic or health crises involving information management or providing the right information to the right people in the right community, it is also important to examine how message dissemination is being perceived especially in the age of digital information. For instance, AI studies of misinformation such as vaccine safety and trust suggest that natural language AI (e.g., Google Nest, Siri, and Alexa) provide users a free way for accessing information that allows individuals to self-select information and know fully that the information or sources have not been verified or proven to be true. In essence, such media platforms have a contributing effect on embodied interaction (Eysenbach, 2020). More importantly, breaking through how people see themselves in tandem with how they seek or interpret information may be more problematic than originally thought and more importantly impact the notion of "control" as indicated in AMCM. In other words, overabundance of information leads to both accurate and inaccurate information, which makes it difficult to discern trustworthy sources and reliable guidance when needed.

A major issue that dominated the election circle and the 2020 U.S. election was COVID-19, which is also a major source of disinformation. For instance, in March 2020, a poll conducted by YouGov and the Economist reported that 13% of Americans believed the virus was a hoax, 49% thought it was manmade, and another 44% thought it was exaggerated for political reasons (Economist,

2020; Motta et al., 2020). Popular-leaning media and pundits like Sean Hannity claimed that the virus was a fraud by the deep state to create panic. Rush Limbaugh implied that it was a plot by the Chinese to harm the U.S. economy while Trish Regan of Fox Business News thought it was another attempt to impeach the president (Motta et al., 2020; Peters & Grynbaum, 2020; Warzel, 2020). Unfortunately, this innocuous denial and misinformation may lead people to a false sense of security to ignore government recommendations and appropriate authority such as the Center for Disease Control and Prevention (CDC) and the Food and Drug Administration (FDA) guidelines (Motta et al., 2020). Furthermore, the highly partisan nature of media coverage of the COVID-19 pandemic had severe and serious health consequences. For example, some believed that COVID-19 was lab-created and that a vaccine already existed—none of which is true (Motta et al., 2020).

In general, more than one-third of individuals (8,914) surveyed in a study (Pew Research Center, 2020), of whom 17% mostly consume left-leaning media thought COVID-19 was lab made, while twice that amount (i.e., 34%) of those consuming right-leaning media thought the same (Motta et al., 2010; Pew Research Center, 2020). In general, of the people surveyed between March 10 and 16 in 2020 that endorsed COVID-19-related disinformation, 22% or one in five Americans believed that COVID-19 was deliberately lab-created, and one in four (24%) believed that a vaccine currently existed or would exist shortly (Pew Research Center, 2020). Unfortunately, the danger with misinformation and misinformed people about COVID-19, especially with people who believe that COVID-19 was lab-created, is that they also think that the CDC is exaggerating COVID-19's health risk danger more than those who do not (Motta et al., 2020). As far as embodiment is concerned, people depending on their political and social beliefs tend to self-seek information that aligns with such beliefs. As a result, they seek information or misinformation that justifies their viewpoint. Google, Facebook, and Twitter for example provide such channels.

While controlling the spread of COVID-19 and bringing an end to the pandemic is the ultimate outcome (Baylor & Dewey, 2021), it also requires behavioral changes. Although COVID-19 vaccines are now available to everyone, convincing people that it is safe to take presents challenges given that agencies (the WHO, CDC, pharmacies, and health departments) use real-time data such as geoposition information to communicate effectively during outbreaks. The information is also used by community pharmacies to sell products and add medicines to the general population (Bahlol & Dewey, 2021). All these point to the issue of planning and preparedness as indicated in the AMCM. Prior to the availability of COVID-19 vaccines, some research suggested that chloroquine, an anti-malaria drug, was effective in treating COVID-19. The consequence was panic-buying, in which only 39.1% of pharmacies had availability (Bahlol & Dewey, 2021, pp. 31–36).

Crisis Communication Center's Role in Research Translation

The CCC's capacity to assist community stakeholders depends a great deal on the ability to turn scientific findings from crisis communication research into useful tools that users are willing to adopt and implement. CCC personnel (e.g., communication and support staff) and experts associated with the CCC (e.g., university researchers) can play an important role in transforming knowledge and research findings of community risks and vulnerabilities into resources that local decision makers, educators, and community members can use to engage in crisis anticipation and management. Such tools can be online videos (e.g., YouTube), infographics, web pages, reading materials, fact sheets, and other resources. This research-to-product transformation is called *research translation* (Grimshaw et al., 2012; Schulte et al., 2017).

The translation process is typically nonlinear and involves: (a) identifying a research finding that could be turned into a solution; (b) using science and observation to test the effectiveness of that solution; (c) taking the tested solution into the field to assess its application with a broader audience; and (d) evaluating the outcomes of the solution when it is adopted in the larger community. To illustrate, the CDC (2014) provides a knowledge-to-action (K2A) framework for turning scientific knowledge into policies, programs, and best practices that are evidence-based (Olaniran & Williams, 2001). K2A categorizes three components of the process: research (e.g., testing the viability of a technology monitoring approach), translation (e.g., incorporating that approach into a training curriculum to be tested on a group of engineers), and institutionalization (e.g., streamlining the tested curriculum to be made available to engineers across various industries, while continuing to test its effectiveness).

Implications for Technical Communicators

As mentioned, Americans tend to gravitate toward news sources that carry their own specific set of views, with their own specific agenda. Heath and O'Hair (2009) state that "if humans were all scientists of the same training, discussions of risk and crisis would be substantially different. But people are not and therefore neither are risk and crises straight forward matters" (p. 2). Brand loyalty is fierce whether the audience listens to Laura Ingraham or Anderson Cooper for guidance and direction, which inevitably pits talking head against talking head. The current polarization of politics in the United States creates a ripple effect that politicizes innocuous acts such as wearing a face mask to protect oneself from COVID-19 in crowds, classrooms, grocery stores, and fitness centers nationwide. Consequently, technical communicators have a specific set of challenges, namely to both inform and persuade the public regardless of bias. This bias, however, constitutes the Herculean task of informing those who are skeptical of taking the vaccine, since it is the job of the technical communicator to present the facts and facilitate compliance. In short, implications for technical communicators are formidable.

Johnson-Sheehan (2015) defines *technical communication* as managing and processing information in ways that allow people to take action (p. 9). Right-wing news sources make this task infinitely more challenging, since we know that there are many conspiracy theories and a great deal of skepticism about both the virus (the "plandemic") as well as the vaccines, which some believe were manufactured before the public learned of the virus. To educate and persuade the various public audiences as effectively as possible, technical communicators must take a multi-pronged approach and learn how to present COVID-19 vaccine information in ways that are ethical, approachable, non-threatening, informative, and persuasive so that as many members of the general public can feel as though it is the right thing to do for the sake of herd immunity.

According to Markel (2018), the technical communicator has ethical obligations to the public: to inform them with clarity and with the truth (p. 22). The truth is that according to the CDC's website, there is currently a virus, with numerous variants, that has killed well over half a million Americans. The truth is also that there are currently three FDA-approved vaccines that are safe and effective against the spread of COVID-19. Unfortunately, the truth is also that some people want to believe conspiracies and are oftentimes willingly obtuse, hence, it will be no small feat to persuade these populations to take action for the public's greater good.

One of the basic tenets of technical communication is audience analysis. Johnson-Sheehan (2015) mentions several guidelines that assist the technical communicator in identifying and catering to audience so that message can be tailored to their specific needs, three of which pertain to the topic of embodied risk. Guideline 1 states that "readers are raiders for information" (p. 19). Most people who read documents in the workplace are interested in skimming and scanning their documents for the specific information that pertains to them, as most readers have very little time for reviewing extraneous content (unless they are interested in learning more about a specific topic). To that end, it is important for the technical communicator to create concise, succinct documentation that is easy to review via subheadings, use of white space, and bulleted or numbered points. By following this advice, the reader is more apt to skim and scan at will and digest those parts of the document that the reader wants to learn more about. Guideline 2 states that "readers are wholly responsible for interpreting your text" (p. 19). In short, the writer(s) of the document will most likely not be around as the reader reviews their document; therefore, it is crucial to provide transparency and clarity of message. It is in a sense a goal of the technical communicator to be invisible in their documentation, which means that the document is so user-friendly that language and content are pored over and the author is unnoticed. A third guideline relevant to the subject at hand is that "readers prefer documents with graphics and effective page design" (p. 20).

Some readers prefer to learn from written content while others learn content faster by reviewing graphs, charts, and tables that accompany the text; others like both. By using relevant, approachable, and easily discernible graphics that accompany and complement content, different kinds of readers are addressed in terms of visual learners and verbal learners.

To tailor messages effectively for different audiences, Johnson-Sheehan proposes to address the needs, values, and attitudes of these various audiences (p. 21). In attempting to address these three components, the technical communicator is required to consider readers' familiarity with the subject, professional experience, educational level, reading and comprehension level, and skill level. Furthermore, it is important to consider the attitude that readers have toward the topic to facilitate stakeholder buy-in of message; without these considerations, message may very well be lost, and the kind of action taken may be no action at all. For example, as our numbers increase in terms of vaccination, our needs change. Although we exist in a state of flux, those who have been vaccinated have mostly been grouped with those who have not; however, the former requires their own guidelines for travel and other behavior, as the latter's guidelines do not pertain to them.

Needs, values, and attitudes are stark reminders that we must focus on differences in intercultural communication that are found throughout the United States. Ting-Toomey and Chung (2005) discuss high- and low-context patterns of communication, which can be somewhat generalized to eastern cultures as more high-context and western cultures as more low-context (p. 170). High-context patterns reflect more of an indirect, passive, read-between-the-lines approach ("It is cold in here") versus a more direct, overt communicative approach ("Please close the window"). While we are more familiar with overt risk messages here in the United States, a great portion of our audience members would nevertheless prefer a more high-context approach. The meta-message is that the more technical communicators know about their audiences, the more effective their risk documentation can facilitate them to take action. To that end, a sensitivity to diversity in intercultural communication, religion, and ethnic background would address audience needs, values, and attitudes and therefore further improve chances for stakeholder buy-in. As technical communication is not an exact science, the technical communicator needs to use as many tools as possible to effect audience action.

Ultimately, as it is far too much of a challenge to enforce conformity in frames of thought about risk, convergence is the goal for the technical communicator. Audiences will see and hear different viewpoints on the issue, and it is up to them to decide how they will take action. Sellnow et al. (2009) discuss this convergence as the primary objective:

> The uncertainty in risk situations gives rise to competing claims about the levels of danger and about the appropriate means for responding.

Thus, diverse arguments emerge. As the public observes these arguments, it is unlikely to fully accept one line of reasoning and totally reject another. Convergence occurs when distinct bodies of knowledge overlap, resulting in some capacity of agreement.

(p. 12)

As opposing talking heads debate the issue at hand, such as COVID-19 vaccinations, audiences will not necessarily stand idly to the side; rather, they will actively think critically about supporting claims from each side and make up their own mind on how they should take action.

Additionally, it would behoove the technical communicator to provide visual rhetoric to accompany text in order for members of various audiences to create new knowledge. Visual and spatial learners learn better by examining visual concepts like maps, graphs, and charts, while verbal learners prefer to glean messages from written content. Beverly A. Sauer (2010) notes that "rhetorical knowledge is both epistemic and inventional. When we individuals represent knowledge in new rhetorical forms, they see the world differently through the lens of new representations; they produce new knowledge in the transformation" (p. 83). Sometimes discovering the big picture, such as trends in pandemic cases, requires more than one mode to fully grasp what is occurring. The use of graphics that complement (not replace) text allows a larger majority of readers to learn the importance of vaccination—oftentimes despite what their selected talking heads say. Including meaningful graphics would be another effective strategy in attempting to persuade skeptical audiences of the intended message.

How much is too much? Another tenet of technical communication involves an overall conciseness with information. Overwhelming audiences with content may be as counterproductive as not providing enough information, as both could generate a deadly combination of confusion mixed with anxiety. Sauer (2010) notes that:

The sheer volume of information creates difficult rhetorical problems when individuals must determine a course of action. Individuals who limit their documentation to a single viewpoint may face less anxiety, but they will not have sufficient resources to operate in conditions of profound uncertainty. Decision makers must have the tools to sort through complexity to isolate and weigh how each individual factor contributes to risk.

(p. 138)

Focus groups and surveys could assist in providing the "right" amount of information for various populations, and their power should not be underestimated. By employing these approaches, technical communicators would be able to essentially see what their different audience members see and act according to participant feedback. As technical communicators take cues from those demographics, they will gain a better understanding of how they can serve various populations to the fullest extent and thereby encourage further buy-in of messages.

For those who identify as left-wing or liberal, the technical communicator's task is a simple one: inform and inform well. Sellnow et al. (2009) discuss this self-efficacy: "Individuals who have a strong sense of risk about a given subject are more likely to respond to messages they believe provide a reasonable strategy for personally reducing their level of risk" (p. 9). The job of the technical communicator is to provide clarity of message to the point that the writer is invisible; there is no ambiguity in the message, and readers wholly understand what to do and how to do it after they read or listen to the message. After all, the writer will not be sitting next to audience members as they peruse materials; therefore, the document must stand well on its own in terms of clarity of information. Graphics must be clear and relevant to the document; language and terminology must be relevant to those who will read or listen; and directions in how to act must be spelled out. We encourage technical communicators to take the reader by the proverbial hand and literally guide them methodically through the entire document. Although it does not take a great deal of persuasion, it *does* take a considerable amount of thought so that formatting is professional and inviting, and information is clear. And of course, as with all groups of audiences, accuracy of message (facts about the virus, how/where to get tested, and how/where to receive the vaccine) is key.

The second group that technical communicators would have to address consists of the ones who are decidedly on the fence to take the vaccine. Perhaps this group consists of those who listen to both sides of the argument and would be receptive to taking the vaccine or disregarding the vaccine, depending on the strength of persuasive measures. In cases such as this section of the public, technical communicators would need to both inform as well as persuade. Along with supplying consistency of message and presentation of facts, it would also assist the technical communicator to employ examples of famous people to sway those who are undecided. Music, film, podcast, radio, TV, and sports celebrities are obvious choices, as are well-known people within religious factions. Baptist, Catholic, Islam, Jewish, pagan, Buddhist, Hindu: Positive representations of all of these groups *from* all of these groups would assist in delivery of the message, especially with images that show the person/people receiving the vaccine. Even if these audience members are skeptical and their selected news sources disregard the vaccine campaign, their discovering well-known figures getting vaccinated can encourage them to follow suit.

The third and final group that the technical communicator would need to convince to take action is the hardest and most complicated: those who do not want to take the vaccine at all. Renn (2010) mentions the need to address the needs of the user; otherwise, the user will not respond. "Unless risk communication explicitly addresses aspects of potential benefits and social needs, it will not correspond to the expressed and revealed preferences of the people it is supposed to serve" (p. 81). Perhaps this group is the most complicated of all, since this is the group that would require the most work and would be the most difficult to convince. To that end, the technical communicator would be required to take a multi-pronged approach for this group alone, as they often do not listen to facts or reason. Conspiracies, such as Antifa and Q-Anon, run rampant, particularly with this third group, to the point that government representatives are elected who support these platforms. First of all, it is crucial to state and restate that ultimately the choice is an individual choice, as a lot of people are more fearful of loss of independence than they are about the virus itself. Nevertheless, the most influential component for the technical communicator to influence this final group is to bring local representatives from numerous life facets into the conversation. CNN recently featured a news story about a Texas pastor who has a disproportionately high number of his congregation taking the vaccine, living in a red town of a red state. The pastor was previously a hospital chaplain in the military, and he said that he sees why it is so important to receive the vaccine. He has the same needs, beliefs, and values of those in his congregation, further allowing them as stockholders buying into the idea of receiving a vaccine. Technical communicators should network with locals in every region of the country—policemen, farmers, pastors, truck drivers—thereby reinforcing validity of message and solidarity. Perhaps even if audience members are not fully onboard, the influence of seeing respectable people in local communities are taking the vaccine, the scales could be tipped so that they would follow suit.

Use of media outlets—and the connection to specific ones—would behoove the technical communicator in broadcasting messages and canvassing en masse. Carmichael and Brulle (2017) comment on the effect that the media can have on public opinion.

> An extensive body of literature has shown that aggregate public opinion is significantly impacted by the extent and prominence of media coverage. Many individuals do not possess integrated true attitudes on most issues that are relatively peripheral to their everyday concerns.
>
> *(p. 6)*

To canvas messages to the public, the technical communicator can approach radio, television, and social media; when combined, a great many listeners and observers would be able to be reached.

Talk radio is disproportionate to a conservative listening base. Mort (2020) mentions that "talk radio has been shown to exert influence specifically during primary elections and to increase conservative and moderate listeners' political efficacy while producing a dissuasive effect on more liberal listeners" (p. 21).

With personalities like the late Rush Limbaugh, dittoheads oftentimes hang on every word and do not question any of them. We suspect that this is the toughest subgroup to deal with, since listeners may not only disagree with taking a vaccination but also be openly hostile with the messenger. Researching and combing through various regions for examples of local heroes who have taken the vaccine—found in local government, the military, the medical community, and by all means the church community—is crucial in bringing others with different views onboard with the idea of herd immunity. Once these valuable players in message are procured, deploying soundbites, posters, and PSAs featuring these local heroes could be created … with the stipulation that ultimately, the decision to be vaccinated lies within the rights of the individual. After all, just because people are able to do what they want to do does not necessarily mean it is the *right* thing to do. Randolph and Barreiro (2020) discuss the effects of herd immunity. "If a fraction of the population has immunity to the same pathogen, the likelihood of an effective contact between infected and susceptible hosts is reduced, and susceptible individuals benefit from indirect protection from infection" (p. 737). Herd immunity threshold is what the country strives for—and what the technical communicator should encourage—as soon as humanly possible.

CNN, The Young Turks, Fox News, MSNBC: Television appearances by local heroes, featured on local and national news networks would also assist in further stakeholder buy-in. The more frequent messages are conveyed, the more the public will see these messages and have time to think for themselves. Competition is indeed formidable, as there are Sean Hannitys and Tucker Carlsons who undermine and outright ridicule the notion of vaccination. This option would most likely be a far more difficult task to accomplish than other media, since one would need money, influence, or both to procure television time, especially on a frequent basis.

Social media is a strong source to not only influence younger audiences but also their friends, too. Facebook, Tik Tok, Instagram, YouTube, Google, Pinterest: The majority of these outlets are frequented by younger demographics who devote a great deal of time on them and their respective apps. One way to reach audiences and tailor messages would be to create competitions within these outlets. Facebook features an older demographic of users who still want to belong and maintain relevance, record a video clip to upload for their friends to see—but other outlets would feature fertile grounds of diversity. Targeting different demographics, younger people could run with their ideas at no cost. The more prevalent the social media messages and competitions, the more publicity would be generated to those in further reaches.

Each group contains multiple audiences from different social, cultural, and economic backgrounds that causes members to perceive messages differently. Therefore, it is imperative for technical communicators to get their messaging right the first time. If posters are translated into Spanish, then the technical communicator must make sure that they are translated effectively. Additionally, they must ensure that differently-abled members of audiences such as those in wheelchairs or living in assisted living know how to take action effectively. Images in the media should accurately portray those we want to reach with a message, too; otherwise, these audience members will feel othered and may perhaps think twice before they proceed to take action.

Compliance is not always such a straightforward response from audiences—*any* audiences. Sauer (2010) discusses the difficulty of compliance. "When the likelihood of a bad outcome seems remote, we develop a sense of complacency, testing the limits of compliance until we once again produce disaster" (p. 51). With the first year of the pandemic well within our rearview mirror, we have seen this behavior play out time and again: States and communities tighten up on COVID-19 restrictive behaviors, and numbers of cases descend; those same states and communities loosen up on behaviors, and numbers plateau then steadily climb. This sense of complacency that Sauer mentions is a deadly response to accrued apathy: "I am wealthy/Caucasian/working from home; therefore, I am not at risk and do not need to be so strict with my daily routines." To be effective, technical communicators should address this complacency, which is oftentimes what lawmakers do as we attempt to round the corner with vanquishing the pandemic. As messages vary from technical communicators—informing, persuading, translating—they must also address that we are not over COVID-19-era safety protocols as of yet.

Risk communication is a process in attempting to get audience members to take action. Influencing others certainly does not happen overnight, especially when it is a highly polarizing topic such as COVID-19 vaccinations. Communication itself is influenced by contextual dynamics, sender and receiver features, message attributes, and elements of "noise" that might interfere with shared meaning. All of these aspects are ever-morphing in terms of substance and relationships to one another (Sellnow et al., 2009, p. 22).

Whatever the case, the technical communicator is bound by the truth. Both honesty and accuracy are fundamental values of human communication that every member of our audiences deserves to know (Seeger et al., 1998). Without them, the public is once again manipulated by powers-that-be, and they become pawns in a sinister power grab rather than having their own interests addressed. In fact, Kunreuther et al. (1996) note that successful risk decisions result from "an atmosphere of trust between the proponent and the host community" (p.111). Additionally, Renn (2010) adds that "trust in control institutions is able to compensate for even a negative risk perception and distrust may lead people to oppose risks even if they are perceived as small" (p. 89). To facilitate trust, stakeholders should have a voice in risk documentation, too. Different models vary in terms of stakeholder participation, but ultimately they should get the attention and recognition that they deserve. After discussing components of trust (perceived competence, objectivity, fairness, consistency, sincerity, and faith), Renn (2010) states that "there is only one general rule for building trust: listening to public concerns and, if demanded, getting involved in responsive communication" (pp. 89–90). One great way to facilitate these dialogues with the public, especially within those regions that are skeptical about risk factors, is through town hall meetings with a panel of experts. This arrangement is no small feat. The technical communicator could liaise with those people in local communities with power and/or influence—those who are trusted and address issues side-by-side with other trusted officials. Questions could be answered, important points could be raised and reviewed, and most importantly, people could feel heard.

Technical communication is not an exact science, and its experts could very well perform every action correctly and still not budge members of various communities to take action in the preferred method. Additionally, technical communicators' attempts at these forms of engagement with the public could foster other challenges. Nevertheless, by strategizing carefully laid plans and paying attention to detail, two attributes associated with the craft, these experts could encourage not only thinking seriously about a topic but also actually changing opinion.

Discussion Questions

1 How can you apply the AMCM to a subset within a specific community to which you belong (online or otherwise)?
2 Why do audiences feel that it is important to glean current news from sources that reinforce their own personal beliefs?
3 Name a specific example in which technical communication can streamline and improve a process, risk, crisis, or otherwise, within your town, city, or state.

Assignments

1 Every day for one week, watch an hour of current news programming on a television channel that you believe opposes your own values. Take notes of the visual and verbal content you notice as pushing an agenda. Were you surprised with your findings? Compare your notes with your classmates' findings.
2 Individually or with a partner, create a document that informs and persuades about a topic of your choice. Be sure to employ effective formatting for professional documents, such as meaningful headings, single-spaces within paragraphs, double-spaces between paragraphs, numbered or bulleted points, meaningful use of graphics, and meaningful use of white space.

References

Adler, N. J., & Bartholomew, S. (1992). Academic and professional communities of discourse: Generating knowledge on transnational human resource management. *Journal of International Business Studies, 23*, 551–569. http://dx.doi.org/10.1057/palgrave.jibs.8490279.

Bahlol, M., & Dewey, R. S. (2021). Pandemic preparedness of community pharmacies for COVID-19. *Research in Social and Administrative Pharmacy, 17*(1), 1888–1896. http://dx.doi.org/10.1016/j.sapharm.2020.05.009.

Carmichael, J. T., & Brulle, R. J. (2017). Elite cues, media coverage, and public concern: An integrated path analysis of public opinion on climate change, 2001–2013. *Environmental Politics, 26*(2), 232–252.

Centers for Disease Control and Prevention (2021, June 14). *COVID data tracker.* https://covid.cdc.gov/covid-data-tracker/#datatracker-home.

Eysenbach, G. (2013). CONSORT-EHEALTH: Implementation of a checklist for authors and editors to improve reporting of web-based and mobile randomized controlled trials. *Studies in Health Technology and Informatics, 192,* 657–661.

Ferrer-Garcia, M., Gutiérrez-Maldonado, J., & Riva, G. (2013). Virtual reality based treatments in eating disorders and obesity: A review. *Journal of Contemporary Psychotherapy, 43*(4), 207–221. http://dx.doi.org/10.1007/s10879-013-9240-1.

Goh, O. S., Fung, C. C., Wong, K. W., & Depickere, A. (2006). An embodied conversational agent for intelligent web interaction on pandemic crisis communication. In *2006 EEE/WIC/ACM International Conference on Web Intelligence and Intelligent Agent Technology Workshops* (pp. 397–400). IEEE. http://dx.doi.org/10.1109/WI-IATW.2006.37.

Gregory, A. (2005). Communication dimensions of the UK foot and mouth disease crisis 2001. *Journal of Public Affairs, 5,* 312–328.

Grimshaw, J. M., Eccles, M. P., Lavis, J. N., Hill, S. J., & Squires, J. E. (2012). Knowledge translation of research findings. *Implementation Science, 7*(1), 1–17. http://dx.doi.org/10.1186/1748-5908-7-50.

Heath, R. L., & O'Hair, H. D. (2010). Introduction. In R. L. Heath and H. D. O'Hair (Eds.), *Handbook of risk and crisis communication* (pp. 1–3). Routledge.

IEEE/WIC/ACM International Conference on Web Intelligence and Intelligent Agent Technology Workshops (pp. 397–400). IEEE.

Jaques, T. (2006). Issue management: Process versus progress. *Journal of Public Affairs, 6*(1), 69–74. ttps://doi.org/10.1002/pa.36.

Johnson-Sheehan, R. (2015). *Technical communication today* (5th ed). Pearson.

Kilduff, M. (1996). Making sense of sense-making: Into the jungle with Karl Weick. *Journal of Management Inquiry, 5,* 246–249. https://doi.org/10.1177/105649269653009.

Konsynski, B., & McFarlan, W. (1990, September-October). Information partnerships-shared data, shared scale. *Harvard Business Review, 68,* 14–120.

Kruzan, K. P & Won—A. S. (2019). Embodied well-being through two media technologies: Virtual reality and social media. *New Media & Society, 21*(8), 1734–1749. https://doi.org/10.1177/1461444819829873.

Kunreuther, H., Slovic, P., & MacGregor, D. (1996). Risk perception and trust: Challenges for facility siting. *RISK: Health, Safety & Environment (1990–2002), 7*(2), 109–118.

Lerbinger, O. (2012). *The crisis manager* (2nd ed.). Routledge.

Lukaszewski, J. E. (2013). *On crisis communication: What your CEO needs to know about reputation, risk and crisis management.* Rothstein Associates, Inc. https://doi.org/10.33077/uw.24511617.ms.2015.62.536.

Markel, M., & Selber, S. A. (2018). *Technical communication* (12th ed). Bedford/St. Martin's.

Merriam-Webster. (n.d.). Retrieved June 15, 2021, from merriam-webster.com.

Mitroff, I. I., Harrington, K., & Gai, E. (1996). Thinking about the unthinkable. *Across the Board, 33*(8), 44–48.

Mitroff, I. I., Puchant, T., & Shrivastava, P. (1989). Can your company handle a crisis? *Business & Health, 7*(5), 41–45.

Mort, S. (2020). Harnessing the potential of the "demotic turn" to authoritarian ends: Caller participation and weaponized communication on US conservative talk radio programs. In Céline Ségur (Ed.), *French perspectives on media, participation and audiences* (pp. 19–44). Palgrave Macmillan. https://doi.org/10.1007/978-3-030-33346-1_2.

Motta, M., Stecula, D., & Farhart, C. (2020). How right-leaning media coverage of COVID-19 facilitated the spread of misinformation in the early stages of the pandemic in the US. *Canadian Journal of Political Science/Revue canadienne de science politique, 53*(2), 335–342. 10.31235/osf.io/a8r3p.

Olaniran, B. A. (2007). The role of perception in crisis management: A tale of two hurricanes. *Multicultural Education, 15*(2), 13–16.

Olaniran, B. A. (2016). ICTs, E-health, and multidisciplinary healthcare teams: Promises and challenges. *International Journal of Privacy and Health Information Management (IJPHIM), 4*(2), 62–75. DOI: 10.4018/978-1-5225–9863-3.ch022.

Olaniran, B. A., & Scholl, J. C. (2014). New England compounding center meningitis outbreak: A compounding public health crisis. *Journal of Risk Analysis and Crisis Response, 4*(1), 34–42. https://doi.org/10.2991/jrarc.2014.4.1.4.

Olaniran, B. A., & Scholl, J. C. (2020). Anticipatory model of crisis management and crisis communication center (CCC): The need to transfer new knowledge to resources. *The Handbook of Applied Communication Research*, 297–311. https://doi.org/10.1002/9781119399926.ch18.

Olaniran, B. A., & Williams, D. E. (2001). Anticipatory model of crisis management: A vigilant response to technological crises. In R. L. & G. Vasquez (Eds.), *Handbook of Public Relations* (pp. 487–500). Sage.

Olaniran, B. A., & Zhang, Y. (2016). Rethinking ICTs and E-Health: A focus on issues and challenges. In M. Cruz-Cunha, I. Miranda., R. Martinho, and R. Rijo (Eds.), *Encyclopedia of E-Health and Telemedicine* (pp. 998–1012). IGI Global. DOI: 10.4018/978-1-4666–9978-6.ch078.

Palttala, P., Boano, C., Lund, R., & Vos, M. (2012). Communication gaps in disaster management: Perceptions by experts from governmental and non-governmental organizations. *Journal of Contingencies and Crisis Management, 20*(1), 1-12. doi: 10.1111/j.1468-5973.2011.00656.x.

Pauchant, T. C., & Mitroff, I. I. (1992). *Transforming the crisis-prone organization: Preventing individual, organizational, and environmental tragedies.* Josey-Bass.

Pearson, C. M., & Clair, J. A. (1998) Reframing crisis management. *Academy of Management Review, 23*, 59–76. https://doi.org/10.2307/259099.

Peters, J. W., & Grynbaum, M. M. (2020, March 11). How right-wing pundits are covering coronavirus. *New York Times.*

Randolph, H. E., & Barreiro, L. B. (2020). Herd immunity: Understanding COVID-19. *Immunity, 52*(5), 737–741. https://doi.org/10.1016/j.immuni.2020.04.012.

Renn, O. (2010). Risk communication: Insights and requirements for designing successful communication programs on health and environmental hazards. In R. L. Heath and H. D. O'Hair (Eds.), *Handbook of risk and crisis communication* (pp. 80–98). Routledge.

Riva, G. (2008). From virtual to real body: Virtual reality as embodied technology. *Journal of Cyber Therapy and Rehabilitation, 1*, 7–22.

Riva, G. (2011). The key to unlocking the virtual body: Virtual reality in the treatment of obesity and eating disorders. *Journal of Diabetes Science and Technology, 5*(2), 283–292. https://doi.org/10.1177/193229681100500213.

Sauer, B. A. (2010). *The rhetoric of risk*. Routledge.

Scholl, J. C., Williams, D. E., & Olaniran, B. A. (2005). Preparing for terrorism: A rationale for the crisis communication center. In D. O'Hair, R. Heath, & G. Ledlow (Eds.), *Community preparedness and response to terrorism: Communication and the media* (Vol. 3, pp. 243–268). Praeger.

Schulte-Schrepping, J., Reusch, N., Paclik, D., Baßler, K., Schlickeiser, S., Zhang, B., & Ziebuhr, J. (2020). Severe COVID-19 is marked by a dysregulated myeloid cell compartment. *Cell, 182*(6), 1419–1440.

Seeger, M. W., Sellnow, T. L., & Ulmer, R. R. (1998). Communication, organization, and crisis. *Annals of the International Communication Association, 21*(1), 231–276. https://doi.org/10.1080/23808985.1998.11678952.

Sellnow, T. L., Ulmer, R. R., Seeger, M. W., & Littlefield, R. S. (2009). *Effective risk communication: A message-centered approach*. Springer.

Sellnow, T. L., Ulmer, R. R., & Snider, M. (1998). The compatibility of corrective action in organizational crisis communication. *Communication Quarterly, 46*, 60–74. https://doi.org/10.1080/01463379809370084.

Serino, S., Pedroli, E., Keizer, A., Triberti, S., Dakanalis, A., Pallavicini, F., & Riva, G. (2016). Virtual reality body swapping: A tool for modifying the allocentric memory of the body. *Cyberpsychology, Behavior, and Social Networking, 19*(2), 127–133. https://doi.org/10.1089/cyber.2015.0229.

Sethi, S. P. (1987). Inhuman errors and industrial crisis. *Columbia Journal of World Business, 22*, 101–107.

Shrivastava, P. (1988). Industrial crisis management: Learning from organizational failures. *Journal of Management Studies, 25*(4), 283–284. https://doi.org/10.1111/j.1467-6486.1988.tb00037.x.

Shrivastava, P, Mitroff, I. I., Miller, D., & Miglani, A. (1988). Understanding industrial crises. *Journal of Management Study, 25*, 285–303. https://doi.org/10.1111/j.1467-6486.1988.tb00038.x.

Simmons, R., & Shiffman, J. (2007). Scaling up health service innovations: A framework for action. In R. F. P. Simmons, & L. Ghiron (Eds.), *Scaling up health service delivery* (pp. 1–30). World Health Organization.

Slovic, P. (1996). Risk perception and trust. *Risk, 7*, 110–118. https://doi.org/10.1201/9781439821978.sec3.

Smircich, L., & Stubbart, C. (1985). Strategic management in an enacted world. *Academy of Management Review, 10*, 724–736. https://doi.org/10.2307/258041.

Ting-Toomey, S., & Chung, L. C. (2005). *Understanding intercultural communication*. Oxford University Press.

Warzel, C. I. (2020). Need you to care that our country is on gire. *New York Times, 9*, A23.

Weick, K. E. (1988). Enacted sense-making in crisis situations. *Journal of Management Study, 25*, 305–317. https://doi.org/10.1111/j.1467-6486.1988.tb00039.x.

Weick, K., & Sutcliffe, K. (2007). *Managing the unexpected: Assuring high performance in an age of complexity* (2nd ed.). Jossey-Bass. https://doi.org/10.1108/ws.2002.07951dae.003.

Yarrow, E., & Pagan, V. (2020). Reflections on front-line medical work during COVID-19 and the embodiment of risk. *Gender, Work & Organization*. https://doi.org/10.1111/gwao.12505.

4

JUDGING THE UNPRECEDENTED

Common Sense and Risk During COVID-19

Scott Weedon

Problem: Technical, science, and risk communicators are concerned with researching and producing documentation and forums to help people make decisions about science and risk. They want them to use their judgment and common sense to understand the stakes of a scientific problem, or the degree to which they are at risk. However, when there is an unprecedented situation, like the one the world faces during the COVID-19 crisis, what counts as good judgment or common sense is revised and contested by the public. Therefore, science, technical, and risk communicators should understand the ways common sense is shaped by crises through attention to the ways the public describe and define common sense.

Solution: Risk communication can be improved by understanding the way common sense is framed in public discourse.

Science and risk communication have offered several frameworks for conceptualizing risk and how the public constructs risk (Jensen, 2015). These perspectives question the relationship between experts and publics and the place for and representation of deliberation about risk (Danisch, 2010; Grabill & Simmons, 1998; Sauer, 2002; Wynne, 1989). The capacities for deliberation, the judgment that we use to orient ourselves to the options presented in any deliberative situation, receive less attention in the literature; meaning, science and risk communicators are particularly interested in making science a matter of democratic judgment (e.g., Gross, 1994), but inquire less into just what sort of judgment they are presupposing. Deliberation is a crucial component for science and technical communication to consider, but it often rests on already settled notions of what judgment or common sense are and how they operate. If asked to define common sense and/or judgment, one might say something like common sense is the "ability to use experience and information to decide

DOI: 10.4324/9781003266549-5

how to act." In this conventional perspective, common sense and judgment are kinds of decision-making devices that can be optimized by education in critical thinking. However, the ongoing COVID-19 crisis reveals that judgment and its cognates such as common sense and reason become highly contested in unprecedented situations. What counts as good judgment or common sense becomes not just a resource to make decisions, but a topic that is evaluated, revised, and transformed through the circulation of opinions and arguments. When you read or hear appeals to good judgment and common sense in talk such as, "Wearing masks is just common sense," or in complaints where someone might take issue with restrictions or social-distancing measures because they "defy common sense," you get a sense that people are framing common sense differently. A quick sample of media opinions during the COVID-19 crisis shows that common sense can be what is done cautiously (Glick, 2020), it can be a resource to detect the overreach of precipitous bureaucrats, and thus be conceived as a conservative force (Goldberg, 2020), or it can be a faculty for dispassionate and apolitical decision-making (Moffit & Johnson, 2020). In times of crises, our cognitive capacity of judgment on which we rely to make decisions in unprecedented situations is itself contested. In other words, common sense is not static nor is it held in common.

Technical, science, and risk communicators are concerned with researching and producing documentation and forums to help the public make decisions about science and risk (Davies, 2013; Ding, 2013; Grabill & Simmons, 1998; Horst & Irwin, 2010; Simmons & Grabill, 2007; Walker, 2016). This chapter argues that science and risk communicators need to understand what kinds of judgments they are appealing to and how they are presented and argued for in the course of a crisis.

The chapter uses news articles and opinion pieces to look at the ways common sense and the capacity to judge are framed. By framing, I mean how common sense is presented through language from a particular perspective and entailing particular consequences. When we look at the ways people talk about common sense and the ways people use common sense, we see that they are embodying different notions of common sense. By the term *embodying*, I intend to call to mind several things: the way we give form or expression to a thing (such as common sense), the way we internalize an idea or mode of being, and the way we are embodied, how we are creatures with bodies that have biological and cultural dimensions.

Discourse about what common sense is shows that the embodiment of common sense is unsettled and contested in the midst of COVID-19. The chapter starts from the foundational assumption that judgment is embodied and situational as established in the technical and science communication literature (Fountain, 2014; Sauer, 2002; Weedon, 2019; Wynne, 1989, 1993). Through rhetorical analysis, paying attention to the way people use language to present a perspective or a set of facts, the chapter describes how popular news and opinion pieces invoke competing embodiments of judgment as a resource for developing a COVID-19-specific capacity to judge. The embodiment of judgment is invoked to understand social distancing and mask wearing, and in other rhetoric, judgment is framed as that which looks beyond local embodiment and takes up bird's-eye views offered by epidemiological and economical frameworks, based upon modeling and statistics. These variations in how judgment and common sense are framed suggest that science and risk communicators should reflect on citizens' capacity for judgment that they take for granted; not to dismiss laypeople's judgments as too variable, and thus justifying a top-down perspective on science communication, but instead to better understand how to shape and appeal to judgment. A quick point about my terms: while judgment and common sense can be defined separately, they are often used interchangeably in the data I gathered, and I adopt that interchangeability in my descriptions to mirror the way the terms are used in popular discourse.

Questioning taken-for-granted notions about judgment and examining its contestation in a crisis opens up a space for technical communication to talk about the perspectives we hold of judgment and risk, and what perspectives we should take up for the future. The chapter will show the different ways common sense can be embodied and why it matters for risk communication. The chapter will then reflect on integrating these findings into social justice research.

Judgment and Common Sense in Science, Technical, and Risk Communication

Research in science and risk communication and allied fields explore the problem of presenting and negotiating differing professional judgments. Beverly A. Sauer's work (2003) is foundational for understanding the problems of translating embodied judgment into textual genres and the difficulty in translating between different professional judgments in the same workplace. She notes that in mining enterprises, managing risk often means coordinating between three senses of judgment: the miner's embodied pit sense that is gathered through experience in the practice of mining, the engineer's judgment that stems from the built environment of the mine (the structure and materials of its supports and space), and scientific judgment, which relies on mathematical understandings of the physical properties of the mine. Coordinating these various perspectives requires rhetorical sophistication to match situation to judgment (Sauer, 2003, p. 199).

Technical communication has also examined how discipline-specific common sense and judgment are formed in educational settings. T. Kenny Fountain's work (2014) in anatomical education describes skills medical students use to develop what he calls "trained vision," a disciplined way of perceiving a cadaver that transitions it into the anatomical body of medical practice. Fountain contends that this type of disciplining of embodied perception scaffolds the clinical judgment healthcare workers will deploy with actual patients (2014, pp. 98–99). Scott Weedon's (2019) research into the ways engineering students make judgments in the design process argues for a perspective on judgment that is not only personal, but emergent from group debate. Engineering students are able to make design decisions by discussing their feelings about a design when the mathematics fail to make sense for them. Engineering judgment is not simply a matter of cognitive processing and mathematical reasoning as it involves bodily feeling and language as well as mathematics and science.

Importantly, the kind of common sense and judgment one develops is affected by the kind of body they have and the way the culture at large labels that body. In a discussion of risk communication pedagogy, Erin A. Frost (2018) notes that the embodiment of risk is not simply a biological or cognitive fact, it is a cultural-historical experience of "being in a particular kind of body that must navigate a variety of cultural contexts" (p. 25). Implicit in Frost's point is that common sense becomes subject to the ways bodies are positioned by discourses and how embodied experiences are inextricable from time and place.

The research in the embodiment of judgment demonstrates that judgment forms through experience and can be marshaled to warrant decisions about risk and technical problems. This is not only a result of professional training, but also a general process of growing up in a particular culture and at a particular time. For instance, Derek G. Ross (2012) examines the way common sense is invoked by the public when asked about environmental policy. He concludes that common sense is informed not just by personal and private experience, but also by "a complex construction of culturally acquired values that facilitates day-to-day decision-making, and is tied inexorably to our politico-culturally and socio-culturally mediated understanding of how the world works" (Ross, 2012, p. 123). In other words, common sense is informed by various values and worldviews that are tied to our cultural affiliations and politics in a particular time and place. Ross finds that appeals to common sense are a necessary resource for dealing with situations where extended analysis or expert reflection cannot be deployed. However, Ross worries that the handiness of appeals to common sense can give short shrift to complex problems. Rather than deal with the complexity of scientific policy, one falls back on common sense, and this concerns Ross because appeals to common sense can rest on outdated conceptions, they can be intensely parochial and personal, and they can be uncritical. Thus, while common sense is necessary, it may also be reductive.

Zoltan Majdik (2012) points out that while common sense can be all too personal, it can also name those attempts where the requirements of deliberation are shared among a group or community, what might be called an emergent sense in common. Common sense in Majdik's view is a common sense that comes from the intersubjective understanding of a group able to mobilize their personal experience for a collective endeavor. People talk with each other, forming relationships and understandings from different perspectives, and adopt compatible stances on topics important to everyone. This is an emergent common sense.

Damien Pfister (2019) points to a similar way of understanding common sense: common sense can be a sensing in common, one where people share and articulate personal sensations in democratic forums. Pfister advocates for technologies that create better opportunities for people to fashion common experiences, rather than technologies that tend to isolate people. Pfister is working with a kind of common sense that seeks to overcome a strictly private and personal kind of common sense by focusing on ways to bring embodied subjects together to share their sensations. In summary, Majdik and Pfister, like Sauer, Fountain, Ross, and others describe judgment and common sense as an embodied, cultural, and experiential resource for knowledge and decision-making in technical, political, and private spheres of deliberation (see Walsh & Walker, 2016).

In addition to carving out the contours of judgment in scientific and technical situations, from both expert and lay perspectives, research in science and technical communication provides a glimpse of how people contest conceptions of judgment and common sense. Sauer shows that miners, engineers, and scientists are all potentially drawing from different frameworks when making judgments. Pfister contends that the technologies we use to experience the world necessarily entail a particular approach to what it means to sense in common. This chapter follows these insights to analyze how judgment and common sense undergo differing conceptions during a unique crisis. The work presented here is preliminary, modest, and satisfied with simply initiating research into how judgment and common sense are defined and mobilized in calls for action to mitigate the effects of COVID-19.

Science and risk communication are thoroughly involved in understanding how judgment can be applied to situations, but what needs more investigation is how situations change what judgment can be. The following section looks at rhetoric in the popular media that define and advocate for judgment and common sense in the midst of the pandemic.

Data and Analysis

To examine what notions of common sense and judgment circulated throughout the first year of the ongoing COVID-19 crises in the United States, I searched the internet for keywords "common sense," "judgment," "COVID-19," and "coronavirus." I used two different browsers, Google and TOR, to cast a wide net for search results. I selected 38 entries from the first several pages of results, contenting myself with this number when the search engines started to note the absence of keywords in the results. The results included news stories, podcasts, interviews, and blogs. I excluded entries from international sources because I did not feel familiar enough with contexts outside of the United States.

In my initial analysis, I read the selections and excluded those that in actuality did not address common sense or judgment in relation to COVID-19 and those that focused on international contexts. With 35 selections, I then analyzed the samples' articulations of common sense. For my first round of coding, I noted the date, title, and genre of the selection and then excerpted pertinent quotes mentioning, exemplifying, or defining common sense and judgment's relation to COVID-19. In a second round of coding, I summarized these excerpts in short descriptions, starting each summary with "common sense/judgment is...." In a third round of coding, I identified 12 general themes for conceptions of common sense/judgment (see Table 4.1). Within these themes, there was much overlap and many internal distinctions, with some samples articulating several characteristics of common sense. For instance, the category of "common sense is personal responsibility" contained three subcategories: 11 instances of "common sense is taking precautions," three instances of "common sense is personal responsibility, but formulated by experts," and one instance of "common sense needs to be discounted for expert knowledge." To start to draw some conclusions from the categories, I charted when they appeared during the COVID-19 crisis in the United States, from February 2020 to February 2021. As will be discussed below, I found particular articulations cluster around particular points of the history of the virus. For instance, understandings of common sense as taking precautions proliferated in the first three months (February–April) of the pandemic, whereas common sense as justifying a relaxation of COVID-19 restrictions appeared throughout the summer (May–September). This mapping allowed me to notice a temporal dimension of the pragmatics of common sense.

These categories' and subcategories' boundaries are fuzzy, and of course, subject to my own interpretative framing. Nonetheless, I argue that they still can provide evidence of how common sense and judgment are contested and

TABLE 4.1 Themes of common sense/judgment across sample from 2/20 to 2/21

Themes of common sense/judgment and their instances across the sample.	Personal responsibility (15)
	Blanket regulations vs. individual situation (8)
	Common sense is not overreacting (4)
	Individual vs. community (3)
	Drawing simple conclusions from numbers (3)
	Risk analysis (3)
	Common sense vs. medical science/experts (2)
	Mathematics vs. common sense (2)
	Decision-making device (2)
	Leadership (1)
	Common sense opposes interestedness (1)
	Balance (1)

formulated in response to crises. Below, I provide a rhetorical reading (characterized by attention to the way arguments are put forth and concepts are framed) of these expressions and evocations of common sense and judgment to understand the particular characteristics these terms take on in response to the pandemic. My goal is to better understand just what kinds of common sense and judgment science and risk communicators might be appealing to in times of crises similar to COVID-19.

Common Sense and COVID-19

My reading of the data led me to conclude that the different framings of common sense oscillate between an understanding of common sense as personal and experiential and an understanding of common sense as guided by entities beyond a person's immediate environment. Such entities may refer to health experts, politicians, media figures, and other types of formulators of opinion. As I discuss below, these two aspects were not always mutually exclusive, for they often blended or were creatively arranged to make appeals to common sense.

Common Sense as Embodying Expertise

The most common meaning of common sense was framed as personal responsibility. It amounted to 15 instances across the dataset. Writers of interviews, news reports, and opinion pieces understood common sense as personal responsibility, and personal responsibility as taking the proper precautions to prevent infection. These proper precautions could range from a kind of underspecified wariness to adhere to established guidelines. The association of common sense with taking personal responsibility through adhering to the proper precautions clustered in the first three months of the pandemic (February–April of 2020). As the possible impact of COVID-19 was still being assessed and there remained much uncertainty, government officials and health experts tied common sense to early versions of guidelines for washing hands, not touching your face, and recognizing COVID-19 symptoms. Concurrently, there were invocations of common sense as an antidote to overreaction or panic. As people heard conflicting reports on precautions, tragedy in Italy and Spain, and the possibilities of lockdowns, healthcare leaders, politicians, and journalists invoked common sense to make people adequately prepared for the future and to implore them to keep necessities like toilet paper and N95 masks on the shelves. Experts provided common sense to the public as a resource to manage the unprecedented and evolving situation they and their community were thrust into.

Early invocations positioned common sense as what could be reasonably accomplished in one's own circumstances. One did not have the expertise to make useful prognostications and had no control of health policy, but they could protect themselves and not overreact. The movement of common sense to the immediate domain of one's personal environment, the micro-domain we might say, evacuated the macro- and meso-domains from the deployment of common sense except when articulated by experts. Social, political, and economic policy and affairs were now under the purview of epidemiologists. Common sense became doing what the health experts recommended. In some cases, common sense was cast in the terminology of health experts, as common sense was equated with performing personal and vigilant "risk analysis" ("Speaking of Psychology: Coronavirus anxiety w/ Baruch Fischoff," 2020).

As the purview of common sense became circumscribed to the local, personal, and micro-domains, implicitly common sense was about embodying the best advice and using the best judgment in a restricted sphere of action. Health officials noted the struggle of turning what they termed "common sense" into routine practice, getting people to keep their distance, wash hands, and eventually wear masks ("Common sense approach is best health advice," 2020). Health experts cast a formulized common sense as a reasonable, mundane way of acting that needed to become habitual. Pleas for exercising common sense when people started to become less vigilant appealed to this connection between common sense and expert advice. In this logic, common sense was derived from health guidelines and that relationship was a mutual reinforcing appeal: the health guidelines were just common sense backed by expert knowledge. However, it was not only the advice of officials, but the data and models that provided the sense of common sense. Numbers provided stark evidence for probabilities and trends, and served as the basis of good judgment. In fact, some expressions of common sense and judgment highlighted the ability to gather and organize data and information to inform decisions about the future, marking a similarity to the capacities of epidemiological models ("'Superforecasters' Are Making Eerily Accurate Predictions About COVID-19," 2020).

Framed in these ways, embodying common sense meant adhering to guidelines that were meant for all situations. U.S. residents were now seemingly equipped with the common sense to safely navigate their environments, no matter where they were. In an unprecedented and disorienting situation of the pandemic, people had a conceptual, if sketchy and provisional, grasp of how COVID-19 would change their lives and how they would have to adapt.[1]

1 It remains to be determined how well people were able to embody this common sense. Were the guidelines presented in a way that facilitated uptake? Did common sense guidelines fit the common situations of many differently situated people? Beverly A. Sauer (2003) describes the struggles of capturing the embodied experience of miners in written safety guidelines and protocols. During the pandemic, reports from hospitals, warehouses, and processing plants often mentioned the impossibility of embodying the common sense of CDC guidelines, since social distancing and staying outdoors was an impossibility due to the nature of their work. It is not hard to imagine that the embodied situations many people faced could not be accounted for by a common sense.

However, this framing of common sense allowed for people to notice that broad, catch-all guidelines could not describe their particular situation, nor did people necessarily want to cede their own judgment to experts.

Common Sense vs. the Experts

Concurrent with this articulation of common sense and judgment as the embodiment of expert guidelines was a pushback on those same expert guidelines in the name of common sense. Common sense began to be mobilized as a safeguard against expert's overreach and nonsensical restrictions. This mobilization of common sense started to gather around the late spring and summer of 2020 as early recommendations were revised, the crisis became political theater for President Trump and many in the United States tired of COVID-19 restrictions. While common sense was positioned in various ways, it often traded on the ability of individual experience and simplification to be the best guides for action. In other words, overanalysis and generalization should give way to people's wisdom ("Bradley County commissioner says officials 'rushed to judgment' in response to coronavirus pandemic," 2020).

Historian of science Steven Shapin (2001) notes that experts see common sense as the naïve opposite of science and technical knowledge. Common sense is empirical and parochial, whereas scientific and technical knowledge is skeptical and exacting. This opposition was used to rhetorical effect as people who were chafing from COVID-19 contrasted common sense knowledge to what experts promulgated. This contrast has many forms: common sense should be prized over blanket regulations ("Common sense shouldn't be a casualty of the coronavirus," 2020); common sense is about not overreacting to the doom saying of experts ("Take common sense over the coronavirus experts," 2020); common sense is clear-eyed and logical, rather than beholden to needless procedures ("Where is the common sense in coronavirus restrictions?", 2020).

In the late spring of 2020, President Trump epitomized the attitude of clear-eyed logical connection when advocating for hydroxychloroquine over health experts' hesitations, saying "I'm not a doctor. But I have common sense." He had heard empirical, but anecdotal reports of the treatments effectiveness, and reasoned from cause to effect. This was in very public contradistinction to the ways epidemiologists viewed the same data. They saw a claim not verified by methods that they understood to provide the most reliable information.

Another permutation of common sense is found in a letter to the Baker City, Oregon city council, where a couple proposed the city become a "common sense sanctuary," one that would allow businesses to stay open at the capacity they could manage and with the restrictions they saw fit ("Baker City business owners urge 'common sense sanctuary,'" 2021). Here and in similar stories, deployments of common sense created antitheses to "one size fits all" approaches handed down by unaccountable bureaucrats and overcautious experts.

These articulations of common sense seemed to create tidy binaries, but in some cases, common sense was formed through creative arrangements of the boundaries between individual and community, expertise and lay knowledge, and complexity and simplicity. For instance, some advocated for self-determination in the name of communities' need to survive. Business and restaurant owners foregrounded their local, professional, and workplace communities as the ones pressured by safety precautions ("Minneapolis bar owner says coronavirus restrictions lack 'common sense,'" 2021). In their eyes, the measures to protect the minority who were at greatest risk from getting sick were hurting the larger community.

As we have already seen, experts occasionally position their advice as common sense action. While President Trump's statement insisted on an opposition between expertise and common sense, and some experts bolstered that opposition by casting common sense as naïve ("'Common Sense' Is No Substitute for Science," 2020), other actors deployed common sense as the product of expertise. In late January of 2021, amid a high number of coronavirus cases and deaths, Dr. Anthony Fauci characterized double masking as common sense, using a simple explanation of how overlapping masks create extra barriers for protection against potentially viral respiratory droplets ("'Common sense': Dr. Fauci says double-masking is 'likely' more effective against coronavirus," 2021). Earlier in summer 2020, a healthcare provider advocated closing up businesses again because any look at the numbers would lead common sense to conclude that the virus was not under control and more social contact meant more transmission ("Doctor urges governors to 'use some common sense' as coronavirus cases surge," 2020). For the provider, common sense was about acceding to medicine's advice, which itself was a simple extrapolation from cause to effect. The reliance on common sense explanations from both experts and laypeople show that both technical and non-technical people use common sense to make decisions (Button, 1991; Lynch, 1985; Shapin, 2001). Science, technology, and medicine are often operating within degrees of uncertainty, and so they must rely on approximate reasoning. While common sense is often deployed as a wedge to separate experts and laypeople, acknowledging that it is a capacity they share may be an important first step in overcoming the binary.

The role of numbers, and by extension, models, figured prominently in how common sense was framed. For some people, common sense was opposed to models since the models were imperfect and made inaccurate predictions, thus highlighting their distance from the reality people were experiencing. In some other cases, common sense was distilled from the complexity of numbers, as the healthcare provider advocating for a return to lockdowns cited above. In another opinion, a Noble Laureate argued for the virus to subside more quickly than prominent models had predicted because his was a common sense perspective on the available data ("Nobel laureate calls for 'common sense' to prevail in fight against coronavirus," 2020). He argued that foregoing models and simply looking at the numbers could provide a basis for judgment. His own account of "looking at the numbers," as he called it, seemed to consist of accusations of rampant overcounting of coronavirus fatalities by those who would claim it as a cause of death when that determination might be disputed. Thus, common sense was figured as that which looks behind the numbers to the real practices of quantifying. This conception of common sense resonated with a columnist who in April of 2020 took the governor of New York to task for failing to do more to prevent nursing homes deaths from COVID-19 ("State lacked common sense in nursing homes coronavirus approach," 2020). He used common sense as a "B.S. detector" to see beyond justifications for the failure to keep infection out of nursing homes. In these accounts, common sense is a kind of device to identify obfuscation. It is a tool to get beyond carelessness, needless complexity, and fishy statistics to the real issues.

This last sense of common sense, the expression of common sense as disinterested and attuned to "what is really going on," is one that seems to have an exceptional amount of currency, though there is just one explicit instance in the data I gathered. I argue for its prevalence because in many instances, it describes the motivation for invoking common sense in the first place. Common sense is that capacity that helps to arbitrate between ideologies, to differentiate what *they think* they know and what *we do* know. It can draw a distinction between one group's decorum and the failure to adapt to the situation by another group. In this capacity, it acts like a *shibboleth*, a distinguishing concept or belief: a group can have a monopoly on common sense and use it to distinguish the excesses or stubbornness of other groups. Common sense is a grounding of politics on an issue that is supposedly free of politics. Saying, "well it's just common sense" is a way to create a faction by designating an issue as beyond factionalism. When someone appeals to common sense, the phrases common sense or good judgment may not even be explicitly invoked; in some cases, it may be enough to cite evidence emblematic or evocative of common sense. One need only consider the way people use the huge number of COVID-19 deaths or the relatively small percentile of infections that end in death to marshal common sense to their arguments for or against restrictions and lockdowns. Common sense, then, is flexible, responsive to different events, and a vehicle for politics and values.

What to Make of Common Sense

The analysis thus far has shown that common sense and good judgment are contested. Common sense is something we fall back on in the face of complexity or unprecedented change. But it is not a value-neutral, cognitive decision-making device that allows us to parse information like a computer. Common sense is many things, and it responds to changes in the world. Even though common sense seems to only create dichotomies of expert vs. layperson, personal vs.

general, and situated vs. abstract, it is quite flexible because it can be applied to many different situations and take many shapes. Common sense is embodied and communal, accumulated through experience and culture, but it can also be a distillation of abstract data from models and statistics.

Common sense's flexibility and contradictoriness pose a challenge for science and risk communicators because they make it hard to know the character of what we are appealing to when we provide documentation and forums for people to make decisions and invent new knowledge. Science and risk communicators seek "to cultivate common sense in citizens" and facilitate the rendering of judgments (Danisch, 2010, p. 188; see also Walker, 2016). From that prerogative, the question arises as to how to best achieve that aim. Robert Danisch (2010) recognizes these issues when he calls for fostering in people a specifically "science and technical prudence" that would "require a familiarity with working through cases of scientific controversy. In other words, understanding the ways in which scientific work is itself deliberative and conditioned by uncertainty and controversy is necessary" (pp. 188–189). A complementary way to dig into this issue is to look at what common sense's flexibility and contradictoriness affords for persuasive action. In other words, how can science, technical, and risk communicators use the flexibility inherent in common sense for the ends of helping the public make good decisions about science and risk.

Social psychologist Michael Billig (1987/1996) found that common sense's contradictory nature contained the means through which attitudes and opinions could be shifted. Common sense, according to Billig, "is not unitary, but is composed of contrary aspects. If we make this assumption, then we can expect to find logos and anti-logos, or contrary common-places, in the minds of the audience" (p. 234). We see this as the case with COVID-19 common sense, especially in the commonplace notion of personal responsibility. This commonplace is found in arguments for adopting health expert's guidelines by some and for resenting their imposition by others. For Billig, this is to be expected: "[b]oth logoi and anti-logoi are presumed to co-exist within the common-sense, and we can assume this state to be quite normal and not associated with the presumed *angst* of cognitive dissonance" (p. 234). For those tasked with persuading an audience to open up their common sense ideas to other aspects, this contrary state provides the seeds for a shift in perspective. The fuzzy boundaries of common sense notions allow for possibilities of reversal or re-inscription according to timing and situation. This means that while common sense can be resistant to more complex forms of knowledge (Ross, 2012), preferring ready-to-hand explanations and concrete experience over abstractions, it is not incorrigible. Common sense may be reshaped if we recognize it as a resource for persuasion rather than an impediment. The final section of the chapter suggests some possible tactics to account for common sense in science and risk communication.

Tracing the Senses of Common Sense

A quick search for the terms "common sense," "judgment," "swine flu," and "avian flu" will turn up results that show that the proliferation of common sense talk is not uncommon to health crises. In fact, many of these framings of common sense and judgment precede health crises, but in unprecedented events like COVID-19, we should expect these framings to be amplified and appealed to in opposition to each other. This development will likely lead, covertly or overtly, to debates over common sense itself: consider the way Anthony Fauci and Donald Trump both appeal to common sense to not so subtly undercut each other. Science and risk communication should view such cases as both challenges and opportunities. The challenge is how to appeal to people whose very capacity to judge is developing, shifting, and coalescing into factions through the course of a long crisis. The opportunity is for science and risk communicators to craft what common sense can be.

While this chapter can only make modest suggestions from a modest data set, one thing that is worth science and risk communication's consideration is how common sense shifts throughout the extent of a crisis. As people try to grapple with how to order their lives, experts and pundits constantly invoke and advocate for common sense and good judgment. Yet, rather than consensus arising on how common sense should operate in the developing situation, common sense fractures along lines of different living environments, race and class, competing ideologies, and politics, among others. This development should lead science and risk communicators to reconsider their position between experts and publics.

Risk and the capacity to judge risk are shaped by the greater media ecology (Grabill & Simmons, 1998). The messages that risk communicators make and the forums they foster are taken up by people in the midst of many perspectives and framings that impact how an issue is understood or recirculated into discourse (Edbauer, 2005). Thus, science and risk communicators need to be aware of the discourse that precedes their own discourse. By that I mean, science and risk communicators' standpoint as translators between experts and publics obscures the more complex circulation of information about science and risk. It excludes participants on social media, punditry, politics, and other interlocutors that frame common sense through competing arguments and visual communication. Science and risk communicators can start to grasp the circulation of common sense(s) in a media ecology by mapping the emergence and transformation of common sense topics, themes and places of argument that gain traction and force in public discourse. Such an approach might make use of what Casey Boyle and Lynda Walsh (2017) call "topological mapping," which allows rhetors to "trace the contours of a discourse" while providing the potential to "fold it into a new configuration" (p. 4). In other words, by following the ways topics emerge and serve as grounds for arguments, science and risk communicators can better assess their own appeals to people's situated reasoning, making more targeted messaging that accounts for how common sense is being configured in the moment. Such an approach maps not only the way common sense is variously framed, but the timing and circumstances that give it shape as it is confronted with new developments and unforeseen contingencies (Walsh & Boyle, 2017, p. 5). Topological mapping may provide science and risk communicators a guide for understanding just what kind of ideological landscape they will have to cover through strategies of documentation and public forums.

The opportunity my analysis points to is the fact that common sense is corrigible rather than stubborn, flexible rather than fixed, an emerging facility rather than one mired in past experience. Following Billig (1987), Shapin (2001), and others, I suggest that these characteristics of common sense can be useful for science and risk communicators. As the above analysis shows, common sense becomes a topic for the invention of arguments, a way to signal that an opinion is grounded in our collective experience. Yet, the sheer variety of common senses show that this is a thoroughly rhetorical move, one that brings into being a common sense as much as it invokes a pre-existing one. I suggest that science and risk communicators take up common sense as a topic to shape themselves, not just as a resource for appeals. Common sense arbitrates whether a recommendation makes sense, accords with good judgment, and should ground decision-making. Thinking about common sense rhetorically means recognizing that the content of our cognitive capacities are subject to contingencies. Science and risk communicators can think about not only how risk is deliberated upon, but also how deliberation itself is called forth through discourse and manifested in various environments.

Finally, a focus on the development and variability of common sense prods us to take a more granular approach and attend to the ways marginalized communities frame common sense. If we grant that common sense is not held in common, that it is drawn from particular relationships to events and power, science, and risk communicators must be aware of how their communication design will appeal to people who hold less power than others. How is common sense being framed in communities that have troubled histories with medicine, as Black and Indigenous populations do in the United States? What is the common sense of an essential worker who must be in contact with people daily and cannot simply embody the health recommendations to stay home or distance themselves from others? How do they navigate the pandemic situated as they are? Science and risk communicators can start to address these questions by not taking for granted a general common sense and not assuming common sense is universally represented in the media. Science and risk communicators need to meet underrepresented communities where they are to get a sense of where they are coming from when they make decisions.

An interesting instance of medicine meeting people where they are to come to a common understanding together is the YouTube series "Greater than COVID," where many different kinds of healthcare workers of Black and Latinx descent respond to common questions about coronavirus vaccines. The series attempts to build trust between marginalized communities and medicine by addressing those communities with their own members. One can read these videos as tacitly negotiating common sense between two communities by using representatives that span both. While the video series should be applauded for addressing marginalized communities, science and risk communicators will note that the videos are less places of dialogue to establish a common sense among communities, and more a platform for medical workers to address marginalized communities. The questions the healthcare workers respond to are not articulated by anyone, but instead are canned renditions of concerns such as "How did we get a COVID vaccine if there's not one for HIV?", "Will the vaccine give me COVID?", and "How were Black people involved in the COVID vaccine development?" Again, the gesture here is exceedingly important, but there is no dialogue in these videos, only healthcare workers paraphrasing common concerns and answering them. Even the comment section is disabled.

While one can easily understand why a YouTube video series would try to circumvent trolls and others spreading misinformation, the lack of dialogue means the series fails to be a place where medicine and Black and Latinx communities can respond to and acknowledge each other's concerns, and in this space of recognition and interrogation, potentially enroll each other in striving for goods each finds worthwhile and mutually beneficial. A more fruitful platform would be one where common sense emerges from intersubjective discourse where different viewpoints are manifested and accommodated (Majdik, 2012). Furthermore, Natasha N. Jones (2016) might characterize such a space as having the potential for moving toward social justice, insofar as marginalized communities are ensured a place to interact and present their own understandings unabashedly rather than be passively addressed by experts. Facilitating better forums is one thing science and risk communicators who are interested in the potential of common sense-making to aid social justice can do.

Conclusion

The art of rhetoric, which many science and risk communicators rely on to think through communication problems, is an art for topicalizing common sense, where the appeal to values and opinions is both the starting point and endpoint for discourse. A speaker works to show an audience how one set of values entails this other set of values, or how one deeply held opinion connects to this other, perhaps novel opinion. One may read Dr. Fauci's belief that double masking "makes common sense" as an attempt to topicalize common sense, to say that what may seem to be overly cautious is commonsensical when accounting for the behavior of aerosols. While I am not suggesting that topicalizing common sense will dissolve disagreements (and I would not want it to), I suggest that by making it a topic for science and risk communication, common sense may be a place to start to understand what is sensible for different groups and how that sense changes in response to different events and environments. Thinking of common sense as contested and emergent, having a past, but subject to revision in the face of novelty, helps science and risk communicators forefront the various needs of different audiences and the development of those needs through the course of a crisis. Using common sense as a topic brings in "a rhetorical component that facilitates finding a shared understanding of the common stakes we have" (Majdik, 2012, p. 4) and the common stakes we are developing in response to a crisis we do not yet fully grasp.

Discussion Questions

1 What are some of the ways you hear and see common sense expressed? Is common sense the same in all of these cases? What differences and similarities are there and what do they tell you about how people relate to scientific and technical information?
2 In the minds of many, common sense and scientific thinking are opposites. Common sense is gained from personal experience and from the way things appear to be and scientific thinking is objective and skeptical of appearances. Given what you know about both, describe how they are similar and how they differ? Does common sense have a place in scientific thinking, or does common sense have elements of scientific thinking?
3 The chapter has focused on common sense and judgment, but what role do the emotions play in framing how people react to risk? Are feelings our primary reactions to risk that then influence our common sense? How can science, technical, and risk communication do a better job of accounting for emotion in decision-making about risk?

Assignments

1 Find a message or visualization communicating risk. Make a short recommendation memo for incorporating appeals to common sense. First, define common sense for the purposes of the message and then think of ways to incorporate that meaning into the content of the message.
2 Watch several episodes of the YouTube series "Greater than COVID" and read Natasha Jones's (2016) "The Technical Communicator as Advocate: Integrating a Social Justice Approach in Technical Communication." In groups of two or three, write a proposal to redesign and/or reformat the series to meet the goals of social justice technical communication that Jones outlines. Consider ways of making forums where people can feel that they are heard and their concerns are adequately addressed by experts.

References

Allain, R. (2020, April 19). Common sense is no substitute for science. *Wired.com*. https://www.wired.com/story/common-sense-is-no-substitute-for-science.

American Psychological Association (2020). Coronavirus anxiety w/ Baruch Fischoff. *Speaking of Psychology*. https://www.youtube.com/watch?v=xrxMDan_5Fc&t=1053s.

Benton, B. (2021, May 23). Bradley County commissioner says officials 'rushed to judgment' in response to coronavirus pandemic. *Timesfreepress.com*. https://www.timesfreepress.com/news/local/story/2020/may/23/bradley-county-commissioner-says-officials-ru/523715.

Billig, M. (1996). *Arguing and thinking: A rhetorical approach to social psychology*. Cambridge University Press.

Button, G. (Ed.). (1991). *Ethnomethodology and the human sciences*. Cambridge University Press.

Danisch, R. (2010). Political rhetoric in a world risk society. *Rhetoric Society Quarterly, 40*(2), 172–192.

Edbauer, J. (2005). Unframing models of public distribution: From rhetorical situation to rhetorical ecologies. *Rhetoric Society Quarterly, 35*(4), 5–24.

Editorial Board of the Orange County Register (2020, April 15). Common sense shouldn't be a casualty of the coronavirus. *Orange County Register*. https://www.ocregister.com/2020/04/15/common-sense-shouldnt-be-a-casualty-of-the-coronavirus.

Elkind, E. (2020, July 10). Doctor urges governors to 'use some common sense' as coronavirus cases surge. *CBS.com*. https://www.cbsnews.com/news/coronavirus-cases-surge-doctor-urges-governors.

Fountain, T. K. (2014). *Rhetoric in the flesh: Trained vision, technical expertise, and the gross anatomy lab*. Routledge.

Frost, E. A. (2018). Apparent feminism and risk communication: Hazard, outrage, environment, and embodiment. In A. Haas and M. Elbe (Eds.), *Key theoretical frameworks: Teaching technical communication in the twenty-first century* (pp. 23–45). Utah State Press.

Garfinkel, H., Lynch, M., & Livingston, E. (1981). The work of a discovering science construed with materials from the optically discovered pulsar. *Philosophy of the Social Sciences, 11*(2), 131–158.

Glick, A. (2020, March). A common sense approach to coronavirus crisis. *CNN.com*.

Goldberg, B. (2020, May 17). Add the death of common sense to the coronavirus's toll. *The Hill*. https://thehill.com/opinion/finance/497901-add-the-death-of-common-sense-to-the-coronaviruss-toll

Goodwin, M. (2020, April 25). State lacked common sense in nursing homes coronavirus approach. *New York Post*. https://nypost.com/2020/04/25/new-york-lacked-common-sense-in-nursing-homes-coronavirus-approach.

Grabill, J. T., & Simmons, W. M. (1998). Toward a critical rhetoric of risk communication: Producing citizens and the role of technical communicators. *Technical Communication Quarterly, 7*(4), 415–441.

Gross, A. G. (1994). The roles of rhetoric in the public understanding of science. *Public Understanding of Science, 3*, 3–23.

Hilder, A. (2020, September). Fauci asks Americans to "double down" on "common sense" public health measures ahead of winter. *Thedenverchannel.com*. https://www.thedenverchannel.com/news/national/coronavirus/fauci-asks-americans-to-double-down-on-common-sense-public-health-measures-ahead-of-winter.

Hindenach, C. (2021, January 12). Minneapolis bar owner says coronavirus restrictions lack "common sense." *Foxbusiness.com*. https://www.foxbusiness.com/small-business/minneapolis-bar-coronavirus-restrictions-common-sense.

Jacoby, J., & O'Connor, S. (2021, February 4). Baker City business owners urge "common sense sanctuary." *The Observer*. https://www.lagrandeobserver.com/coronavirus/baker-city-business-owners-urge-common-sense-sanctuary/article_cd6b55d4-6667-11eb-bab7-a34fefd7d07b.html.

Jensen, R. (2015). Rhetoric of risk. In Hyunyi Cho, Tursten Reimer, and Katherine McComas (Eds.), *The SAGE Handbook of Risk Communication* (pp. 86–97). Sage.

Jones, N. N. (2016). The technical communicator as advocate: Integrating a social justice approach in technical communication. *Journal of Technical Writing and Communication, 46*(3), 342–361.

Law, T. (2020, June 11). Superforecasters' are making eerily accurate predictions about COVID-19. Our leaders could learn from their approach. *Time*. https://time.com/5848271/superforecasters-covid-19.

Loreno, D. (2021, January 6). "Common sense": Dr. Fauci says double-masking is "likely" more effective against coronavirus. Fox8 News. https://fox8.com/news/common-sense-dr-fauci-says-double-masking-is-likely-more-effective-against-coronavirus.

Lynch, M. (1985). *Art and artifact in laboratory science: A study of shop work and shop talk in a research laboratory*. Routledge.

Majdik, Z. P. (2009). Judging direct-to-consumer genetics: Negotiating expertise and agency in public biotechnological practice. *Rhetoric & Public Affairs, 12*(4), 571–605.

Majdik, Z. (2012). Reply to Derek Ross's "Ambiguous weighting and nonsensical sense." *Social Epistemology Review and Reply Collective*, 1–5.

Marshall, K. (2020, March 15). Common sense approach is best health advice for fighting coronavirus, Legacy medical director says. *Click2Houston.com*. https://www.click2houston.com/news/local/2020/03/15/common-sense-approach-is-best-health-advice-for-fighting-coronavirus-legacy-medical-director-says.

Moffit, R. E., & Johnson, D. (2020, June). Common sense on COVID-19 beats second-guessing every time. *The Heritage Foundation*. https://www.heritage.org/public-health/commentary/common-sense-covid-19-beats-second-guessing-every-time

Pfister, D. S. (2019). Technoliberal rhetoric, civic attention, and common sensation in Sergey Brin's "Why Google Glass?". *Quarterly Journal of Speech, 105*(2), 182–203.

Ross, D. G. (2012a). Ambiguous weighting and nonsensical sense: The problems of "balance" and "common sense" as commonplace concepts and decision-making heuristics in environmental rhetoric. *Social Epistemology, 26*(1), 115–144.

Ross, D. G. (2012b). Reply to Zoltan Majdik on "Ambiguous weighting and nonsensical sense". *Social Epistemology Review and Reply Collective*, 1–3.

Sauer, B. A. (2003). *The rhetoric of risk: Technical documentation in hazardous environments*. Taylor & Francis.

Shapin, S. (2001). Proverbial economies: how an understanding of some linguistic and social features of common sense can throw light on more prestigious bodies of knowledge, science for example. *Social Studies of Science, 31*(5), 731–769.

Strawn, S. (2020, May 17). Where is the common sense in coronavirus restrictions? *Capitalgazette.com*. https://www.capitalgazette.com/opinion/columns/ac-ce-column-steve-strawn-20200517-ynzw5emqdved3c2wfpolralsfy-story.html.

Vaughan, D. (2020, April 23). Take common sense over the coronavirus experts. *Thereporter.com*. https://www.thereporter.com/2020/04/23/david-vaughn-take-common-sense-over-the-coronavirus-experts.

Walker, K. C. (2016). Mapping the contours of translation: Visualized un/certainties in the ozone hole controversy. *Technical Communication Quarterly, 25*(2), 104–120.

Walsh, L., & Boyle, C. (2017). From intervention to invention: Introducing topological techniques. In L. Walsh and C. Boyle (Eds.), *Topologies as techniques for a post-critical rhetoric* (pp. 1–16). Palgrave Macmillan.

Walsh, L., & Walker, K. C. (2016). Perspectives on uncertainty for technical communication scholars. *Technical Communication Quarterly, 25*(2), 71–86.

Weedon, S. (2019). The role of rhetoric in engineering judgment. *IEEE Transactions on Professional Communication, 62*(2), 165–177.

Wynne, B. (1989). Sheepfarming after Chernobyl: A case study in communicating scientific information. *Environment: Science and Policy for Sustainable Development, 31*(2), 10–39.

Wynne, B. (1993). Public uptake of science: a case for institutional reflexivity. *Public Understanding of Science, 2*, 321–337.

Yahoo News. (2020, March 30). Nobel laureate calls for "common sense" to prevail in fight against coronavirus. https://www.yahoo.com/entertainment/nobel-laureate-calls-common-sense-025457829.html.

5

FIRST-YEAR COLLEGE STUDENTS CHALLENGING EMBODIED ENVIRONMENTAL RISK

Uma S. Krishnan

Problem: Risks are generally considered a taboo subject in some cultures as they bring into the forefront the question: What if something goes wrong? In some communities, the moment this question is raised by anyone who is concerned about the flip side of an action or plan, the general reaction from the other members of the group is that the person voicing this sentiment is thinking negatively even before the action has started and they are bringing bad vibrations and negative energy to the whole situation. In addition, voicing "in case something goes wrong" is viewed as "fostering unhappy associations," and suggesting that the process or action will fail even before it begins (student interviews). Given such connotations about discussing risks in general with family members and friends or other members of their communities, how do first-year students assess embodied risks in their lives and address them?

Solution: In this chapter, I show how students consider embodied risks and understand them, and why they should address them in college while acquiring a higher education. I also discuss how students exposed to the concept of Vygotsky's Zone of Proximal Development (ZPD) and Maslow's Hierarchy of Needs understand that situated risks are a part of life based on self-actualization.

Embodied Risks and Student Perceptions

Instructors in colleges often receive emails from academic services about player athletes' performance in their classes. The emails are sent to professors as a means of assessing "at-risk" students, so that the academic staff can help and coach these players to stay focused on their studies and manage their sports activities. Interestingly, this criteria of identifying students who are failing in their classes or having retention issues has now expanded and is being used

DOI: 10.4324/9781003266549-6

across academia to monitor and target "first-year students, student-athletes and students with demonstrated academic difficulties" under a system called the "Early-Alert Program" (Hanover Research, 2014, p. 3). This system has been put in place as a means to identify students who are at-risk on three levels: "academic, social, or otherwise," and to intervene in a timely fashion and provide the much-needed "intrusive and individualized interventions to students in need" (Hanover Research, 2014, p. 5). But questions remain: How and why does this happen in college, when students were successful in graduating out of high school? At what point does the risk factor enter their life so that early alerts have to be sent out to monitor them?

In response to these questions, students in first-year Honors Colloquium and College Writing, both core classes, suggested that risk factors in academia begin when they fall or are trapped into temptations after entering college.[1] The temptations are usually "considered as taboos in society," but are accessible on campus, "from underage drinking, drugs, food to even being promiscuous," and they find that it is very easy

> to lose control of what we want, leading to health issues, depression, eating disorder, missing classes and assignments, gaining weight and even, feeling guilty as we are not following what is expected of us. This leads to early alerts and warnings to shape up or you will be....

Further, many of the participants felt they were unable to discuss these problems with their family or siblings as "they all have their own baggage to deal with and we don't want to add to their life problems" (student interviews).

Although they are "technically considered as adults, being 18 and 19," they were still unsure how to comprehend and address some of these issues and felt that there were serious gaps in the educational system that needed to be addressed. For example, a student in the first-year writing class wrote that the pandemic has caused him to become sleepless, that insomnia has set in, "as I am not sure how my future is going to be. This is just May 2020. I have three more years of college and I am unable to even visit the health center." Sami claimed that she was unsure if she had chosen the right major and felt that there was a risk in continuing the current path as she was not happy in any of the classes she was taking. She suggested, "Because I don't like any of the courses I am taking, I don't feel like going to class and ... attendance will soon be a problem. This slowly leads to missing assignments and then, it is a domino effect." One of the students suggested that losing a member of the family has already made him miss so many classes that he is all poised to fail his courses: "I am not sure how

---

1 Some of the students suggested that although this starts at high school, it is "living independently" in college that accelerates these temptations.

to cope with the present situation." A freshman in another writing class was concerned about the future and wrote,

> What will my major look like given that we, as Science majors, can't attend labs. I feel everything is closing in; what will our future look like; what if I fail in my classes as I am unable to attend lab? How will we be graded? I am on scholarships and need to maintain my GPA and can't fail, but with no lab I am taking a big risk in my major?

Cory expressed this anxiety as, "The embodiment of outside changes can be seen and felt by all of us and we are all anxious about our lives, our family members, friends, and everything. This is scary."

On the other hand, some of the students were more concerned about their social skills, in case the pandemic continued, and they were unable to attend college in person. Students referred to their peers in current classes as "my tech or collaborate friends" or "my Zoom friends" and some of them were concerned about their psychological growth and peer interaction. This was reflected in their writing: "I am not sure how to interact with my peers, although we are in a group chat. I feel awkward and feel we are all losing out on real time, like the physical interaction and communication." One of them expressed this concern as:

> I feel our generation will not know what will be the 'freshman experience' that my siblings and older peers talk about. If this continues, we will only know how to interact through technology filters and maybe, we will all become introverts—less talk, more texts—I am so unsure of everything that it is depressing.

An added response from another student was,

> I want to change my majors as I have no interest in math and failing in this class. This is risky as I am not considering the consequence of my current actions and not being proactive about it. But, if I had friends on campus, I could seek guidance in so many different ways.

Geri suggested that her parents will be upset if she were to inform them about the early-alert signal. Her response was, "I am a little skeptical about these systems helping me, does it really work? Do people really see what I am undergoing? I feel like a caged bird." Another student wrote,

> We have great scientists in our country, rather the BEST in the world, why is it that this type of risk and planning wasn't taken into consideration. Did people know about it? When I first heard about COVID, I imagined this was something out of a dystopian novel.

The students during class discussion summed up the situation as,

> Although these measures are good and intended to help students, they are like a 'band-aid' being placed on our 'heads and bodies with no proper cure from within' to see and measure what students consider as 'embodied risks in their everyday lives' and how they address them 'physically and psychologically'.
>
> *(Student interview)*

In some ways, this process of interaction that students consider as environmental risks, Bazerman (2019) considers an "interactant." He suggests that the individual's "interactant experiences" and "perception of the semiotic environment from an individual perspective" leads to a development of a "unique individual," which in turn contributes and transforms "the environment for themselves and others" (p. 3). When students in Honors Colloquium were questioned on what they considered as embodied risks, their responses ranged from "anything that is contrary to our previous way of life" to "what we were engaged in that is contrary to what we see around us before joining college." They viewed many aspects of their college life as risk factors, such as— changing opinions on race and color to underage drinking, to majoring in a subject matter of their choice, to natural disasters happening around the world, including the current pandemic. The uncertainty of the future, especially the Covid situation, was like a "cave with no light." One of them pointed out, "I didn't know I could add to the national average of Covid patients." Further, many of them mention that "although these situated environmental risks" were there for them to see and recognize before they joined college—they were never clearly articulated by family and friends—so that they could recognize the risk factors that they would be involved in and be better equipped to renegotiate these spaces.

The solution for this situation, according to many scholars in the field, is to provide students the tools to negotiate the risk spaces on their own, through interpersonal and intrapersonal ways, and by providing students opportunities to represent their own cultural products and problem-solving skills in the environment (Howard Gardener, 1999; Marcus & Kitayama, 1991). Given these perceptions of embodied risks, in the following section, I define the term embodied risk and how students interpret and translate them physically and psychologically in their lives.

Embodied Risk: Definition and Student Perceptions

Kavanagh and Bloom (1998) suggest that *embodied risk* can be defined as a probable health issue located in the body of the person said to be "at-risk." They explain this more in terms of sickness that is embodied or manifested within an individual, and how they need to "cope with uncertainty" that "they must consider what it means to be in danger of developing an illness even though most have no symptoms; and they must mobilize appropriate surveillance and perhaps risk reduction" (1998, p. 437). Kriger (2005), from an *epidemiology* (study of the determinants of health) perspective, suggests, "Embodiment is a verb-like noun that expresses an abstract idea, a process, and concrete reality. Whether used literally or figuratively, it insists on bodies as active and engaged entities" (p. 351). Kriger believes that embodiment can come into existence only when bodily characteristics engage with the outside environment to express their fears or insecurities in two ways: as biological organisms that "reproduce; develop; grow; interact; exist in time and space; and evolve" or as social beings existing within a "societal context; social position; social production; social consumption; and social reproduction" (p. 351).

Interestingly, all these definitions showcase that embodied risks have been associated with human psyche and body for a long time. Kriger argues that this word:

> ... advances three claims: 1) bodies tell stories about—and cannot be studied divorced from—the conditions of our existence; 2) bodies tell stories that often—but not always—match people's stated accounts; and 3) bodies tell stories that people cannot or will not tell, either because they are unable, forbidden, or choose not to tell.
>
> *(2005, p. 350)*

These definitions further reveal that the external environment plays a huge role in how we, as humans, understand, analyze, embody, and react to the risks internally—within ourselves. It also reveals how we ignore them as we grow older and recognize that every one of the decisions or choices we make in life are inherently risky. We also know that "replacing one potential action with something different also means replacing certain risks with other risks" and this in turn allows us to move on with our lives (Igoe, 2018).

Critical analysis of this concept reveals that from the time we are born till the time we die, we are constantly facing challenges and undertaking risks, consciously or unconsciously. And, despite all the challenges in the environment, external and internal, we survive, develop, live, and grow. As adults, we consciously never address, every moment of the day, the dangers lurking behind us and events that could and might happen to put us in mortal fear of living. We address them as conditional conjectures and never pay heed to them. We watch television news or shows or read the newspaper to recognize there are international issues that affect us directly and indirectly, and we ignore them. We hear about acts of terrorism on foreign soil or catastrophic climate effects in other parts of the world or about deforestation or genocide or political negotiations among countries, and we accept them as normal. But what we fail to address often as parents, as guardians, and as instructors is to discuss "these issues with our children and students." We assume that these are global or national issues, and they will learn to deal with it "when the time comes." We never spend time discussing "how these issues trickle down into their own lives" and provide coping mechanisms. "We pass on our anxieties to them but never bother to discuss it with them" (student interviews). A student used all these situations to showcase what they were all referring to as embodied risks on many levels. He said that they had to attend freshmen lectures on what they termed as "Be warned about temptations in college environment" and how:

> ...parents and siblings never sat me down and said if you find yourself in a sticky situation like you are drunk or get into drugs or oversleep—leading you to losing interest in studies and failing in a lot of ways—seek help or call us. So the question is who should I talk to as it is hard to come out of this vortex without help. These are risky situations for us. We deal with a different set of issues, that are all trickled down from the top into our lives. We didn't ask for it, but it was bestowed upon us. The global warming—cyclones, hurricanes, storms—the immigration issues, educational issues like loans, debts; health issues—obesity, diabetes, cancer—so much more that I can keep ranting but it all boils down to what did we do ... but we suffer with all of you, right? We didn't create these problems—you all did, and we pay the price? So, we are all affected in this and you are right, Dr. K, it all is about the ripple effect and concentric cycles. Yes, we are all in a symbiotic relationship.

Students felt that embodied risks were constantly changing forms and in turn leading to a form of physical anxiety or psychological sickness to deal with issues they had never faced before in their lives. They categorized embodied risks on four levels: international, national, local, and oneself and why it needs to be addressed in high school or in college classes.

Methods

In 2017–18, when I wanted to teach a course on self-actualization (or the realization of one's full potential) to my college writing and freshmen colloquium class, there were quite a few aspects that I needed to consider before teaching it. Class lessons and materials had to be scaffolded in such a way that they addressed the learning outcomes and yet, addressed the risk factors in students' lives and the process of self-actualization. I wanted to experiment with this concept, as there was enough research to prove that students taking courses on self-actualization were well equipped to handle risks in their lives (Barocas, 1967; Bordages, 1989; Lefcourt, 1966; Ryff, 2014). Although unsure of how students will react to such a course, like Bracher (1993), I had "little sympathy with the assumption that the basic functions of the humanities should be to indoctrinate students into a monolithic system of ideals, values, knowledge, and belief" (ix). Like Bracher, my sentiments and purpose were to enable students "analyze literature and other discourses which will make our scholarly and pedagogical activities more valuable—for ourselves, our students, and society in general—without turning literary and cultural studies into indoctrination" (p. ix).

I felt the most appropriate method to study and measure the growth of approximately 32 students, in terms of self-actualization and acquired confidence, in my colloquium classes of 2017–2020, was through a qualitative case study. This was mainly due to having no control over the participants' way of thinking and views (Yin, 2014). Although this was a very small sampling, I felt it was still worth viewing what my students had to say, as the demographics of this population was in many ways a representation of the larger student population across campuses, as students were from different parts of the United States with different majors and from different school districts. I considered fall 2017 to spring 2018 as a pilot study and repeated the same course methods, text, discussions, and assignments to the new batch, from fall 2018 to spring 2019, then from fall 2019 to spring 2020. I mostly used the writing samples only from my Honors Colloquium class as they were with me for 2 semesters, 30 weeks, as opposed to College Writing class students who were with me for 1 semester or 15 weeks. All students were informed that this course would be an experimental study and to maintain anonymity and they were referred to as student participants in general and pseudonyms were provided while referring to their writing samples.

Cole et al. (1978) explain Vygotsky's use of the experimental method:

> As every student of an introductory experimental course knows, the purpose of an experiment as conventionally presented is to determine the conditions controlling behavior. Methodology follows from this objective: the experimental hypothesis predicts aspects of the stimulus materials or tasks that will determine particular aspects of the response; the experimenter seeks maximum control over materials, tasks, and response in order to test the predictions. Quantification of responses provides the basis for comparison across experiments and for drawing inferences about cause-and-effect relationships. The experiments, in short, are designed to produce a certain performance under conditions that maximize its interpretability ... To serve as an effective means of studying "the course development of process" ...include a variety of materials that could be used in different ways to satisfy the demands of the task.
>
> *(pp. 11–12)*

Students' thoughts and reactions to risks and self-actualization, from the beginning till the end of the semester, were recorded based on: (a) the diagnostic essay they wrote during the first few weeks of the semester, toward the middle of the second semester based on class discussions on Maslow's Hierarchy (Figure 5.2), and then, at the end of the second semester, (b) class assignment writings as the semester progressed, (c) final portfolios, and (d) class presentation. Student participation was voluntary. The class lessons were all scaffolded to make students: (a) read articles, (b) take notes in three columns—before the reading, during, and after the readings, (c) annotate paragraphs that made them think critically or appeared controversial, (d) have class discussion about their readings, and (e) write short journals leading to six short papers and a final portfolio and presentation. The final presentation in class was about their own self-actualization journey; provided they were comfortable discussing their personality growth with others in class.

In terms of text, I used *Ways of Reading* by Bartholomae, Petrosky, and Waite (2016) to expose students to rich scholarship by authors from different fields. Each author's purpose, approach, thoughts, sentiments, beliefs, attitudes, and appeals were discussed and thoroughly understood before asking students to write about their perceptions after the reading. Students were then asked to internalize and see what these authors were reacting to in terms of risks in their lives, and what types of discussion and solutions they were offering. Using the text as a model, the students were asked to construct their papers on different levels.

Each paper they were going to be writing was to be based on the readings for the week and then, exploring one specific concentric cycle in their life and how every circle impacted them in unique ways. As Figure 5.1 reveals, each circle represents relationships, issues, perceptions, beliefs, communications between: (1) I/Me—myself—who am I, (2) I/Me—my family, siblings, and friends, (3) I/Me—my major and Local University Community, (4) I/Me—State/National Community, and (5) I/Me—Global Community, leading to articulating them in their final writing portfolio and presentation. For example, each student would begin their exploration of oneself by asking myriad questions related to them, such as: Who am I? Why am I here in this universe? What is my purpose in life? Once they explored this cycle from many angles, they would move to the next cycle—to explore their relationship with their family, siblings, and friends and to see how these bonds were created, how they remain strong or weak over a period of time, and what can the self/person do to become better at communicating their thoughts to others. After they explore the second concentric circle fully, they progress to explore the other circles—from their education/majors to national policies and issues to global situations and policies. During this process of exploration from one circle to another, over a period of one or two semesters, two things materialize: (1) students understand how they have formed certain opinions or beliefs or ideologies in their lives and (2) students question some of these beliefs to see if they are ethical, applicable, and valid given the current global, national, and local circumstances. This in turn provides them with a deeper understanding of themselves, of who they

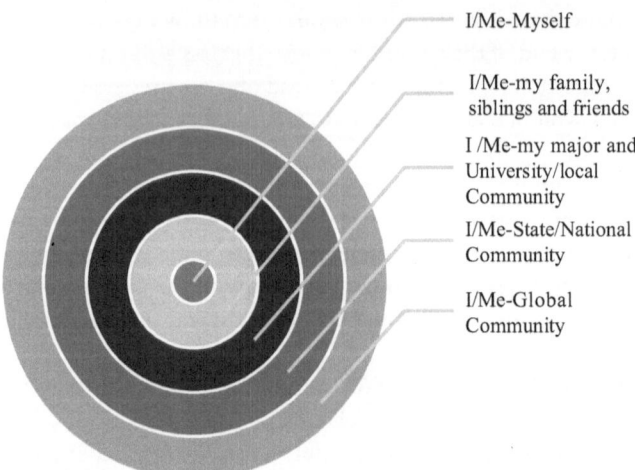

I/Me-Myself

I/Me-my family, siblings and friends

I /Me-my major and University/local Community

I/Me-State/National Community

I/Me-Global Community

FIGURE 5.1 Concentric cycles and ripple effects represent the relationship between the nucleus "I/Me" and the different environmental circles one is a part of in life.

are and what they want from life. It is at this point they start valuing certain personal beliefs and rejecting some to remain true to their core identity. Thus, scaffolding each lesson using ZPD enabled students to see that identity plays an important role in every level of this concentric cycle.

The reason I wanted to use the ZPD, developed by Vygotsky, a Soviet psychologist and social constructivist, was to provide an understanding of the human psychological processes that exist between thought and language (Cole et al., 1978). Vygotsky's (1978) research on the impact of social context in which the human mind, cognitive thought process, and behavior developed—led him to define ZPD as "the distance between the (child's) actual developmental level as determined by independent problem-solving and the level of potential development as determined through problem-solving under adult guidance, or in collaboration with more capable peers" (p. 86). Based on his scientific experiments, he believed that *scaffolding* (or temporarily supporting) a particular task or lesson for a child, under the direction and supervision of a knowledgeable adult or peer, allows the child to think a certain way leading him/her to better understand a concept, greater learning, knowledge building, and new experiences. He believed that these experiences lead to new layers of higher thinking and concept formation that are superimposed upon the earlier lower layers.

Vygotsky using the earth analogy explained the growth as a "structure of behavioral development to some degree resembling the geological structure of the earth's core. Research has established the presence of genetically differentiated layers in human behavior" (Kozulin, 1986, p. xxix). He further suggested that these layers, when interconnected, produced complex thinking processes and concept development, where at the center remained the nucleus of thought and around it developed other developmental layers of association, collection, chains, and diffusion, leading to pseudo concepts. Vygotsky believed that with the growth of complexity of psychological processes, the preceding development stages also grow along, and this is only possible under the supervision of an adult who guides the child (Kouzlin, p. 124). Thus, the adult because of his/her experience is able to shift the focus for the child from a situational position to a conceptual situation leading to self-understanding (Kouzlin, p. 133), and what can be considered as a shift taking place from an egocentric awareness to theory-of-mind, or TOM (Callaghan et al., 2005; Dennett, 1987).

This shift happens over stages from childhood to adolescence, and at every stage, the child recognizes his/her role in society and learns to express it in a cogent way through the use of language and concept of the self, and this is mainly due to what he/she observes in the environment and due to parenting, school, and peer interaction (Erikson, 1968). In addition, concrete thinking changes to more complex and abstract thinking only during the teenage years due to the expansion of existing skills and thoughts, leading to cognitive empathy, an integral aspect required for problem-solving and conflict resolution (Shamay-Tsoory et al., 2007). Erikson (1968) further suggests that it is at this stage of their life, these young adults start recognizing their sense of self, leading to identity and identity crisis, where "the man … experiences it as something that 'comes upon you' as a recognition, almost as a surprise rather than as something strenuously 'quested' after" (p. 20). This further allows the young adult to view things in a mature way and make them understand a situation in its proper light and come to conclusions that are beneficial for them and for the society as a whole.

Over the years, Vygotsky's ZPD has been used in many areas of education from elementary to high school and used in teacher training (ZPTD) to provide a holistic approach to teaching and learning (Warford, 2011). Thus, scaffolding all materials for the class and introducing students to Maslow's Hierarchy of Needs (see Figure 5.2) led to rich writing samples, where my students could see their own individual growth and identity development from the beginning to the end of the semester, over a period of one year. I introduced Maslow's Hierarchy during the latter part of the second semester as students: (a) would have been exposed to different texts, ways of thinking, discussions, and writing, and (b) would be able to understand that the physiological needs, safety needs, love and belonging needs, and esteem needs need to be addressed before focusing on self-actualization (Maslow, 1943). Maslow's (1954) explanation for self-actualization was, "What a man can be, he must be. This need we may call as self-actualization" (p. 93). Couture et al. (2007) refer to it as the psychological process aimed at maximizing the use of a person's abilities and resources and how this process varies from person to person. It is the recognition by the individual of their own creative, intellectual, and social potential through the

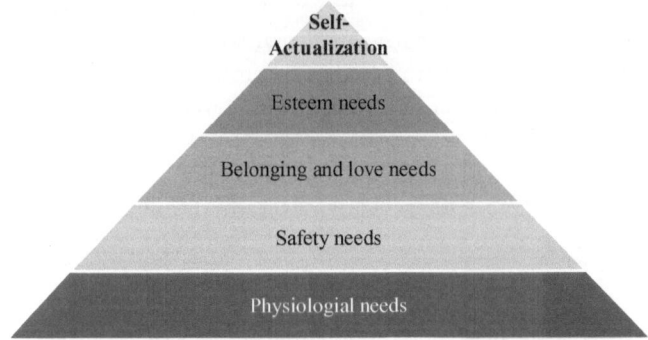

FIGURE 5.2 Maslow's Hierarchy of Needs.

process of internalization, which begins by asking questions like, who am I? Why am I here in this universe? What is my purpose in life? Why am I in this university, and so on.

Further, this concept when introduced and discussed promoted students' "synergetic self-developing abilities to improve and form professional knowledge, to initiate new creative ways of working abilities and innovative decision-making methods in the modern educational environment" (Dmitrienko et al., 2017, p. 162). But for this development to take place, Maslow suggests, the teacher must appeal to students' systems of values and support their thinking pedagogically. When students are exposed to the concept of hierarchical needs and self-determination, there is a personal growth trajectory of self-development and a shift toward "overcoming inner hesitations and personal problems" (p. 164). The process of self-determination, as defined by Deci and Ryan (1986), is that it is an independent activity performed by a human and that person's ability to choose independently or autonomously the direction for their own self-development. Students recognize that given the risk factors in the environment, it is only through self-determination, they can overcome the challenges they face. They recognize that the act of self-determination is an intricate and delicate process, giving them the freedom to choose the direction of their "own development and ability to act regardless of external circumstances in a way." Individuals observe small changes happening within themselves, yet outsiders do not witness the changes within the individual, "but consistent with its innate logics and beliefs of the individual," the changes happen (p. 341)—what Vygotsky considered as higher-order thinking.

Results and Discussion

The final results based on students' final portfolios and presentation revealed that 29 out of the 32 students (90%) were highly motivated and determined to overcome their personal challenges. A total of 2 out of 32 (6%) were still worried that they would be unable to perform and without fail the courses they were in currently and had already changed majors twice. They were still unsure "about how their majors will help them and worried what if they fail in their current major?" (student interview). One student left the program as he felt that there was "no support from his parents, and felt society didn't care to address the issues and innate risks students undertake on college campuses." It was sad to see him drop the program, a reminder that more courses on self-actualization need to be developed sooner than later.

Based on students' final portfolios, which consisted of all their papers and writing, developed through the process of concentric circles, they reflected and referred to their growth and attitude toward life at various levels as: "innate growth," "a new sense of liberation," "I can do it attitude," "what happens within me determines what I see outside," and "change is within me." In the following paragraphs, I showcase a few samples of students' growth, perception changes, beliefs, willingness to exceed, and live up to their full potential, and how they interrelate and connect global, national, local and personal identities to see the change within themselves. Students have all been provided pseudonyms.[2]

Evan writes about the process of self-actualization as "Walking by Moonlight: A Journey." His title page begins with a poem:

> From charcoal gray to color untold
> From chains of steel to open sky
> From dusk to dawn's gentle glow
> From impenetrable darkness: To Light.

The student describes the risks in life as a journey of misconceptions and how self-realization and education helped him to understand that recognizing one's identity and being true to the core allows for creativity and greater understanding of human problems, issues, and perspectives on every level. He writes:

> ...the cracks in the foundation of my identity was to deny the truth that I've learned. I am flawed, I am human, there is no escaping that. Nonetheless, I believed I had ... changed; ...and to amend my misconceptions, I must redefine who I thought I was with the clarity of the present. I was and continue to be the remnant of a "Scholarship Boy" ... always successful, I was always confident (Rodriguez; p. 534). I feared failure in every facet of my life, even though I'd never really experienced it. In high school, I suffered a "B" in physics, which dropped my rank from

2 I am truly grateful to my students for being part of this study and allowing me the opportunity to showcase their work. Some of their work has been edited for this study.

first to fifth. I was not cast in a lead role ... and I labeled these memories as *failures*. Along with these came misguided delusions that I even resisted coming to Kent as it did not sound prestigious enough. But now, I absolutely love it.

He continues to express these thoughts in another section as,

> ...Yet, during those moments, and as little as a few months ago, the outcomes felt devastating ... yet I realize now that being vulnerable has made me stronger as Behar suggests, "to write vulnerable is to open a Pandora's Box" (p. 123) ...I have created the identity I need, an identity that can "recognize both the centrality of difference within human identity and the fundamental moral unity of humanity" (Appiah p. 63) ...Ah, yes. Solace. Solace at last. And night turns to dawn.

This student is a leader in many student organizations and a role model for many first-year students.

Another student suggested that growing up fatherless makes a person very vulnerable and fear a lot of things in life, but reading John Berger's *Ways of Seeing* and understanding the process of self-actualization helps in the way we view life, such as:

> Society and culture heavily influence us, and we learn to look at things—"understand the images that surround us and on the culture that teaches us to see things as we do" (p. 140) ... But I think it is necessary to find a balance between individuality and existence in society ... One cannot isolate oneself and worry about the risks one will face in society ... there is a difference between being a cookie cutter human and being able to pull inspiration and creativity from others.
>
> Said showcases that point in his writing about the Palestinians and global strife for justice. Said shows how the identity of this nation has been taken away, they simply exist with nothing to show ... opening my eyes to the fact that *I am blessed with my own identity* ... despite the fact that as I look at the universe, the university, my major, my existence, they make feel tinier and tinier ... but I know what I want and will achieve it.

Randi in her preface refers to one of my class lectures and begins:

> Through these essays, I have dragged my insecurities, my fears, ... my regret into the light and watched them burn in the sun. From these ashes grew my strength, my individualism, my beauty and my power. Through the text, our discussions and writing—I have been able to address many of the complex insecurities that exist in our world and understand the

relationship between society and me. This excavation of the self has been beneficial and eye opening as I continue to uncover the truths under the layers of complex ideas and uncertainties. I have begun to uncover the beauty in myself ... I have found the source of my passion for my goal is to pursue foreign policy. I am a feminist. I am an advocate. I am a voice for the voiceless ... for the first time in my life, I will fight for myself, creating my true parallel identity.

One of the students from the writing class sums up the class teaching, learning, and the whole development process. This student claims that using Maslow's approach to needs was very helpful in understanding how in every stage of our life, we have certain requirements and when they are fulfilled, we want to move to the next. The student elaborates on this by suggesting:

> Now that all the physiological, safety, belonging, esteem needs have been fulfilled or I know where and how to acquire it, now what? I am in college, now what? What should be my course of action? Where should I be heading?
>
> At the beginning of the semester, when you asked us to take a journey, I was perplexed and fought the idea and hated—introspection—as all the deep fears and anxiety that I had nicely kept under covers would loom and make me showcase what I consider as embodied risks. For example, I considered even walking alone in the night with my partner as a risk, as I was fearful someone will attack us. I would get hives thinking about it. The mortal fear of someone finding out about us and who I was. I was ashamed and scared. You were right about referring us to Krieger (2005) as bodies do tell stories. Coming from a very conservative religious family—having partners and (underage) drinking—is a taboo and not permitted. I felt I was failing myself, expectations of my family and people around me ... but what happened over two semesters is a tale that must be told.

Further comments include:

> From the beginning of the semester till now, I see how my energy has been channelized to recognize who am I? I am comfortable with who I am. I have realized what is my role in life, based on these different concentric cycles. I am meant to serve humanity and be a field officer. I know my calling and I will achieve it. People like me are hurt in many ways and misunderstood as we are not the way others see us. We are different, and people see us among their family members, in friends circle, in local communities, in national communities and in global communities—the message is the same—we are misunderstood, so the fear is not just within

me. It is everywhere across boundaries. We are slowly being accepted but we still aren't there. There is a long journey ahead of us, but I know my identity, personal and professional. And, I will prevail.

Recommendations

The above student samples reveal what Gaddis (2002) suggests, "It's like taking off in an airplane: the establishment of identity recognizing our relative insignificance in the larger scheme of things ... each event diminishes your authority at just the moment at which you think you've become an authority" (p. 5). While students, on one hand, recognized that their authority at every level was reduced when they recognized they are a speck in the universe, they also recognized on the other, that they have a certain place in this universe and "it isn't a large one" (p. 5). This change in thought provided them an important perspective that "instead of relating the world to oneself they relate oneself to the world" (p. 5). It is this change in positioning that allowed students to speculate and accept the vulnerabilities within themselves and address it in a bold way in their writing.

The interesting aspect of teaching this course was that although, throughout the semester, students were given choices to respond to either prompts that related to the texts or explore the path of concentric cycles, they all volunteered to take this route and write about their journey in their final portfolio. The only thing I reminded them of was C.D. Wright's quote, "All writing is a risk and a trust; all writing is critical as it pertains to consciousness, and every word of it is for the record." Another interesting aspect is that almost all students from these two classes are in touch with me and send me emails or texts to inform me of their progress and future plans. Last year, when Black Lives Matter movement started, I received an email from one of the students with the subject line "Long overdue message from a former student." Perry in his email refers to the class and writes about how he is proud of his race and being who he is with his identity. In his email, he writes about the class and the takeaway:

> I have had many great teachers both in high school and here at the university. I will never be as blessed as I was to step into our classroom in Johnson (I believe that was the first semester location) and learn from you. You did not teach us what you thought was right and wrong, but instead gave us the tools to guide our own awakening and produce our own ideologies. I will never forget the first day of class when I was offended by you telling us that our ideologies will change. Obviously, you were right about that. And that makes me laugh....
>
> The course on self-actualization has created a lasting impact on me ... While ancient literature and other works are important to read about, there is nothing more critical than creating your own thoughts in this

world (as college students). You could have been like any other teacher and made us write the same papers on the same works of literature. You chose to take a burden upon yourself and challenge us to be better and truly look at ourselves. For that, I am forever thankful to you.

Such emails only validate why we need to teach more courses on self-actualization to first-year students. What students recognize by taking a scaffolded course on self-actualization are: (a) situated risks are a part of our life, (b) addressing embodied risks allows for personal growth, (c) underestimating or overestimating risks only leads to anxiety and fears, and (d) measuring and taking preventive steps allows for a positive attitude toward achieving personal and professional goals and allows for healthy and confident living. The key takeaway being that only by changing the perception of embodied risks from a taboo to a constructive terminology will students feel comfortable to address and overcome their fear factors—to find the right solutions for themselves and for their communities—in a micro- and macroscale.

Discussion Questions

1 How would you define the terms *risk* and *embodied risk*? Using three students' responses, provide a detailed description of why you think there should be more steps taken to address embodied risks in our environment.
2 What would you consider as embodied risks in your life, and how would you describe them?
3 Do you think it is important to explore your identity in terms of the concentric circles of Vygotsky's Zone of Proximal Development and why? How will it help you in your life?

Assignments

1 Based on reading this chapter, using Maslow's Hierarchy chart, have you ever wondered and explored questions that relate to you: Who am I? Why am I here in this universe? What is my purpose? Have all my needs been met—of food, shelter, clothing, etc.? Now that I am in college, what should I be doing? I am in search of what—define. If you have not explored these questions, then it is time to explore them, by using the concentric cycle chart, from inward to outward. At every stage of exploration, ask yourself, what is my opinion on this subject matter or why do I feel this way? Why do I have this belief system on this subject matter? Who are the people who have influenced me? Where are they gaining their insight/information from? Is this what I truly believe in? If not, why? Then, what are my beliefs and why? Your objective should be to understand yourself as much as possible and to assess your strength and weakness; addressing them will allow you to become more … (complete the blanks after you have explored all the circles).
2 Do you feel the students' explanation about embodied risk is true? How do you see their transformation from the beginning to the end of semester as a sign of growth and maturity, and as a sign of developing confidence-in-the-self?

References

Barocas, R., & Gorlow, L. (1967). Self-report personality measurement and conformity behavior. *Journal of Social Psychology, 71*(2), 227–234. https://doi.org/10.1080/00224 545.1967.9919785.

Bartholomae, D., Petrosky, T., & Waite, S. (Eds). (2017). *Ways of reading: An anthology for writers.* Bedford/ St. Martin's, Macmillan Learning.

Bazerman, C. (2019). Development makes history, where inside meets outside. In *Didactique du français et construction d'une discipline scientifique* (pp. 83–92). OpenEdition Books. https://doi.org/10.4000/books.septentrion.78401.

Bordages, J. W. (1989). Self-actualization and personal autonomy. *Psychological Reports, 64*(3), 1263–1266. https://doi.org/10.2466/pr0.1989.64.3c.1263.

Bracher, M. (1993). *Lacan, discourse, and social change: A psychoanalytic cultural criticism.* Cornell University Press.

Callaghan, T., Rochat, P., Lillard, A., Claux, M. L., Odden, H., Itakura, S., Tapanya, S., & Singh, S. (2005). Synchrony in the onset of mental-state reasoning: Evidence from five cultures. *Psychological Science, 16*(5), 378–384. https://doi.org/10.1111/j.0956-7976.2005.01544.x.

Couture, M., Desrosiers, J., & Leclerc, G. (2007). Self-actualization and poststroke rehabilitation. *International Journal of Rehabilitation Research, 30*(2), 111–117.

Deci, E. L., & Ryan, R. M. (1986). The dynamics of self-determination in personality and development. In R. Schwarzer (Ed.), *Self-related cognitions in anxiety and motivation* (pp. 171–194). Erlbaum.

Dennett, D. (1987). *The intentional stance.* MIT Press.

Dmitrienko, N. A., Gorbina, M. A., Porozhnyak, N. F., Trusova, T. V., & Konovalenko, T. G. (2017). Formation of students' professional self-actualization in modern educational environment. *Journal of Social Studies Education Research, 8*(2), 161–177. https://doi.org/10.17499/jsser.360871.

Erikson, E. H. (1968). *Identity youth and crisis.* W. W. Norton & Co.

Gaddis. J. L. (2002). *The landscape of history: How historians map the past.* Oxford University Press.

Gardner, H. (2006). *Changing minds. The art and science of changing our own and other people's minds.* Harvard Business School Press.

Hanover Research (2014). *Early alert systems in higher education.* Academy Administration, Practice.

Igoe, K. J. (2018, September 26). *Critical analysis of our daily lives.* Harvard T. H. Chan School of Public Health. https://www.hsph.harvard.edu/ecpe/critical-risk-analysis-daily-lives.

Kavanagh, A. M., & Broom, D. H. (1998). Embodied risk: My body, myself? *Social Science & Medicine, 46*(3), 437–444. https://doi.org/10.1016/s0277-9536(97)00188-3.

Krieger N. (2005). Embodiment: A conceptual glossary for epidemiology. *Journal of Epidemiology and Community Health, 59*(5), 350–355. https://doi.org/10.1136/jech.2004.024562.

Lefcourt, H. M. (1966). Internal versus external control of reinforcement: A review. *Psychological Bulletin, 65*(4), 206–220. https://doi.org/10.1037/h0023116.

Markus, H. R., & Kitayama, S. (1991). Culture and the self: Implications for cognition, emotion, and motivation. *Psychological Review, 98*(2), 224–253. https://doi.org/10.1037/0033-295X.98.2.224.

Maslow, A. H. (1943). A theory of human motivation. *Psychological Review, 50*(4), 370–396. https://doi.org/10.1037/h0054346.

Maslow, A. H. (1954). *Motivation and personality*. Harper & Row.

Ryff, C. D. (2014). Self-realization and meaning making in the face of adversity: A eudaimonic approach to human resilience. *Journal of Psychology in Africa (South of the Sahara, the Caribbean, and Afro-Latin America), 24*(1), 1–12. https://doi.org/10.1080/14330237.2014.904098.

Shamay-Tsoory, S. G., & Aharon-Peretz, J. (2007). Dissociable prefrontal networks for cognitive and affective theory of mind: a lesion study. *Neuropsychologia, 45*(13), 3054–3067. https://doi.org/10.1016/j.neuropsychologia.2007.

Vygotsky, L. S. (1978). *Mind in society: Development of higher psychological processes*. M. Cole, V. John-Steiner, S. Scribner, & E. Souberman (Eds.). Harvard University Press.

Vygotsky, L. S. (1986). *Thought and language* (Alex Kozulin, Trans.). MIT Press.

Warford, M. K. (2011). The zone of proximal teacher development. *Teaching and Teacher Education, 27*, 252–258. https://doi.org/10.1016/j.tate.2010.08.008.

Yin, R. K. (2014). *Case study research design and Methods*. (5th ed.). Sage.

II

Representations of the Earth's Body

Part Four

6

THE OHIO RIVER

Re-Imagining Water Risk through Embodied Deliberation

Barbara George and Heather Manzo

In understanding the need for technical communication to more fully represent intersectional risks within environmental communication, this chapter explores representations of the earth's "body," particularly in the ways waterways are discursively framed. It focuses on communication surrounding water risk related to the Ohio River headwaters, part of the Ohio River Basin and Pittsburgh's famed "Three Rivers" that converge in the city and are often referred to as "arteries" of the region (Heinz Endowments, 2019). Any body of water is a web of tributaries leading to and mingling with other bodies. In this case, the Ohio emerges from many tributaries leading southward from Lake Erie in Pennsylvania, and the Ohio River exists as a tributary itself, traveling through several states before eventually joining the Mississippi River. Bodies of water can impact other bodies, human and nonhuman. Stacey Alaimo (2010) discusses the "interconnections, interchanges, and transits between human bodies and nonhuman natures" before centering her work on the idea that natural bodies, or ecosystems, and human bodies have profound interactions. This chapter discusses the problems of the river's pollution and the difficulties in changing regulations and explores solutions for reimaging the river as a healthy ecosystem. It focuses on the way language can contribute to both the problems and solutions in terms of river health.

There are many immediate examples of the tensions that exist in balancing the needs of human communities with the ability for ecosystems to provide the natural resources requested of them. Sometimes this is referred to as *ecosystem services*, the positive benefits ecosystems provide to people and animals. Balancing what can be asked of or taken from an ecosystem before that ecosystem fails is known as *carrying capacity*. It is important to consider that ecosystems should not be viewed in a way that puts focus on what can be taken for human

DOI: 10.4324/9781003266549-8

consumption and recognize that nature provides much more than raw materials to be extracted and used for human consumption or economic gain.

This balance, and tension when there is an imbalance, can be seen at the intersection of human communities demanding too much of an ecosystem and can be seen in the way the Ohio River is currently being framed. Access to nature has been proven to have positive mental and physical health benefits; the Japanese call this *shinrin-yoku*. People desire a closeness to nature, and it is our responsibility to make choices that leave wild spaces for animals to migrate, aquifers to recharge, perennial ecosystems to thrive, and rainwater to infiltrate and recharge aquifers rather than cause flooding.

This chapter attends to social justice concerns in technical communication by exploring how waterways are framed through language. Water has been a flashpoint for environmental protections and social justice concerns; however, there are important questions about the variations of protections afforded to different bodies of water, and by extension, bodies living by that water, which can often be observed through language patterns that describe and entrench policies for different waterways. This space is repeatedly influenced by a variety of intersectional factors: both nonhuman and human. Human activity impacting the river is particularly problematic in terms of layered and accumulating or *intersectional* risks: raw sewage, outlets, non-point pollution, including agricultural runoff and toxic discharge, acid mine drainage (AMD) from past mining industrial practices, lack of water quality standards for "bacteria and pathogens, PCBs, lead, mercury, metals, organics and other pollutants," and the system of dams that impact natural ecosystems, among other issues (Ohio River Foundation, n.d.).

Problem: The headwaters of the Ohio River in Pittsburgh have been designated a "working river," which narrowly defines the river as a waterway to move commodities.

Solution: Networked communication and policy efforts that expand the definition of this body of water beyond the rigid framing of the river as a commodity are needed. For example, various stakeholders, people with an interest in a given ecosystem, seek to portray this waterway as an embodied, or tangible space. By framing water language through an *intersectional lens* (a way to understand the environment through layered risks and environmental justice) and through an *environmental justice lens* (the meaningful involvement of all who might be impacted in an ecosystem), more just protections will be afforded in spaces like Appalachia where environmental protections have lagged.

This research applies a discourse analysis lens to explore how particular *ideologies*, or a system of fundamental beliefs about the world, found and reiterated in language, frame the Ohio River (Dryzek, 2013; Stibbe, 2015). *Discourse analysis* studies language as related to its social use. The Ohio River is a waterway that intersects multi-state boundaries, but its framing through policy and definition as a "working river" leads to a lapse in possible water protections and contributes to the marginalization of the river as a "body" and the bodies that live near the river. According to the EPA, the Ohio River is currently one of the most polluted rivers in the United States (Johnston, 2019). This brings us to the question: How and why do certain bodies of water attain protected status with language framing that leads to policy, while others do not? By exploring language or discourse patterns about risk, we can advocate for more equitable water rights, bringing bodies of water and the organisms that exist in this water to more parity.

To understand how discursive framing can lead to protections, this research explores how other bodies of water in the northeastern United States have been framed linguistically or through discourse through an intersectional environmental justice lens that has allowed for protections in places like the Chesapeake Bay and the Great Lakes, both of which are protected, federally, by interstate agreements and federal funding. Discourse analysis is also used to understand how various Ohio River stakeholders, from current institutional regulatory, or managing stakeholders, to quasi-regulatory stakeholders currently use discursive frames for the Ohio River, including complications in terms of how its working river designation solidifies industrial use as an Appalachian and Midwest waterway that makes protections more difficult. Finally, this research analyzes intersectional environmental justice discourses used by nonprofit and activist groups who seek the same federal designations as the Chesapeake Bay and the Great Lakes for the Ohio River. Here, we explore how groups are seeking to redefine and reframe the Ohio River as a natural asset through a variety of practices, including networking and resource pooling from various Ohio River stakeholders. Through networking, these groups seek to highlight both storylines of embodied, situated risk inherent in current industrial practices, and alternatives to that storyline by expanding stakeholder input, collecting data, publishing research, and offering education about the Ohio River.

There are important questions about the variations of protections afforded to different bodies of water, and, by extension, bodies living by that water, which can often be observed by analyzing language patterns that describe different waterways. By framing water discourse as through an intersectional environmental justice lens, more just protections will be afforded in spaces like Appalachia where environmental protections have lagged.

Regulating a Working River

The Ohio River's name comes from the Iroquois word meaning "The Great River." The headwaters begin with the joining of the Allegheny and the Monongahela in Pittsburgh, Pennsylvania, and end as a tributary to the Mississippi in Illinois. This 981-mile stretch of water flows through or touches on the border of 6 states and its drainage basin includes 14 states (National Park Service, 2019). Since pre-colonial times, the Ohio River has been used to move goods and people, but it was during colonial times that the river became a political dividing line between European land claims of the French and British, then between the British Empire and the new world settlements (NSP, 2019). That very body of the river, like many rivers, was used for land boundaries suggests a colonizing mindset, where intricate socio–ecological relationships from indigenous cultures become "erased" (Whyte, 2016).

Initially, the Ohio River was used to move agricultural products domestically, but by the early 1800s, the Ohio River was increasingly used to move commercial commodities toward seaports for export, confirming the working nature of the river as key infrastructure facilitating the extraction of natural resources, which continues to this day. According to the National Park Service, the Ohio River carried more than 230 million tons of commercial traffic in 2019 (NPS, 2019). This shaping of the Ohio River continues. A contested Shell ethylene cracker plant was recently constructed on the banks of the Ohio River not far from Pittsburgh, which creates plastic pellets from the controversial practice of HVHF or fracking (Corkery, 2019).

The Ohio River's work in service to natural resource harvesting in Pennsylvania is inextricably tied to generations of laborers and speculators fueling pre- and post-colonial growth during the Industrial Revolution, and into today's continued push for natural gas, timber products, and development projects. A chemical analysis of Ohio River sediment near its origins in Pittsburgh, Pennsylvania, at the confluence of the Allegheny and Monongahela Rivers, shows this legacy. This legacy sediment is attributable to human disturbances from sources such as commercial, residential, and infrastructure development; stormwater, industrial, agricultural and mining runoff; and raw sewage discharged into the rivers during heavy rain events through combined sewage overflow pipes.

This story of pollution is not unique to the Ohio River Basin, and in an effort to alleviate and mitigate pollution to the "Navigable Waters of the US," the first U.S. law to consider water pollution, the Federal Water Pollution Control Act (FWPCA) of 1948, was enacted in 1948. The period following FWPCA included mounting evidence that the way natural resources were managed were resulting in a toxic environment, disruption of ecosystems, and negative consequences to human health. The idea that humans needed highly functioning ecosystem services to have the human species survive grew over the next decade and a half and by the 1960s, the modern environmental movement was a part of America's growing awareness of environmental concerns (Cox, 2006; Kroll, 2001).

As such, water has been a driving force for policy making about environmental risk. Numerous landmark regulatory protections, including the development of NEPA (National Environmental Policy Act) and the EPA (Environmental Protection Agency) in the 1970s stem from public outrage as related to the Santa Barbara oil spill and a fire on the Cuyahoga River in 1969 (Cox, 2006). After the first Earth Day, organized efforts to strengthen laws related to pollution pushed legislators to amend the 1948 FWPCA. In 1972, sweeping changes were made and the Clean Water Act was born (EPA, 2020). The EPA gives states oversight of definitions and regulations delegated to states. In Pennsylvania, this is the responsibility of the Department of Environmental Protection. The Pennsylvania DEP delegates regulatory enforcement of Chapters 102 and 105 of the Clean Water Act to Conservations Districts, which operate on the county level.

However, due to the Ohio River being defined as part of the "Navigable Waters of the US," the DEP works in tandem with the Army Corps of Engineers to manage the Ohio River, so protection of waterways is complicated by various language and policy frames of waterways and various stakeholders with various interests (33 CFR Part 329, n.d.). Because the Ohio River is used for commerce, it is subject to management through dam construction and other engineering projects that alter the Ohio River's body, most notably through a series of locks and dams. The Army Corps of Engineers can make decisions over water bodies that "are subject to the ebb and flow of the tide and/or are presently used, or have been used in the past, or may be susceptible for use to transport interstate or foreign commerce" (33 CFR Part 329, n.d.). "Navigable Waters" also include interstate wetlands and all other intrastate bodies of water that could affect interstate or foreign commerce, recreational interests of interstate or foreign travelers, or for interstate industrial purposes. These provisions for the Army Corps to control and potentially alter through civil engineering means the natural ebb and flow of any waterway that has or could be used to move commerce confirms the working nature of a river in terms that do not necessarily include ecosystem services, recreational access, or access by river adjacent communities. There has been movement toward more sustainable collaborations with the Army Corps with groups like the Nature Conservancy, such as the Sustainable River Programs, but these have yet to yield results (Nature Conservancy, 2021).

While there are stakeholders and agencies that push for a wider protection of the Ohio River due to concerns about the health of the Ohio River and its ecology, the research and advocacy surrounding the Ohio River largely does not have the political agency, particularly with federal protections, as compared with bodies of water on the East Coast, for example, either the Chesapeake Bay or the Great Lakes Restoration Initiative. This can be clearly seen by comparing discourses that represent these different bodies of water.

This is despite the "protections" of the Ohio River headwaters that imply protection. For example, The Pennsylvania Constitution Article I § 27 declares (Pennsylvania General Assembly, n.d.):

> The people have a right to clean air, pure water, and to the preservation of the natural, scenic, historic and esthetic values of the environment. Pennsylvania's public natural resources are the common property of all

the people, including generations yet to come. As trustee of these resources, the Commonwealth shall conserve and maintain them for the benefit of all the people.

(May 18, 1971, P.L.769, J.R.3)

While there is a "right" to pure water in language, in actual practice, state environmental regulatory agencies routinely balance Pennsylvanian's right to quality natural resources and the economic opportunity of extraction. Legal challenges often side with industry (Widener School of Law, 2010). This lack of protection for waters is another instance of lack of protection for human and ecological health in the industrial Midwest and Appalachian regions.

The current state of the river is testament to questionable regulation created by discursive frames and attendant policy. The Ohio River, in particular, has been beset with an industrial approach to use that is very much still in effect. The river is largely seen as an industrial service provider—as a waterway and as a commodity. The historical extractive approaches to the environment in southwest Pennsylvania center on the way the river has become a repository for pollution. Layers of industry leave a legacy in the body of the Ohio River. The Clean Water Act was supposed to have eliminated discharge of polluted water in the United States Waterways by 1985 (Congressional Declaration, n.d.). This goal was not met, and there is no anticipated future date where this will be achieved (FelKamp, 2020). By many measures, the Ohio River has become more polluted since then, with little oversight. For example, EPA data gathered by the nonprofit group Eye on Ohio shows 6,900 pollution discharge points in the Ohio River that have associated National Pollutant Discharge Elimination System (NPDES) permits. This large quantity of pollution discharge points is not exhaustive and does not include non-permitted sites or permitted sites that are not reporting as active (FelKamp, 2020).

Despite the wide footprint of the river and the focus on the working river designation, there is little research about environmental justice and intersectional approaches to the bodies of water as riverways in the industrial Appalachian and Midwest regions, including the Ohio River. While there are studies about the impact of attendant industries in the Ohio River's drainage basins, calling for environmental justice lens to address fracking and mountaintop removal, as well as effects on human health (Morrone, 2017; Kolkowski & Perkins, 2016), there is little research about risks inherent in the working river designation that has become status quo, leading to a dearth of protections.

New Models for Deliberating Environmental Policy

The designation of particular waterways having protections is also complicated by how stakeholders are able to communicate and discuss environmental policy.

Finding a deliberative space to examine issues of complex environmental risk is difficult, partly because of the struggle in determining who might have "true" authority or expertise to engage in these deliberations. Environmental risk scholars suggest risk representations are socially situated and socially contested (Beck, 1987; Lash et al., 1996; Sauer, 2003). This is particularly salient in variations of protected water status. Why are some waterways protected and others are not? When environmental risk assessment is linked to science (as it often is), an emerging body of work exploring rhetorics of science reveals nuances of authority, *ethos*, or credibility of the speaker, and the descriptive or prescriptive approaches to science and policy. Historically, there has been a divide in science between knowledge claims and policy. The way in which science claims are used by different stakeholders to establish authority, particularly in environmental studies, makes "science" an important area of rhetorical investigation in understanding environmental risk assessment and policy. Whose "science" speaks for what particular environmental policy, especially as linked to water policy?

This idea of ideologies informing different definitions of protection is highlighted in Frank Fischer's concerns about ideologies that underscore environmental policy:

> The production of knowledge about environmental problems and the subsequent policy responses are more a function of political and economic power than "objective science," and that leaving decisions up to "the experts" is simply a recipe for furthering the interests of corporate and state elites at the expense of both ordinary citizens and the environment.
>
> *(2000, pp. 263–264)*

Of central importance to this chapter is the use of stakeholder networks to address complex, costly, and potentially lethal uncertainties surrounding practices on bodies of water. This scholarship has informed more recent work in sustainability studies, which links knowledge-making across disciplines through complex theories, which increase validity in addressing complicated issues. Sustainable science, in general, is concerned with the emergence of more complex notions of science and decision-making (McGreavy et al., 2012). In response to the emergence of sustainability studies, several researchers promote public deliberation in the interest of a sustainable future. These rhetors acknowledge possible deliberative spaces among varied authorities and ideological frameworks that showcase ways authorities might deliberate (Goggin, 2009; Scholz, 2011).

One of the reasons consensus or agreement is so difficult for complex environmental projects is because varied stakeholders often seek to make decisions based on different values, using different discourses. Dryzek's (2005) notions of discourse as "a shared way of apprehending the world" explains a variety of discourses, often disparate, about the environment that, when combined with political power, frame policies and practice through particular discourse "stories" (pp. 9–10). Similarly, Arran Stibbe (2015) defines environmental ideologies as "belief systems about how the world was, is, will be or should be which are shared by members of particular groups in society" (p. 23). Certain discourses, such as those that normalize industry are often accepted as the "status quo," but alternative discourses based in environmental justice and intersectionality question beliefs that Dryzek outlines as patterns found in environmental discourse: ontologies, assumptions about natural relationships, key players (human or nonhuman), and metaphors that frame particular actions.

Part of water regulation includes policy that is meant to regulate or manage a given waterway. However, not all bodies of water are defined or protected in the same way through the technical language in regulatory agencies. This brings up questions about intersectional environmental justice in approaching a definition of water risk, which might impact environmental and human health, including whether stakeholders are included in the process of definition, and whether their long-term accumulated, intersecting, embodied, and layered risk concerns are acknowledged—an intersectional consideration (Crenshaw, 1989; Holifield et al., 2010; Simmons & Grabill, 2007; Walker, 2010).

Routinely, we see that industrial discourses, ideologies, and storylines permeate descriptions of the Ohio River, leading to policies and actions that reiterate the pollution injustices that occur on the river and minimizing risk to human and nonhuman bodies. By analyzing discourse patterns surrounding new approaches redefining the Ohio River beyond a working river, including intersectional and environmental approaches that understand waters and people and organisms along the river as embodied vs. static entities that do not interact, new ways of interacting with the river might be enacted. This is particularly true in light of other waterways that have been successful in this redefinition.

Discourse Analysis as Applied to Environmental Policy

Both authors of this chapter, Barbara and Heather, grew up and live in the Rust Belt. As such, they have seen, throughout their lives, the way past industrial communities deal with the long-term effects of intersectional pollution from heavy industry. We argue that social justice concerns are many: they exist at the intersection of poverty, public health, access to green space, racist development practices, food apartheid, subpar housing and schools, lack of medical care, and lack of public transportation. These divestment communities that we both have witnessed are often plagued by soils contaminated by lead and other heavy metals and drinking water carried by lead pipes. Lead is a known human toxin, which, when ingested, leads to permanent damage to the brain, kidneys, and can cause death.

This toxic legacy in soils and water often occurs in low-income communities of color where zoning allows businesses and industries to discharge toxins in the air, soil, and water. Working with communities on any project that includes an environmental or social justice goal requires a practical and tactical engagement plan as a first step. Practitioners are cautioned to develop relationships with community members as a first and highest-level outcome, before tackling environmental issues. This ensures respect for the agency of community members and the culture of the community.

Issues related to social diversity in this work are multidimensional and include: research to identify risks in local communities, many of which are communities of color or low-income communities; development of equitable model ordinances on natural resources used by diverse communities; oversight of regulatory enforcement of laws; data mining, collection, and GIS mapping to increase understanding of how variables such as storms, green infrastructure, and best management practices impact the built and natural environment. Research methods referred to in the nonprofit section include feminist technical research methods such as interviews, think-aloud-protocols, participatory methods, community member engagement, surveys, focus groups and listening circles, program delivery and evaluation, fiscal sponsorship for developing projects, and creating scholarship funds for low-income youth to access professional development on natural resource topics. Research is conducted on natural resource quality indicators in soil and water in areas that have experienced legacy industrial pollution.

It is important to make genuine relationships with people and get to know them. This takes time and can be overlooked in the rush to apply solutions to environmental problems. It is important to not act in the role of a savior or overpromise. Respect and active listening, sharing meals, getting to know elders are all critical components to building successful communications. Understanding how, and in what modes (text, cell, social media, in-person meetings) community members and practitioners communicate is necessary to meet people where they are digitally. Finally, using language that is nontechnical is essential to being understood.

The analytical research in this chapter applies a discourse analysis lens to explore how particular ideologies found and reiterated in language frame the Ohio River (Dryzek, 2005; Stibbe, 2015), specifically applying analysis to the problem of lack of risk protection and the possibility of more protections. The Ohio River is a waterway that intersects multi-state boundaries, but its framing through policy and definition as a working river leads to a lapse in possible water protections and contributes to the marginalization of the river as a body and the bodies that live near the river.

Discourse analysis is a powerful analytical tool to explore patterns about the ways in which dominant neoliberal energy discourses become normalized within social contexts. This analysis is applied to the language on websites used by various stakeholders. By comparing languages from different stakeholders with different discursive approaches to the river, one can see various frames or stories that permeate discourse (Dryzek, 2005).

For example, a *green politics discourse* frames the river as an ecological entity, with complex ecosystems and interconnections between humans and nature. Green politics also brings into play the idea of extended stakeholder communities that communicate about complex environmental concerns. By contrast, a *Promethean discourse* frames the river as matter, or part of a commodity system, driven by markets with little to no acknowledgment of the interactions between bodies of water and other bodies. An *administrative rationalism discourse frame* views the river as a resource that can be managed, so while this discourse might acknowledge, there may be interactions between bodies of water and other bodies, the approach to any risk in this relationship is that the risk can be managed. These discourses appear in language on various stakeholder's sites, revealing underlying ideologies that orient humans and the river in particular ways. The problem with Promethean and administrative rationalism discourses, many stakeholders argue, is that they wait until an environmental issue has proven harms (these may occur immediately or over time), and only then are regulations created or enforced. This many point source issue is particularly difficult in terms of many intersecting environmental justice concerns that exist through time.

We apply discourse frames to texts by analyzing language found on various stakeholders' websites to understand how bodies of water and the relationship to other organisms are constructed. Two of the sites are used to compare language and policy to the Ohio River to show the protections for some waterways and the lack of protections for others. Other stakeholders are chosen because they are the primary regulatory institutions making policy about the river or are the primary driving forces behind redefining the river (see Table 6.1). While we are not sampling all language on a website, the samples we choose have potential impact on many species' bodies, and are listed prominently on websites, clarifying current policies that are underscored by particular discourses. Also, we note that many stakeholders often use "green" language at some points, but actual policy and action reveal deeper and sometimes problematic ideologies, so, when possible, we choose examples that highlight language and policy in action versus only language. In addition to analyzing language on sites in assessing green politics discourses as related to the Ohio River, we note the way networked stakeholders communicate strategically and beyond hierarchies to enact policy to promote more just definitions of the how the Ohio River might be envisioned beyond a working river to a river with a complex ecosystem, a body of water that affects other human and nonhuman bodies and that is worthy of protection.

TABLE 6.1 Comparative discourse

Stakeholder name	Discourse ideology	Example
Chesapeake Bay Foundation Great Lakes Restoration Initiative	Green Politics Here, examples show complex ecosystems and a willingness to work across varied agencies.	"As the saying goes, 'everything flows downstream.' If we are to Save the Bay we must also save the hundreds of waterways that flow into it." (Chesapeake Bay Foundation, 2021) "Since 2010 the multi-agency GLRI has provided funding to 16 federal organizations to strategically target the biggest threats to the Great Lakes ecosystem." (Great Lakes Restoration Project, 2021)
Army Corps of Engineers ALCOSAN (Allegheny County Sewer Authority)	Promethean Here, nature is seen as "brute matter" that is controlled by human	• Water infrastructure • Environmental management and restoration • Response to natural and man-made disasters • Engineering and technical services (Army Corp of Engineers, n.d.)
Pennsylvania DEP ORSANCO	Administrative Rationalism Nature is protected through human expert management	"The Ohio River Valley Water Sanitation Commission (ORSANCO), was established on June 30, 1948 to control and abate pollution in the Ohio River Basin." (ORSANCO, 2021)

A Tale of Three Waterways and Language

As outlined earlier, the Ohio River's current discourse frame is that of a working river. However, the frame of a working river does not create space for dialogue or recognition of ecological and public health realities associated with this body of water. While a working river serves industry, it often does not include ecosystem services or public health needs in that definition of *working*. Residents of six states, roughly 10% of the U.S. population, use the Ohio River for drinking water, food, and recreation despite the fact that large sections of the river have biological, sewage, and heavy metal levels that are above regulatory thresholds. Aquatic life faces pressure from toxins and physical barriers from dams that disrupt species health, migration, and reproductive patterns. As the river works to convey freight, effluent from the oil, gas, and coal industries act as a sink for unmitigated sedimentation and legacy sediments all impact the communities and species that depend on this river for drinking water, habitat, food, and recreation (Ohio River Foundation, 2021).

By contrast, bodies of water like the Chesapeake Bay and the Great Lakes Restoration Initiative have numerous interstate stakeholders collaborating through nonprofit, legal, and environmental designations that protect the water and the ecosystems surrounding them.

Green Politics Discourse: Chesapeake Bay Foundation and Great Lakes Restoration Initiative

Stakeholder collaborations have been gaining agency to preserve the ecology of the Chesapeake since legislation was enacted beginning in the 1970s and 1980s. The efforts have resulted in a plethora of studies that maintain protections for a broad ecosystem: "The Chesapeake Bay is arguably the most studied large body of water on earth. Forty years of intense scientific investigation by leading estuarine scientists have identified why the Chesapeake is degraded and how to fix it." ("The History of Chesapeake Bay Cleanup Efforts" on the Chesapeake Bay Foundation website). Protective policies grew from interstate and federal protections through the Chesapeake Bay Program Partnership (Governance and Management Framework, 2020).

The language about the Chesapeake Bay highlights the way that body of water, and bodies near the water, are protected through policy. For example, green politics discourse is apparent throughout the Chesapeake Bay Foundation site, showing intersectional environmental justice ideologies in the following lines of "Mission Statement" of the Chesapeake Bay Foundation first refers to the bay as a "National Treasure." Later, the mission is described as a group "fighting" pollution collaboratively for the betterment of a larger ecosystem.

> Serving as a watchdog, we fight for effective, science-based solutions to the pollution degrading the Chesapeake Bay and its rivers and streams. Our motto, "Save the Bay," is a regional rallying cry for pollution reduction throughout the Chesapeake's six-state, 64,000-square-mile watershed, which is home to more than 18 million people and 3,000 species of plants and animals.
>
> *(Chesapeake Bay Foundation, 2021)*

Similarly, the Great Lakes have recently been afforded protections through the interstate Great Lakes Restoration Initiative that seeks to restore lake ecosystems as a valuable source of freshwater: The Great Lakes Restoration Initiative accelerates efforts to "protect and restore the largest system of fresh surface water in the world." Again, the protection of this body of water and attendant bodies has been established through powerful stakeholders, including those that offer federal protections, collaborating to recast the lakes and their ecologies as having "value" (Great Lakes Restoration, n.d.). In the "About" section of the Great Lakes Restoration Initiative, intersectional environmental justice ideologies are seen at play by sections on strategic collaborations across stakeholders, ecosystems, and healthy outcomes for a variety of species.

Promethean Discourse, Ohio River Headwaters

While some waterways' ecosystems are valued in complex ways and through protections, the Ohio River struggles to evoke discursive and policy frames of value and protection from important stakeholders responsible for current policy. The forces that frame the Ohio River mainly show the river as a commodity to move commodities versus a complex ecosystem, revealing a Promethean discourse focused on markets, placing humans as actors that approach nature as "brute matter." An acknowledgment that bodies might interact does not appear in official language in this discourse.

For example, the Allegheny County Sanitary Authority (ALCOSAN) sanitary sewer infrastructure in Allegheny County controls combined sewage overflows (CSOs) (ALCOSAN, n.d.), which allow raw sewage to enter the Allegheny, Monongahela, and Ohio Rivers during high rain events. Essentially, at its headwaters in Pittsburgh, the Ohio River is a sewage repository after large rain events. While ALCOSAN is currently under a consent decree from the EPA to reduce the volume of untreated water discharged from, several of ACLOSAN's plans are couched in Promethean discourse ideologies. For example, a new "Clean Water Plan" claims to lower incidents of "sewage overflows into rivers and streams almost every time it rains" (Hopey, 2020). While providing infrastructure to better address the sewage issue is laudable, the deadline to do so has been extended to 2036, the green infrastructure possibilities are optional and not required, and part of the plan includes risk: "It also proposes building more than 15 miles of underground 14-foot-in-diameter storage tunnels along the city's rivers to capture the overflows during storms" (Hopey, 2020). This is an interesting case because sewage becomes divorced from the bodies from which it emerges and from the body of the Ohio River. A potential plan to store raw sewage under drinking water is a public health concern that is not addressed by the proposed "Clean Water Plan." The assumption is that these storage tunnels will somehow be immune to leaks or degradation through time. Here, the river is seen as a storage facility for raw sewage, though the public is largely unaware of this plan.

Additionally, other risks are largely ignored with other projects that view the river as moving commodities without connections to other bodies. The Army Corps of Engineers recently approved a plan for fracked wastewater which will go into place in 2021. Because fracked wastewater has many unknown risks, and the possibilities of spills are quite high. However, this is framed as an economic benefit for the movement of matter through Promethean discourse of a barging facility that suggests benefits for this permit (Comtech Industries, n.d.). Again, in these lines, the river is seen as a commodity—nature is seen as brute matter in the service of markets and prices. However, concerned citizens' reservations about spill risk and radioactive waste that might affect many bodies of many species are not addressed (Hunkler, 2021).

Administrative Rationalism: Ohio River Headwaters

There are many regulatory agencies that frame the Ohio River ecological services that can be managed. While there are some environmental protections with this approach, this ideology is limited in terms of a lack of communicative ability to understand and approach risk as intersectional and impacting bodies in cumulative ways. For example, the EPA has made suggestions for minimizing pollutants in the sewage plan noted in the previous section but lacks the ability to mandate true targets and protections.

Like the Promethean discourse approach, nature is subordinate to humans, in this case, human problem solving. Environmental risk is considered, but the assumption is that experts can manage risk. Another specific example of this that relates to the body of the river, bodies of fish species, and human bodies occurs with fish consumption advisories put out by a regulatory group, the Ohio River Valley Water Sanitation Commission (ORSANCO), which is linked to several other regulatory agencies, including the Pennsylvania Department of Environmental Protection, to provide public health guidance on what contaminants have made their way through the food chain to fish caught in the Ohio River. Contaminants include bacteria from raw sewage overflows, mercury, and polychlorinated biphenyls (PCBs), which have a cumulative effect in the human body and can lead to birth defects and cancer (Ohio River Fish Consumption Advisories, n.d.).

The advisories are listed by fish species, recommending if fish may be eaten at all, and if so, outlining the maximum number of fish that may be consumed per month and per year. The map begins at Pittsburgh, Pennsylvania, at the confluence of the Monongahela and Allegheny Rivers where the Ohio River begins, and is divided into units. The section of the Ohio River is labeled Unit 1, and the following species are recommended to never be eaten in any frequency include bottom feeders such as carp and catfish, and a maximum of six fish per year are recommended (Common Carp, n.d.). What is noteworthy about this approach is the lack of work to understand the root causes of these risks. While bodies are considered in terms of safe consumption, the status quo of the sewage issue is maintained by limiting the number of fish recommended.

Collaborative Stakeholders and Green Politics

Despite the current Promethean and administrative rationalism discourses that are the primary ideologies underlying current policy and practice in terms of the Ohio River as a working river, this research analyzes intersectional environmental justice discourses used by nonprofit and activist groups who seek the same federal designations as the Chesapeake Bay and the Great Lakes for the Ohio River. Here, we explore how groups like ORBA (Ohio River Basin Alliance) and the collaboration of the Heinz Endowment and the UPenn Water Center, the Allegheny Conservation District, and the National Wildlife Foundation are seeking to redefine and reframe the Ohio River as a natural asset through networking and resource pooling from various Ohio River nonprofits. While there are many other groups, these examples highlight the concept of networking green politics intersectional environmental justice concerns. In networking, these groups seek to highlight both storylines of embodied, situated risk inherent in current industrial practices, and alternatives to that storyline by expanding stakeholder input, collecting data, publishing research, and offering education about the Ohio River. These discourse moves are within a green politics discourse pattern that allows for ideologies of intersectional environmental justice to be part of the conversation that more inclusively engage with a wider variety of stakeholders.

For example, the University of Pennsylvania Wallace Water Center (UPENN) and Heinz Water Works, with UPENN acting as a convener and facilitator with local stakeholders allows local actors to create an empowered interorganizational network. This network acts in a way that uses a collective impact model that shifts from organizational level wins on individual projects to planning for and acting on opportunities to improve the water and ecosystems of the Ohio River (Peterson & Zimmerman, 2004). They do so by stacking waterway improvement opportunities through multi-organizational grant-making, improvement projects, and coordination of other activities across nonprofits, holding the accomplishment together as a network rather than siloed organizations. This transformative work is changing culture and relationships between stakeholders in Southwestern Pennsylvania who work on Ohio River Water projects from organizationally controlled and owned into network ownership and may hold the key to shifting the perception of the role of a body of water.

One notable difference between these groups and the regulatory agencies listed in the administrative rationalism section is the way in which these groups network leadership. For example, the Allegheny Conservation District (2021) outlines communicative acts that move from hierarchical discourses where information is siloed to a networked communication possibility where leadership is collective, diverse, collaborative, and recursive. These communication patterns open up opportunities across stakeholders from different agencies to more effectively enter into deliberative spaces about the health of the body of the Ohio River, and finding these deliberative platforms is an environmental justice concern.

In Figure 6.1, the Allegheny Watershed Alliance (n.d.) shows the value of networking in the initial statement "partnerships and supporters are powerful." Different groups are framed as having different types of expertise from various environmental nonprofits, showing wide concerns about the environment and an understanding of a complicated ecosystem that can impact bodies in various ways.

The potential to transform discourses within regulatory agencies who, currently and in the past held to Promethean and administrative rationalism discourses and policies is also clear. The ORBA is a nonprofit arm of ORSANCO, which works with the Army Corps of Engineers. This group is interesting in that it emerges from traditional regulators. In 2020, the organization refreshed its working committees and for the first time, it invited voting members from

Partnerships and supporters are powerful. We all share a common goal of improving the environment and our communities. We can make incredible strides if we team up with the organization that has the capacity and expertise for a particular job. Our region is rich with innovative and capable organizations!

PARTNERS

Engages and leads through partnerships, innovation, and implementation to conserve, promote, and improve Allegheny County's natural resources.

To protect and restore Pittsburgh's urban forest through community tree planting and care, education, and advocacy.

Engaging and empowering residents to eliminate illegal dumping in their community.

To protect the water quality of the Monongahela, Allegheny, and Ohio Rivers, and their respective watersheds.

FIGURE 6.1 Partnerships.

Southwestern PA. The re-envisioned working groups are centering issues around equity, environmental justice, and community impact of industry and pollution that affect human and nonhuman bodies. The new model of communication has led to broader discussions of the river beyond that of a working river, but one where goals including water, ecosystems, stakeholder education, risk management, recreation, and economics are more clearly explored (Ohio River Basin Goals).

What is noteworthy about this collaboration is the way in which the collaborative effort offers a counter to the working river narrative, even when offered in collaboration with the Army Corps of Engineers. Additionally, a networked discourse approach allows wider discussions of risk to bodies—the river as a body and human and nonhuman life as impacted by the body of the river, acknowledging connections between bodies. Statements from key stakeholders from the UPENN and Heinz Water Works show a green politics discourse approach that underscores intersectional environmental justice ideologies that offer wider perceptions of the river (The Heinz Endowments). Selected statements from the 2019 "Accelerating Transformational Change in Pittsburgh's Three Rivers" Phase I Report from this group details a green politics discourse where specific risks to the river body and attendant risk to human and nonhuman bodies are outlined. In addition, models for a more just approach to water protections are shared.

This intersectional environmental justice work through a green politics discourse extends to policy. A group intent on aligning federal protections with new frames for the river is through the National Wildlife Federation's efforts. Murray and Rau (2020) outline the National Wildlife Federation's efforts to have the Ohio River classified as a federal waterway, with the goal of science-based ecosystem restoration and equity and justice language that considers ecosystems as benefiting a variety of stakeholders. Similarly, selections detail a variety of indicators for the health of bodies in complex ecosystems, including "healthy species and habitats, hydrology, water quality, aquatic invasive species, climate change, and healthy communities" (Murray & Rau, 2020, p. 7). Further, this group explicitly calls for restoration of the Ohio River Basin, using collaboration and justice as a goal for future projects, again, linking complex understandings of ecosystems and bodies in them for an understanding of the Ohio River beyond the working river designation it currently has:

> In summary, we are proposing a general restoration framework for the Ohio River Basin that will consider the conditions of species and habitats and stressors affecting them, use conceptual models to inform targets and strategies, identify additional research, monitoring and other information needs, and consider equity and justice and restoration economy principles and information in producing a comprehensive ecosystem restoration plan.
> *(Murray & Rau, 2020, p. 9)*

This approach challenges the status quo of the Ohio River as a working river and offers a wider view of a complex ecosystem with many bodies.

Conclusion

The Ohio River has been beset by issues of intersecting pollution for over a century. While other waterways enjoy protected status through federal programs, the Ohio River has a current designation as a working river and current plans from key regulatory agencies reiterate that designation, compounding concerns about pollution for the body of the river and the bodies of human and nonhuman organisms that interact with the river. However, many stakeholders from various groups are working to reframe the designation into an understanding of a river as a body that has value in terms of many complex ecosystems within a green politics discourse that allows for intersectional environmental justice concerns. The success of a new frame can happen through collaborative networking toward action: working groups establishing better ways to reimagine the Ohio River. This networking involves new forms of communication with collaborative, networked language. While much of this work is in the early planning stages, intentional networking can reframe the body of the Ohio River, leading to federal protection that might act as a model for other Appalachian and Midwest waterways that have historically been impacted by heavy industrial practices that, to this day, are found in the bodies of the river and the human bodies that interact with the river.

Discussion Questions

1   Typically, ecosystem risks in the United States are designated by agencies like the Environmental Protection Agency; however, there has been a call to widen the way risks are defined. Discuss how regulatory agencies define the Ohio River and how

nonprofit groups define the Ohio River. Who defines the risk? What kinds of words are used to describe the risk? Is the process participatory?

2 Much of the story of environmental risks and possibility for previously marginalized land is changing. The Great Lakes Restoration Initiative took years to develop, but now offers protections for the once industrial Great Lakes. Look at the Great Lakes Restoration Initiative website. How are the Great Lakes depicted through language and images now? What is the new story of the Great Lakes as revealed through this initiative? How is this more inclusive than past uses of the area? Are there any ecosystems in your immediate vicinity that have a changing "story"? How does language or images show that change?

3 Solar farms promise fossil-free electricity. When the embodied energy, which is the total amount of energy needed to make, transport, use, maintain, and dispose of solar panels is calculated, there must be a net gain to make solar a worthwhile proposition. Often, solar farms are installed on agricultural land or in prairie ecosystems. Investigate a solar project. Are those systems impacted? What language discusses those impacts? Discuss how those ecosystems could be impacted. Then, imagine and discuss ways of creating a solar farm that could repurpose antiquated or unprofitable spaces in the built environment. What language could reimagine this land use?

Assignments

1 Find a website of an environmental page devoted to a waterway, land area, or other ecosystem. Using discourse analysis, look at the words used. How do particular words create perceptions of the ecosystem you chose? Who is the speaker (could be a group)? Who is the audience? Are there attempts made to include concerns of a diverse audience?

2 Find a website devoted to an environmental risk in an ecosystem. Using discourse analysis, look at the words used. How do particular words create perceptions of risk? Who is the speaker (could be a group)? Who is the audience? Are there attempts to include concerns of a diverse audience?

References

33CFR 329. (n.d.). https://www.nap.usace.army.mil/Portals/39/docs/regulatory/regs/33cfr329.pdf.

Alaimo, S. (2010). *Bodily natures: Science, environment and the material self.* Indiana University Press.

ALCOSAN (n.d.). https://www.alcosan.org.

Allegheny County Conservation District (2021). Chapter 105 programs. https://www.conservationsolutioncenter.org/solution-center/soils/chapter-105-program-information.

Allegheny Watershed Alliance (n.d.). About. https://www.awapa.org/board-partners.

Army Corps of Engineers (n.d.). About. https://www.lrp.usace.army.mil/About/Mission-and-Vision.

Beck, U. (1987). Risk society and the provident state. In Lash, Szerszynski & Wynne (Eds.), *Risk, environment, & modernity: Towards a new ecology* (pp. 28–43). Sage.

Chesapeake Bay Foundation (2021a). About the bay. https://www.cbf.org/about-the-bay/more-than-just-the-bay.

Chesapeake Bay Foundation (2021b). About. https://www.cbf.org/about-cbf/our-mission.

Common Carp (n.d.). Ohio River fish consumption advisories. http://216.68.102.178/comm/fishconsumption/carp.asp.

Comtech industries (n.d.). Produced water barging. https://comtechindustriesinc.com/our-solutions/barging.

Congressional Declaration (n.d.). USC Title 33. https://www.govinfo.gov/content/pkg/USCODE-2011-title33/html/USCODE-2011-title33-chap26-subchapI-sec1251.htm.

Corkery, M. (2019, August 12). A giant factory rises to make a product filling up the world: Plastic. *New York Times.* https://www.nytimes.com/2019/08/12/business/energy-environment/plastics-shell-pennsylvania-plant.html.

Crenshaw, K. (1989). Demarginalizing the intersection of sex and race. *University of Chicago Legal Forum, 1,* 139–167.

Dryzek, J. (2005). *The Politics of the earth: Environmental discourses.* Oxford University Press.

EPA (2020). History. https://www.epa.gov/laws-regulations/history-clean-water-act.

FeldKamp, B. (2020). We mapped the toxic wastewater discharges along the Ohio River. Here's what we learned. *Allegheny Front.* https://www.alleghenyfront.org/we-mapped-the-toxic-wastewater-discharges-along-the-ohio-river-heres-what-we-learned.

Fischer, F. (2000). *Citizens, experts, and the environment: The politics of local knowledge.* Duke University Press.

Goggin, P. (2009). *Rhetorics, literacies, and narratives of sustainability.* Routledge.

Governance and Management Framework (2020). Chesapeake Bay program. https://www.chesapeakebay.net/documents/CBP_Governance_Document_version_3.1_%28updated_03.31.2020%29.pdf.

Great Lakes Restoration Initiative (2021). About. https://www.glri.us/about.

Heinz Endowments & The Water Center, University of Pennsylvania (2019). *Accelerating Transformational Change.* https://watercenter.sas.upenn.edu/accelerating-transformational-change-in-pittsburghs-three-rivers/

Holifield, R., Porter, M., & Walker, G. (2010). *Spaces of environmental justice.* Wiley-Blackwell.

Hopey, D. (2020). Long-awaited federal approval granted for ALCOSAN wet weather plan. *Pittsburgh Post-Gazette.* https://www.post-gazette.com/news/environment/2020/05/15/Long-awaited-federal-approval-of-Alcosan-wet-weatehr-plan/stories/202005150129.

Hunkler, B. (2021). Barging radioactive frack waste threatens the Ohio River. *Concerned Ohio River Residents.* https://www.concernedohioriverresidents.org/post/barging-radioactive-frack-waste-threatens-the-ohio-river.

Johnston, A. (2019). "That's vinegar." The Ohio River's history of contamination and progress made. *Environmental Health News.* https://www.ehn.org/ohio-river-pollution-cleanup-2641307895.html.

Koslowski, M., & Perkins, H. (2015). Environmental justice in Appalachia Ohio? An expanded consideration of privilege and the role it plays in defending the contaminated status quo in a white, working-class community. *Local Environment, 21*(10). https://doi.org/10.1080/13549839.2015.1111316.

Kroll, G. (2001). The "silent springs" of Rachel Carson: Mass media and the origins of modern environmentalism. *Public Understandings of Science, 10*(4). https://doi.org/10.3109/a036878.

Lash, S., Szerszynski, B., & Wynne, B. (Eds.). (1996). *Risk, environment, & modernity: Towards a new ecology.* SAGE.

McGreavy, D., Silka, B., & Hart, D. (2012). Creating a place for environmental communication research in sustainability science. *Environmental Communication, 6*(1), 23–43.

Morrone, M. (2008). Environmental justice and health disparities in Appalachia, Ohio. In P. H. Liotta, D. A. Mouat, W. G. Kepner, & J. M. Lancaster (Eds.), *Environmental change and human security: Recognizing and acting on hazard impacts.* NATO Science for Peace and Security Series C: Environmental Security. Springer. https://doi.org/10.1007/978-1-4020-8551-2_14.

Murray, M., & Rau, E. (2020). Seizing the day for Ohio River restoration: A vision for science-based ecosystem restoration integrated with equity and justice. *Ohio River Basin Symposium & Summit 2020.* https://www.kyscience.org/docs/OhioRiverBasin_EcoRest_BackgrdDoc_NWF.pdf.

National Park Service (2019). *The Ohio River.* https://www.nps.gov/articles/the-ohio-river.htm#:~:text=Beginning%20at%20Pittsburgh%2C%20Pennsylvania%2C%20the, 130%20feet%20near%20Louisville%2C%20Kentucky.

Nature Conservancy (2021). *What we do.* https://www.nature.org/en-us/what-we-do/our-priorities/protect-water-and-land/land-and-water-stories/sustainable-rivers-project.

Ohio River Basin Goals (n.d.). Army Corps of Engineers. https://www.lrh.usace.army.mil/Missions/Civil-Works/ORBA/ORBA04.

Ohio River Fish Consumption Advisories (n.d.) What are fish consumption advisories? http://216.68.102.178/comm/fishconsumption/default.asp.

Ohio River Foundation (n.d.). About. https://www.ohioriverfdn.org/about_us/about_orf/index.html.

ORSANCO (2021). About. http://www.orsanco.org/about-us.

PA General Assembly (n.d.). *The constitution of Pennsylvania.* https://www.legis.state.pa.us/cfdocs/legis/LI/consCheck.cfm?txtType=HTM&ttl=00&div=0&chpt=1&sctn=27&subsctn=0.

Peterson, N. A., & Zimmerman, M. A. (2004). Beyond the individual: Toward a nomological network of organizational empowerment. *American Journal of Community Psychology, 34*(1–2), 129–145.

Sauer, B. A. (2003). *The rhetoric of risk: Technical documentation in hazardous environments.* Erlbaum.

Scholz, R. W. (2011). *Environmental literacy in science and society: From knowledge to decisions.* Cambridge University Press.

Simmons, M., & Grabill, J. (2007). Towards a civic rhetoric for technologically and scientifically complex places: Invention, performance and participation. *College Composition and Communication, 58*(3), 419–448.

Stibbe, A. (2015). *Ecolinguistics: Language, ecology, and the stories we live by.* Routledge.

Walker, G. (2010). Environmental justice, impact assessment and the politics of knowledge: The implications of assessing the social distribution of environmental outcomes. *Environmental Impact Assessment Review, 30*, 312–318.

Whyte, K. (2016). Indigenous experience, environmental justice and settler colonialism. *SSRN.* http://dx.doi.org/10.2139/ssrn.2770058.

Widener School of Law (2010). *A citizen's guide to Article I, § 27 of the Pennsylvania Constitution.* https://blogs.law.widener.edu/envirolawcenter/files/2010/03/PA_Citizens_Guide_to_Art_I_Sect_27.pdf.

7

PRIVATE GROUNDWATER CONTAMINATION AND INTEGRATED RISK COMMUNICATION

Simon Mooney, Sarah Lavallee, Jean O'Dwyer,
Anna Majury, and Paul Hynds

Water is the basis upon which human development is sustained and thereby a critical focus of modern risk management. The continued availability and consumption of this intrinsic resource stems largely from the withdrawal of groundwater (i.e., water contained within subterranean soils and layers of permeable rock or *aquifers*) from the earth's subsurface. Groundwater is the most extracted raw material on earth and currently serves approximately 2.2 billion people (Gleeson et al., 2016; Murphy et al., 2017). It is further estimated that groundwater supplies 50% of drinking water and 33% of the water used for industry globally (UN-Water, 2018). Where appropriately utilized, groundwater provides a ubiquitous asset in arid and remote regions lacking a perennial surface water source and is uniquely insulated from anthropogenic contamination and climate change impacts (Green, 2016; Lall et al., 2020). Accordingly, the equitable use of groundwater qualifies as a modern global priority—its importance embodied in ongoing recommendations to assign groundwater a priority focus under the UN's Sustainable Development Goals (Velis et al., 2017).

A large volume of historical groundwater abstraction globally is limited to the last half-century owing to advancements in drilling technologies (Gorelick & Zheng, 2015; Margat & van der Gun, 2013). As a consequence of the concurrent, interrelated expansion of rural electrification, land availability, and irrigation requirements, a large quantity of recent abstractions have centered in rural areas (GWP, 2012; Villholth, 2016). Parallel to the increased extraction of groundwater for agriculture has been the proliferation of private groundwater wells for domestic use. Private wells, encompassing boreholes, dug wells, and tubewells, pose an accessible, affordable alternative to public mains and other point-of-use water supplies and have thus become the dominant conduit of local drinking water in many rural regions. While the global reliance on private wells

DOI: 10.4324/9781003266549-9

is unknown, they are believed to contribute significantly to drinking-water intake within rural South Asia and Africa and serve over 10% of the combined Canadian and U.S. population (Chappells et al., 2014; Grönwall & Danert, 2020). The newfound geographical prevalence of these household and community-level supplies has rendered groundwater a coupled socio-ecological system, underscoring the value of effective risk management toward preventing supply degradation and human health ramifications downstream (di Pelino et al., 2019).

Problem: Despite conferring benefits, the global escalation in private groundwater abstractions has placed contamination risk at the forefront of modern groundwater issues alongside resource depletion (Fienen & Arshad, 2016; Lall et al., 2020). The historically uncoordinated installation and maintenance of private wells in rural regions (typified by high geochemical, agro-industrial, and localized wastewater disposal activity) has been central to diminishing groundwater's impenetrability to contaminant ingress and exposed well users to myriad harmful pollutants (Gorelick & Zheng, 2015; Murphy et al., 2017). At present, over 200 million people worldwide are exposed to toxic groundwater concentrations of natural and synthetic chemicals while as many as 59 million annual cases of acute gastrointestinal illness (AGI) are attributable to microbial groundwater contamination (Murphy et al., 2017; Podgorski & Berg, 2020). With private wells susceptible to an ever-increasing number of contamination episodes and mechanisms via climate change-induced extreme weather events (EWEs), the applicability of an integrated, interdisciplinary risk-management response is unequivocal (Andrade et al., 2018; Kløve et al., 2014). However, the global paucity of private well regulation largely prohibits direct government intervention, vesting first-order risk-mitigation responsibilities (e.g., well testing and treatment) with well users themselves. Integrated *risk communication* thus serves as a proxy for both risk assessment and broader risk management, emerging as the most universal, pragmatic means of addressing private groundwater contamination.

Solution: Risk communication can alleviate barriers hindering private well stewardship but cannot eradicate them in and of itself, with supply contamination often remaining a persistent problem irrespective of information conveyance (Jakeman et al., 2016). In addition to elucidating a complex, dynamic risk, communicators must variously address rural isolation, limited communication resources, diffuse stakeholders, divergent risk perceptions, social inequality, and culturally entrenched behavioral impediments (Limaye, 2017; Re, 2015). Formidable though these obstacles are, the past decade of applied and theoretical research into groundwater end-user outreach reflects a gradually intensifying effort to disentangle and optimize communication of private groundwater contamination risk (Colley et al., 2019; Mooney et al., 2020a). A recent body of literature drawing from risk-communication theory and the socio-ecological systems paradigm has sought to promote corrective, feasible risk-communication frameworks. The dawn of the sustainability science of *socio-hydrogeology*, which conceptualizes the dynamic interactions and feedback between hydrogeological and social systems, plots a course to facilitate the bidirectional, multidisciplinary engagement required to accomplish effective, integrated groundwater risk communication (Barthel et al., 2017; Re, 2015).

Private groundwater contamination represents a truly unique contemporary risk—at once global yet predominantly rural and remote; spatially ubiquitous, yet largely invisible. The issue variously underscores urban versus rural inequality, social justice issues, discrepancies between risk-management disciplines, and the increased interconnectedness of environmental health risks, encapsulating many pressing challenges within modern risk communication (Balog-Way et al., 2020). With groundwater management entering a crucial era in reckoning with climate change and increasingly widespread human–water interactions, the moment is opportune to reflect on the instrumental role that risk communication plays in addressing private groundwater contamination (Grönwall & Danert, 2020; Hynds et al., 2018). Building on an emerging body of research, the current chapter presents a comprehensive examination of private groundwater risk communication, outlining:

- Global private groundwater contamination issues and management
- Historical risk-communication responses in developed and developing regions
- Barriers to groundwater end-user risk perception/behavior and associated predictors
- Social justice and diversity concerns within private groundwater risk communication
- Potential future frameworks for effective private groundwater risk communication.

Private Groundwater Contamination and Management: A Global Overview

Sources and Mechanisms of Private Groundwater Contamination

Contamination of private groundwater supplies is conventionally considered a developing region concern due to disparities in economic resources and sanitation services compared with developed nations (Li et al., 2021). However, contemporary epidemiological research affirms that private well contamination is neither a socio-economically nor regionally homogeneous phenomenon (Murphy et al., 2017; Podgorski & Berg, 2020). This fact is testament to the multitude of causal factors such as well depth, local geology, hazard proximity, and, increasingly, climate change. Prevention is heavily contingent on acknowledgment of contamination mechanisms and maximum permitted contaminant thresholds for drinking water quality, non-observance of which may lead to acquisition of fatal respiratory and gastrointestinal illnesses (Lall et al., 2020). Sources and types of private groundwater contamination can be broadly classified as both natural and anthropogenic.

Natural private groundwater contamination is typically a result of the dissolution of geological mineral deposits, with such contaminants referred to as *geogenic* (Figoli et al., 2016; Li et al., 2021). Geogenic contaminants of human health concern have been identified in private wells since the 1960s, the most significant of which are uranium, nitrate, fluoride, and arsenic (Grützmacher et al., 2013; Hug et al., 2020). The health impacts of chronic exposure to such contaminants (among them arrhythmia, cancer, and methemoglobinemia) are extensively documented in geochemical hotspots spanning North America, Central Africa, and South Asia (Hug et al., 2020). While often associated with deep aquifers, geogenic contaminants may also occur at shallow depths—as typified by a national tubewell installation program undertaken in Bangladesh in the late 20th century, which inadvertently exposed millions to elevated levels of shallow arsenic and caused the largest mass poisoning in human history (Kumar et al., 2020). Saltwater intrusion into coastal aquifers poses a further natural threat and has contributed to severe groundwater degradation across the Mediterranean Basin, Mexico, and other arid regions (Alfarrah & Walraevens, 2018; Lall et al., 2020). Consumption of saline-rich groundwater is associated with raised risk of cardiovascular disease and a problem exacerbated by rising sea levels due to climate change (Talukder et al., 2017).

Anthropogenic contamination of private wells arises from sources as varied as manure fields, pesticides, septic tanks, mineshafts, and pharmaceutical industrial facilities (Lall et al., 2020; Lapworth et al., 2012). Contaminants consist of pathogens (i.e., bacteria, protozoa, and viruses) and industrial chemicals ranging from cyanide and lead to emerging chemical contaminant classes such as polyfluoroalkyl substances (Lee & Murphy, 2020; Richardson & Kimura, 2020). As pollutants comprise both microbial and chemical contaminants, anthropogenic contamination of groundwater presents a ubiquitous, double-jeopardy threat in many rural regions. Microbial contamination of private wells may lead to incidents of severe gastrointestinal illnesses while elevated levels of synthetic chemicals are linked with heightened risk of thyroid disease, cancer, and birth defects (Lee & Murphy, 2020). The issue received considerable attention in the early 2000s, following a large-scale outbreak of *E. coli-* and *Campylobacter-*related disease originating from a public groundwater well in Walkerton, Canada, which resulted in over 2,000 local illnesses and six deaths (Chique et al., 2020; Murphy et al., 2017). The event was initiated by ingress of cattle manure containing enteric pathogens subsequent to two days of heavy rainfall, illustrating the ability of EWEs to expedite transport of anthropogenic contaminants into unprotected groundwater supplies (Andrade et al., 2018).

Risk Management

Heightened recognition of the international scale of private well contamination has coincided with the emergence of two dominant contemporary paradigms for water resources risk management—integrated water resources management (IWRM) and risk governance. While these approaches, in principle, promote groundwater management in harmony with related aquatic and terrestrial resources and bottom-up, participatory risk mitigation, their utility pertaining to private groundwater supplies is contentious (Pittock et al., 2016; Simpson & de Loë, 2020). As the fundamental management unit of IWRM is the river basin, IWRM is ultimately incompatible with large-scale groundwater supply management—a reflection of the longstanding disciplinary boundaries between hydrologists and hydrogeologists (Foster & Ait-Kadi, 2012; Staudinger et al., 2019). The viability of risk governance is meanwhile compromised by the ambiguity of the term *governance* and absence of representative bodies for private sector actors (e.g., well drillers) and well users (Grönwall & Danert, 2020; Villholth et al., 2018). Further problems stem from groundwater's innately invisible, dispersed nature, which has rendered implementation of private supply regulation (e.g., well registration) a global challenge due to prospective financial costs and perceptions of private groundwater access as riskless and an inherent human right (Villholth et al., 2018). Although a number of countries such as the United States have been able to enact subnational or, in rare cases, national regulations, the majority of direct government-led risk management measures are limited to groundwater monitoring using purpose-drilled wells (Figure 7.1) (Molle & Closas, 2020; Zheng & Flanagan, 2017). This places the onus on private well owners to undertake the bulk of mitigating actions and

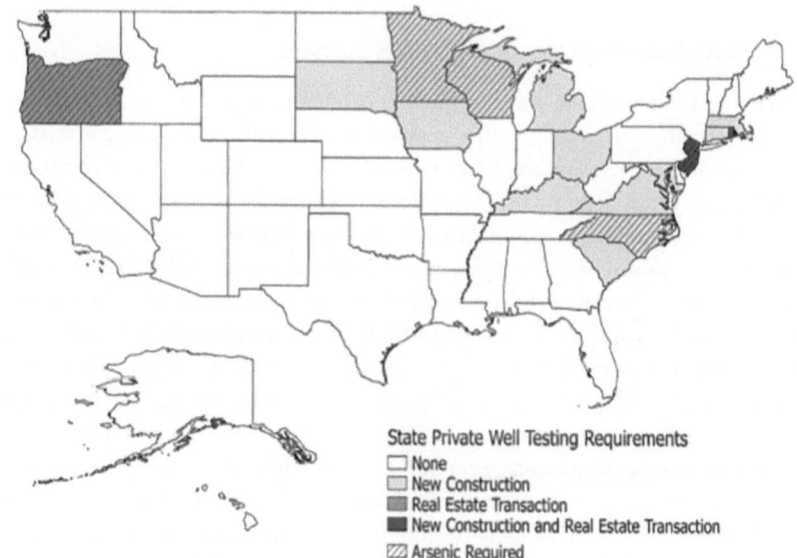

FIGURE 7.1   State-level requirements for private well testing in the United States (Zheng & Flanagan, 2017).

communication of risk information at the uppermost level of importance in managing private groundwater contamination risk.

In light of the growing interlinkage between contemporary risks and number of implicated actors, the definition of risk communication has been expanded to represent integrated, two-way communication between experts and at-risk populations concerning the likelihood and impact of a threat to human health (Gamhewage, 2014; Heath, 2018). This definition stipulates multidimensional conceptions of risk and reciprocal, interdisciplinary knowledge exchange as essential to optimal conveyance of risk (Lundgren & McMakin, 2018; Shreve et al., 2016). While the merits of this approach in the context of a socio-ecological risk such as groundwater contamination are clear, empirical evidence would suggest that categorically *integrated* communication has been scarce (Mitchell et al., 2012; Mooney et al., 2020b). A range of deficiencies have been identified including absence of empirical or theoretical bases, limited target-audience categorization, fragmented depictions of risk factors and absence of climate change adaptation information (MacDonald Gibson & Pieper, 2017; Mooney et al., 2020b). Such shortcomings are magnified when combined with practical difficulties in directly engaging with often remote, undefined populations (private wells are rarely inventoried), and limited financial resources (Cole & Murphy, 2014; Molle & Closas, 2020). In view of the central role played by government authorities in providing risk information to groundwater end-users, the performance of top-down communication interventions warrants analysis.

Historical Risk-Communication Responses

Two broad modes of risk communication can be discerned based on directness of information conveyance: active risk communication and passive risk communication. Although passive information sources (e.g., online information repositories) often accompany or nominally function as risk communication interventions, public communication campaigns represent the most concerted form of risk communication (Fox et al., 2016; Mooney et al., 2020b). Public communication campaigns targeting private well users most closely resemble individual behavior-change campaigns, which seek to change the behavior of a specific segment of the public to minimize human health risks at various geographical scales (Atkin & Rice, 2012). Individual behavior change campaigns attempt to influence knowledge, beliefs, and perceptions of a given threat, often employing social marketing techniques (principles derived from consumer advertising) to stimulate risk prevention behaviors (Lefebvre, 2013).

While it is difficult to establish a timeline for groundwater risk interventions implemented to date due to publication bias and limited geographical coverage globally, a growing number of interventions have been implemented since the 1990s (Mooney et al., 2020a). Constituent activities or communication mechanisms, as defined by Rowe and Frewer (2005), range from electronic and print (entailing one-way information flow from intervention coordinators to target audiences) to interpersonal (entailing two-way information flow between intervention coordinators and target audiences) and convenience (entailing direct financial or practical supports for target audiences). Overarching groundwater risk intervention types (i.e., distinct outreach strategies) identified within the global literature may be summarized as follows:

- Information campaigns: Behavior promotion campaigns utilizing media and/or interpersonal-based communication mechanisms to encourage supply maintenance.
- Well installation programs: Interventions entailing installation of safe, structurally secure private wells and provision of risk information to promote contamination risk mitigation and conversion from unsafe supplies.
- Well remediation programs: Interventions entailing remediation/retrofitting of unsafe wells and provision of risk information to encourage supply maintenance.
- Well testing services: Interventions entailing voluntary microbiological and/or chemical testing of well water and subsequent provision of test results and supply maintenance information to well users.
- Well treatment system distribution programs: Interventions entailing dissemination of well water treatment systems and/or supply disinfectants to encourage supply maintenance.
- Workshops: Interpersonal, practical interventions delivered in an instructional setting to elucidate key processes supporting supply maintenance and contamination risk mitigation.

Intervention Efficacy Toward Behavior Change and Knowledge Gain

The existing literature outlining the efficacy of well user-oriented risk communication is relatively sparse. While a series of scoping reviews (MacDonald Gibson & Pieper, 2017; Morris et al., 2016) have sought to summarize potential determinants of success, a global systematized review and meta-analyses of communication interventions undertaken by Mooney et al. (2020a) represents the first-known attempt to quantify the efficacy of groundwater risk-communication interventions globally (Figure 7.2). Review findings thus merit consideration in examining the performance of historical private groundwater risk-communication interventions. Mooney et al.'s (2020a) study identified a total of 40 interventions within 31 evaluation studies, with knowledge and behavioral attainment outcomes compared between socioeconomically developed and developing regions via quantitative surveys and water sampling rates.

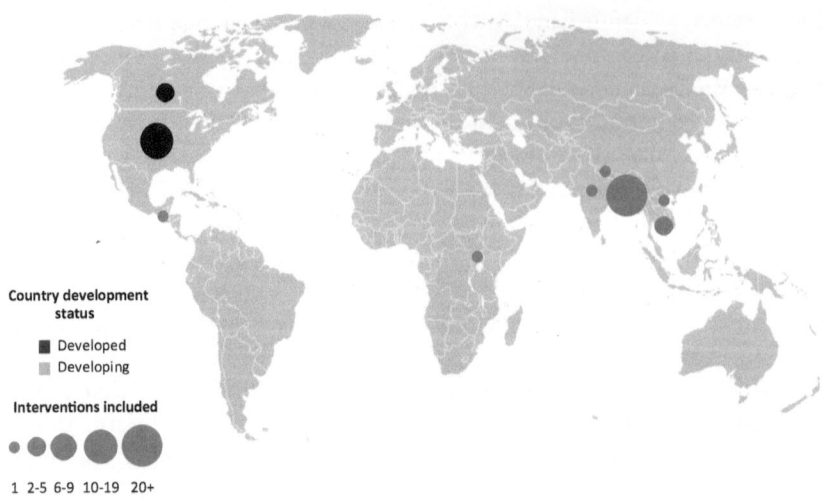

FIGURE 7.2 Global distribution of identified private groundwater risk-communication interventions, 1990–2018 (Mooney et al., 2020a).

Identified interventions attained a mean behavioral attainment of 53% and mean knowledge attainment of 48%, with no significant difference identified between developed and developing regions. While these figures indicate somewhat marginal gains, accompanying study findings denote that intervention outcomes can be attributed to several factors. Presence of an educational coordinating body (i.e., university or research agency) was demonstrated to significantly increase behavioral adoption in both developed and developing regions. As such, involvement of educators in risk-communication campaigns appears to be conducive to risk mitigation; Boholm (2019) suggests that government agencies alone are typically ill-equipped due to internal deficits in financial resources and communications knowledge, while Thornton and Leahy (2012) and Little et al. (2016) highlight the potential efficacy of university-assisted citizen science initiatives as risk-communication apparatuses. As higher behavioral outcomes were also noted in interventions characterized by small geographical reach (i.e., community or local) and use of electronic engagement mechanisms, local-scale interventions paired with both targeted and mass-media activities may represent optimal risk-communication strategies. Higher knowledge outcomes were reported where interventions utilized interpersonal mechanisms—suggesting that practical, direct learning, and information reciprocation may favor knowledge acquisition. Local, interpersonally oriented interventions have been repeatedly recommended in the context of private groundwater as they facilitate trust-building and long-term knowledge gain (Henry & Suk, 2018; Morris et al., 2016).

Barriers to Risk-Communication Interventions

While Mooney et al.'s (2020) review demonstrates potential success factors within prior private well user interventions, several potential shortcomings were identified. Of 40 communicative interventions analyzed, only one intervention (an information campaign in Québec evaluated by Renaud et al., 2011) reported use of a theoretical communicative model—the Precede–Proceed Model. Although message framing was referenced by Paul et al. (2015) in the United States and Bennear et al. (2013) and George et al. (2013) in Bangladesh, no other studies defined whether or not interventions were guided by communication theories or models. Such deficits in reporting indicate that communication science procedures may be underutilized, supporting concerns raised by Mitchell et al. (2012) and Hynds et al. (2018) and empirical data outlining limited adaptation of social science research within groundwater management steps (Barthel & Seidl, 2017; Niu et al., 2014). While policymakers and science communicators have historically opted for an "information-deficit model" approach to promote risk-mitigation actions, the assumed causal relation between this approach and knowledge and behavioral application has been signposted as empirically weak (Simis et al., 2016). Despite recent steps to better align relevant spheres (outlined in the "Potential Future Frameworks for Effective Private Groundwater Risk Communication" section), scholars such as Limaye (2017) note that the hydrogeological community has been historically reluctant to incorporate social sciences research due to disciplinary discrepancies. Absence of collaboration between hydrogeologists, communication scientists, policymakers, and public servants represents a palpable barrier to design of theoretically informed, multidisciplinary-led approach required in contemporary groundwater risk communication.

With respect to risk characteristics, Mooney et al.'s (2020) review identified only one intervention (Lule et al., 2005), explicitly addressing microbial contamination and a limited coverage of local hydrogeological and climatological characteristics. These oversights suggest that microbiological contamination of private wells and the role of climate change are overlooked in existing risk-communication interventions—as evidenced by the obsolescence of current annual supply testing recommendations for microbial contamination in the face of increasing EWEs (Ford et al., 2017; Lim & Prakash, 2020). Khan et al. (2015) note that absence of climate change adaptation in water supply guidelines is globally ubiquitous, with end-users often required to seek such information independently. As public perceptions of climate change impacts may be governed by their own distinct set of parameters (e.g., cultural, experiential, political), impacts of EWEs on private wells represent an integral research focus and topical focus for future groundwater outreach initiatives (Demski et al., 2017; van der Linden, 2015). While communicators must variously navigate organizational barriers, rural isolation, and limited internal resources, they must also contend with public intransigence, distrust, and risk perceptions associated with issues such as climate change. Of crucial importance is understanding the source of such dispositions and the unique challenges and barriers that well owners themselves face in mitigating supply contamination risk.

Risk Perceptions and Health Behavior

Perceptions associated with health and environmental risks represent important parameters in design of effective public health risk-communication strategies and policy implementation. While exceptions may exist where perceived self-efficacy (i.e., confidence in undertaking an action) of perceived applicability of protective actions wield an influence (as documented in the case of flood risk), risk perception generally constitutes an important determinant of health- and risk-related decisions (Bubeck et al., 2013; Ferrer & Klein, 2015; Fox-Rogers et al., 2016). Risk perception refers to the subjective judgment that people make about the severity, consequences, and probability of a risk in relation to their personal susceptibility and often relates to multiple factors including socio-demographics, awareness, and personal experiences (Ferrer & Klein, 2015; Tandi et al., 2018). Examining private well users' perceptions of contamination risk and associated determinants is critical toward understanding decision-making processes underlying well management behaviors. Private well users are regularly faced with a series of decisions (where or when to get water tested, what parameters to test for, etc.), with such decisions often predicated upon both perceptions and knowledge of risks associated with well water contamination. As such, low perceptions of contamination risk can result in failure to undertake necessary protective actions, thus increasing risk of waterborne infection (Chappells et al., 2015; Malecki et al., 2017).

A systematic literature conducted by Munene and Hall (2019) represents a recent attempt to understand the factors that influence perceptions associated with well water quality among private well users in North America. Further contemporary studies outside of North America within developed (Hooks et al., 2019; McDowell et al., 2020; Musacchio et al., 2021; Schuitema et al., 2020) and developing countries have also investigated factors that influence perceptions of risk and health behaviors among private well users. The next section draws heavily from this literature and outlines some of the key socio-demographic, experiential, and psychological factors demonstrated to impact well user risk perceptions and behaviors.

Risk Perceptions and Health Behaviors among Private Groundwater End-Users

Socio-demographics have been repeatedly found to act as potential modifiers to risk perceptions of water quality and the choices well users make regarding supply management (Figure 7.3) (Lavallee et al., 2020; Munene & Hall, 2019). Gender is frequently associated with risk perceptions, with females documented to exhibit higher perceptions of risk regarding well water contamination and health and a greater tendency toward use of well water treatment systems compared with males (Lavallee et al., 2020; Munene & Hall, 2019; Severtson et al., 2006). Length of residence on a property with a well has also been shown to influence risk perceptions and health behavior (Shaw et al., 2005; Ugas et al., 2019). A study conducted by Ugas et al. (2019) found that longer-term residents (reporting residence exceeding 5 years) were more complacent about testing than new occupants (reporting residence of 1–5 years). Newer residents may be less likely to disregard testing as their temporal frame of reference with their well water quality is smaller. In terms of developing regions, several studies report that socio-economic status and risk perceptions were found to be the primary determinants to health behaviors. For example, a study conducted in India by Delaire et al. (2017) reported socio-economic status and perceived likelihood of gastrointestinal illness were the primary determinants to the use of alternatives to untreated shallow groundwater.

With respect to well characteristics, research in Ontario, Canada, found that respondents with dug wells exhibited negative attitudes and higher perceptions of risk associated with groundwater quality, than drilled well users (Lavallee et al., 2020). As dug wells are typically more susceptible to pathogen ingress as water is acquired from relatively shallow groundwater reserves with greater likelihood of contamination from the surface (Hynds et al., 2012), well characteristics may act as potential modifiers to awareness that consequently influence perceptions of risk. The relative distance of supply from household and invisibility of groundwater itself have also been demonstrated to impact risk perceptions; visual and other forms of sensorial imperceptibility (i.e. organoleptic properties) relating to supply risk may significantly impede perceived danger of contamination (Munene et al., 2019). As contamination of private wells often represents a slow-onset risk (the duration of point and nonpoint source contamination ingress into supply may take up to several months) and certain populations may develop immunity to infection overtime, well users may develop a false sense of security (Jones et al., 2006). As contaminant ingress into supplies accelerates in both pace and occurrence through increased occurrence of climate change-induced EWEs, this false sense of security may place well owners who are insufficiently educated about supply maintenance in increased jeopardy (Andrade et al., 2018).

Prior experience has also been shown to act as a predictor of awareness and perceptions of risk (Figure 7.3). Previous studies demonstrate that if well users test once and there is no evidence of contamination, they may believe it unnecessary to submit another sample and place misguided confidence in their water supply quality (Jones et al., 2006; Roche et al., 2013; Ugas et al., 2020). This subjective judgment may be associated with a lack of awareness regarding groundwater transport mechanisms, the fluid nature of groundwater

FIGURE 7.3 Conceptual model ("Well User Risk Perception Model") outlining the potential influential paths pertaining to well users' experience, socio-demographics, well and property characteristics, socio-psychological factors, and health behaviors.

contamination, and the requirement for routine testing (Qayyum et al., 2020). In contrast, well users may perceive their personal risk of waterborne infection to be higher if they or a family/household member have experienced an illness in the past (Imgrund et al., 2011; Jones et al., 2006; McLeod et al., 2015). Of relevance to climate change, a recent survey of Irish well users found that prior experience of an EWE, such as drought and pluvial flooding, is associated with significantly higher-risk perceptions of EWE-induced supply contamination (Mooney et al., 2021). Accordingly, experiences of adverse health effects and/or recurrent problems with or changes in well water quality may influence well users' perceptions of supply contamination and associated health risk and determine the frequency of risk-mitigation actions (Munene & Hall, 2019).

Socio-psychological factors or "risk domains" (i.e., awareness, attitudes, risk perceptions, and beliefs) have been defined as cognitive characteristics potentially affecting personal risk-mitigation behaviors (Figure 7.3) (Lavallee et al., 2020; Musacchio et al., 2021). While these factors are somewhat interrelated, previous research shows that awareness, attitudes, and beliefs exert significant influence on well users' perceptions of risk and consequent health behaviors (Munene & Hall, 2019). A study conducted in the Republic of Ireland (ROI) found that a lack of awareness related to contamination risks, and the relationship between flooding events and potential contamination of well water, decreased perceptions of risk associated with waterborne infections conveyed by flooding (Musacchio et al., 2021). Furthermore, well users' decisions to participate in health behaviors were influenced by whether or not they held a positive attitude toward their water quality. Positive attitudes may influence lower perceptions of risk related to their well water quality, resulting in infrequent well management practices (Summers, 2010). With respect to developing regions, a study conducted in Nicaragua found that positive attitudes toward the use of solar water disinfection were associated with both the intention to adopt and actual adoption of health behaviors (Altherr et al., 2008).

Adoption of Socio-Psychological Models and Frameworks in Behavior-Change Studies

Several socio-psychological models have been developed to understand perceived risks and predict health behaviors in support of effective risk-communication strategies. These models draw from various elements of psychology (i.e., social, health, and motivational) to identify and understand the pathways that influence participation in individual or community health behaviors. Both developed and developing countries have adopted such theoretical models to inform future risk-communication and behavior-change interventions among private well users.

With respect to developing regions, several socio-psychological models (i.e., Risk, Attitudes, Norms Abilities, and Self-regulation [RANAS], Health Action Process Approach [HAPA], Theory of Planned Behavior [TPB], and Protection Motivation Theory [PMT] [Figure 7.4]) have been applied in Bangladesh to systematically review the relative significance of predictors to risk-mitigation behaviors among well users. For example, Inauen and Mosler (2014) found that commitment, descriptive norms, self-efficacy, and perceived vulnerability were the main predictors of health behaviors (i.e., using arsenic-safe wells) using the RANAS approach. Similarly, a combination of two approaches, namely, RANAS and HAPA, was applied in a study conducted by Inauen et al. (2013a). Findings indicated that self-efficacy and descriptive norms were the strongest predictors of arsenic-safe water consumption (e.g., arsenic-safe wells, rainwater harvesting, switching to neighbors' uncontaminated well water). Furthermore, the TPB approach has been implemented in two Bangladeshi studies (Inauen et al., 2013b; Mosler et al., 2010). Specifically, Inauen et al. (2013b) extended the TPB by including the factor of commitment (i.e., reminders, implementation intentions, public-self commitment), and found commitment strength to significantly increase the health behavior associated with switching to arsenic-safe wells. Mosler et al. (2010) combined the PMT and TPB and found that social norms, self-efficacy, and perceived taste of shallow tubewells were all influential factors for the use of arsenic-safe wells.

Socio-psychological models and frameworks, such as RANAS, Health Belief Model (HBM), and the Common Sense Model (CSM) (Figure 7.4), have been applied in developed countries to examine relationships between socio-cognitive factors and health behaviors among private well users. A common theme was found within the literature that used the RANAS approach. Specifically, the "norms" factor was significantly associated with gender (McDowell et al., 2020), well water testing behaviors (Flanagan et al., 2015), and adoption of suitable behaviors when a flooding event occurs (Andrade et al., 2019; Musacchio et al., 2021). The HBM has been used to explain and predict individual changes in health behaviors among private well users by focusing on individuals' perceptions on a number of factors. Both studies that used the HBM alone as a

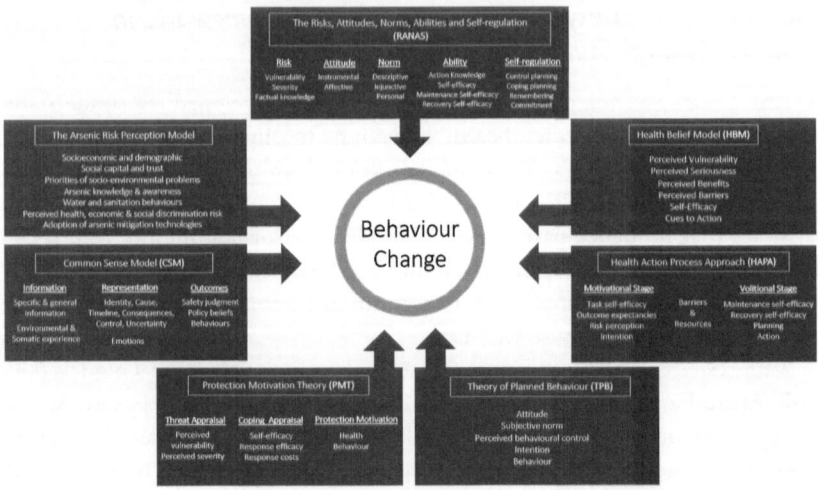

FIGURE 7.4 Overview of the seven socio-psychological models and associated factors used in private well user risk perception literature. Models were adopted to understand perceived risks and predict health behaviors.

framework for predicting well water testing behaviors indicated that reminder cues to action would encourage testing (Munene et al., 2020; Straub & Leahy, 2014). Perceived barriers included concerns related to cost of treatment systems and how a well water issue would influence their property value (Straub & Leahy, 2014), while cost of testing was a barrier in an Irish study using both the HBM and RANAS frameworks (Musacchio et al., 2021). Last, a study framework based on the CSM was applied by an American study to understand how well owners responded to information about arsenic-contaminated well water (Severston et al., 2006). Over half of surveyed well owners perceived their water as "safe" and of "good quality." Additionally, external information and experience (i.e., perceived water quality and health effects) was found to influence health behaviors (i.e., actions used to mitigate personal exposure to arsenic-contaminated wells).

Barriers to Risk Perception and Risk Mitigation

Risk perceptions among well users have been shown to exert a significant influence on the adoption of voluntary health-related behaviors such as well testing, treatment, maintenance (Colley et al., 2019; Morris et al., 2016; Munene & Hall, 2019) (Figure 7.3). Therefore, barriers to risk perceptions and health behaviors must be identified to design effective risk communications that promote targeted protective actions to private groundwater-reliant populations. Several systematic literature reviews have been conducted to understand the barriers to risk perception and health behaviors (Colley et al., 2019; Morris et al., 2016; Munene & Hall, 2019). Numerous factors additional to lack of knowledge have been identified, such as inconvenience and cost, weak social norms, a lack of perceived vulnerability, and feelings of control (Colley et al., 2019; Hooks et al., 2019; Hynds et al., 2013; Morris et al., 2016; Munene & Hall, 2019; Schuitema et al., 2020).

Previous studies have shown that travel requirements and the distance to and operating hours of testing laboratories and health centers can make well water sample collection and drop-off inconvenient, thus representing a signif-icant barrier to well stewardship actions within rural and remote communities (Colley et al., 2019; Munene & Hall, 2019). As such, Morris et al. (2016) rec-ommended expanding public health services in more remote communities to minimize testing inconvenience via provision of drop-off points in publicly accessible areas (e.g., hospitals and municipal offices) and extended hours for sample collection. The cost of purchasing drinking-water treatment systems and submitting a water-quality test may also limit the willingness of well users to engage in health behaviors (Flanagan et al., 2015; Straub & Leahy, 2014). However, it should be noted that well users in North America (some of whom have the opportunity to submit well water tests free-of-charge for contaminant indicators through local public health centers) exhibit low uptake of testing actions, likely due to gaps in knowledge of the importance of testing (Hexemer et al., 2008; Jones et al., 2006; Maier et al., 2014; Ugas et al., 2020). In terms of developing regions, inconvenience and cost have been found to act as barriers to health behaviors. A study conducted in Bangladesh found that a lack of conve-nience (i.e., distance traveled to avoid arsenic exposure) of safe drinking-water behaviors lead people to persist in drinking arsenic-contaminated well water (Aziz et al., 2006). Additionally, monetary costs (i.e., for building or installing arsenic-removing sand filters) were found to act as a barrier to the decision and acquisition of sand filters in a study conducted by Tobias and Berg (2011).

Less obvious barriers associated with perceptions of negative consequences have been noted throughout the literature (Colley et al., 2019). Many well users have been found to perceive a low likelihood of adverse events to their well, and a lack of perceived personal vulnerability (i.e., low perceptions of risk), result-ing in barriers to routine testing and well maintenance behaviors (Colley et al., 2019; Munene & Hall, 2019; Munene et al., 2020). For example, several studies found that most well users perceive low personal vulnerability to a threat in their area, while acknowledging that the local contaminant can pose severe health threats and could be found in water supplies (Chappells et al., 2015; Flanagan et al., 2015). Consequently, low perceptions of risk associated with the quality of their personal well water supplies have resulted in infrequent testing behaviors. Moreover, previous research in the ROI has found that strong feelings of being in control over contamination risks are key moderators that affect the influence of risk perceptions in relation to drinking-water quality and health behavior

(Hooks et al., 2019; Schuitema et al., 2020). For example, Hooks et al. (2019) found that strong feelings of control can lead to an "illusion of control" in well users (i.e., an exaggerated belief in one's capacity to control an independent or external situation [Langer & Roth, 1975]). As a result, well users tend to underestimate contamination risks, which subsequently impact their willingness to engage in well testing behaviors. Similarly, Schuitema et al. (2020) found that strong feelings of control over contamination risks of their drinking water suppress how risk perceptions influence well users' water-quality perceptions.

Social Justice and Diversity: Further Barriers

In 2015, world leaders adopted the 2030 Agenda for Sustainable Development and the Sustainable Development Goals (SDGs), with SDG-6 established to ensure availability and sustainable management of water and sanitation for all (United Nations General Assembly, 2015). This goal is challenged and typically neglected in rural areas that rely on private groundwater supplies within developed and developing countries (Omarova et al., 2018, 2019). Private well owners in rural areas are faced with the task of managing their own water supply and must variously interpret water-quality information, implement routine testing practices, purchase appropriate treatment systems, and retrofit existing supply/treatment systems (Schimpf & Cude, 2020). These challenges are compounded by the often limited educational and monetary resources available to well owners, with rural areas increasingly characterized by geographical and financial isolation from urban population centers serving as important regional/national administrative government hubs (Cole & Murphy, 2014).

Social justice supports the idea that all people, regardless of background and status, deserve equality in social, environmental, and economic benefits within a society. An appreciable body of literature has emerged over the past two decades demonstrating a tangible inequality between urban and rural populations on the basis of water supply, with private domestic wells lying at the center of this disparity with respect to human health (Ford et al., 2017; MacDonald Gibson et al., 2020; ÓhAiseadha et al., 2017). While regulated and monitored treatment systems are in place for urban areas, inadequate surety of clean and safe water occurs in most rural communities reliant on private (unregulated) wells. Private well water supplies are frequently contaminated with microbial pathogens and chemical contaminants from both natural and anthropogenic sources, and poor well maintenance (Charrois, 2010). Additionally, many communities may have limited capacity to withstand EWEs (e.g. drought, flooding) and insufficient infrastructure to support the safety and integrity of their wells (Schimpf & Cude, 2020). The knowledge, attitudes, and practices (KAP) of private well owners serve as an alternative mechanism for water-resource management, as standard governance systems are not implemented for private drinking-water systems. However, individuals may not be aware of their responsibility to maintain their drinking-water supply. This scenario may be more prevalent for those who have moved from an urban to rural setting, where they have previously acquired drinking water from a municipal (city) water system (Simpson, 2004).

In rural areas, diverse populations that represent a wide range of socio-economic statuses, and include all racial, ethnic, and cultural groups, rely upon private wells (Fox et al., 2016). Private well users with low means to ensure adequate supply and quality are disproportionately affected by water insecurity (Schimpf & Cude, 2020). For example, a state-level survey in Wisconsin found that 71% of families with incomes over $75,000 had their supply tested compared with 33% of families with incomes less than $20,000 (Knobeloch, 2013). Similarly, low-income and minority communities can often experience higher risks of exposure to sources of contamination and environmental pollution in drinking-water sources (Flanagan et al., 2016; Schaider et al., 2019). A review by VanDerslice (2011) found that in a low-income Hispanic community in Washington State, more than 10% of private wells exceeded the nitrate maximum contaminant level (MCL) (10 mg/L NO_3^-N). Gender disparities in health risk perception have also been documented within the literature ("Social Justice & Diversity: Further Barriers" section). Special consideration should therefore be given to socio-demographics and underprivileged communities within targeted private groundwater risk-communication interventions and/or cost recovery campaigns.

Potential Future Frameworks for Effective Private Groundwater Risk Communication

Socio-Hydrogeology: A Collaborative Approach to Private Groundwater Management

In 2016, an expert panel convened by the National Centers for Disease Control Prevention (CDC) in the United States formally recognized the lack of evidence-based education, research, and training programs to support users and educators of private water systems (Fox et al., 2016). These concerns highlight the important role that public servants and environmental health experts play in public education and the importance of interdisciplinary collaboration in supporting environmental public health workers and sanitarians as participants in educational outreach programs. Training materials that bridge knowledge gaps between expert scientists (e.g., hydrogeologists, microbiologists) and environmental health professionals and acknowledge the complexities inherent in private well management are integral to the implementation of effective outreach strategies. Without such measures, theoretically and empirically informed interventions, which incorporate target audience characteristics, acknowledge contextually significant local factors and present concise, intelligible information will likely remain scarce.

Understanding the interactions and feedback loops between hydrological and social systems is fundamental to the development of effective risk-communication interventions. The term *socio-hydrology* was developed to describe dynamic feedbacks between the natural, technical, and social dimensions of human–water systems and these dimensions within predictive models to assist in the water resources management (Sivapalan & Bloschl, 2015; Sivapalan et al., 2012). The recent sub-discipline of *socio-hydrogeology* is a sub-disciplinary extension of socio-hydrology intended to understand drivers of human impacts on groundwater systems and represents a considerable step in the development of private groundwater risk communication (Figure 7.5) (Re, 2015; Re et al., 2017). Socio-hydrogeology seeks to integrate socio-hydrological and

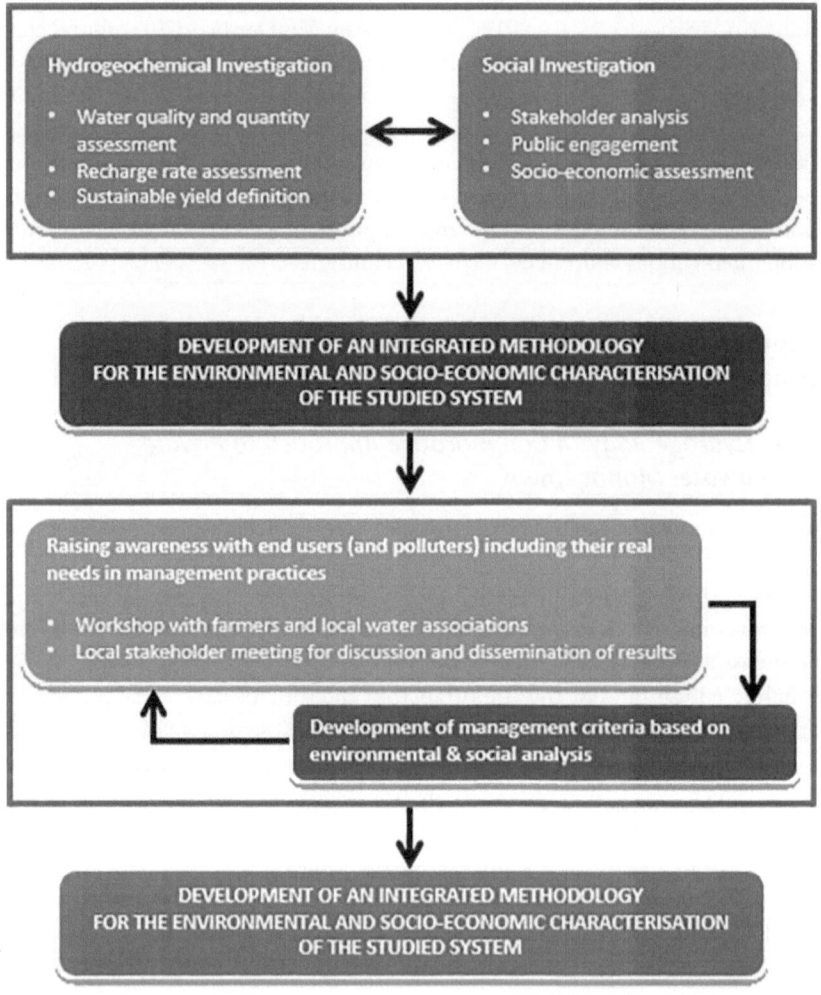

FIGURE 7.5 The Bir-Al Nas Approach for Socio-hydrogeology (Re, 2015).

science-based groundwater management practices through interpretation and exchange of knowledge with and between nonexpert end-users while involving the input of hydrogeological experts (Hynds et al., 2018; Re, 2015). Socio-hydrogeology integrates local knowledge and consultation via stakeholder analysis, public engagement, and socio-economic assessments and can therefore foster increased knowledge transfer and consultation between multiple local/regional/national government, groundwater users, community groups, the water industry, and other professionals (hydrologists, hydrogeologists, social scientists, public health practitioners, etc.) (Limaye, 2017; Re, 2015). Involving a diverse range of stakeholders, it is theorized, allows risk-communication strategies to take place collaboratively. As the articulation of "shared risk experience" (i.e., commonalities in vulnerability to a given risk among analogous populations and stakeholders) has been exhibited to influence risk-mitigation behaviors in both indirect, media-based, and interpersonal settings (Aldoory et al., 2010), incorporation of local stakeholders represents an integral element should this model be adopted.

Increasing Community Engagement and Partnerships

Absence of collaboration, consultation, and consideration of specific needs in a community may result in an inaccurate representation of effective approaches to enhancing private well management. Bottom-up approaches can encourage participatory strategies, whereby local individuals, groups, and communities (i.e., local actors) can work together on locally based environmental issues and potential solutions (Smith, 2008). Since well management behaviors are largely voluntary and self-motivated, increasing community engagement and partnerships is critical to ensure well user engagement and effective risk-communication interventions.

Knowledge and expertise of local actors can increase the capacity of well users to successfully participate in health behaviors at the local level. Licensed well contractors and drillers have been demonstrated to constitute a trusted source of information for well users and wield a positive influence on well maintenance knowledge where well users are present during the well design and construction process (Hynds et al., 2013; Kreutzwiser et al., 2010; Summers, 2010). As well users are likely to come in contact with a well contractor or be involved in the process of well location or construction, this form of engagement represents a critical juncture in the acquisition of supply risk-management information. Given the burden of disease resulting from waterborne illnesses among well users, rural physicians and other healthcare providers also have a role to play through inquiry about water supplies and potential sources of contamination during patient evaluation (Charrois, 2010). Involving community physicians and other healthcare providers in risk-communication and educational strategies after the occurrence of potential waterborne infection within the household can be an effective approach for decreasing adverse human health risks associated with private well water.

As well users may require ongoing motivation to engage in such health be-haviors, social norms can act as a further motivation to behavior change (Morris et al., 2016). Social networks such as neighbors, friends, and acquaintances can be important sources of information and may act as motivators to well manage-ment behaviors (Colley et al., 2019). Previous research in Québec found that well users were 11 times more likely to test their well water if they knew an ac-quaintance who had already tested (Renaud et al., 2011). A review of effective methods for outreach among private well users recommends that community outreach campaigns initiate this social environment by first teaching the new behavior to a few respected members in the target community (Morris et al., 2016). Consequently, local actors can spread the message through a bottom–up approach and motivate individuals in various social groups (Summers, 2010).

Trust within communities represents another essential factor in successful risk-communication strategies. To increase well management practices in rural communities, outreach and educational efforts must establish trust. Many well users lack trust in government officials and exhibit fear of losing autonomy over their private well and groundwater resource (Jones et al., 2005; Morris et al., 2016). Trust within communities can be established when researchers, scientists, and/or government officials work in an open, honest, and transparent manner with the community in which they are collaborating (Minkler, 2004). Government agencies that build partnerships with respected community lead-ers and non-governmental organizations (NGOs) can build trust within a community and can motivate participants of outreach campaigns to become more receptive to the notions of behavior change (Morris et al., 2016). There is growing recognition that sustainability outcomes, particularly in rural areas dependent on private wells, cannot be achieved by top-down management alone but instead must be integrated with human systems, and with trusted community members leading local, grassroots stewardship efforts.

Integrating Social Dimensions of Private Well Users into Quantitative Risk Assessments and Socio-Epidemiological Modeling

Involving local knowledge can embed social uncertainty into quantitative risk assessments and epidemiological modeling. A quantitative microbial risk assessments (QMRA) can provide an accurate assessment of water contamina-tion and likelihood of exposure to a waterborne illness (Peterson & Ashbolt, 2016; WHO, 2016). Similarly, socio-epidemiology focuses on the socio-structural factors on states of health and disease in populations (Honjo, 2004). Integrating quantitative and qualitative results can provide findings that are more meaningful, localized, and, consequently, derive evidence-based inter-ventions and risk-communication strategies that effectively promote health behaviors among private well users. One of the key challenges in quantitative risk assessments is how to address the role of individual perceptions of risk and how these perceptions influence risk-mitigation behaviors (i.e., well management behaviors) (Kleindorfer et al., 1993). The incorporation of knowledge, attitudes, perceptions, and behaviors of private well users, directly into QMRA of private drinking-water systems is an area previously not explored or implemented.

Quantifying and understanding the factors such as previous experiences (e.g., health impacts, past test results, EWEs) and socio-demographics (e.g., age, income, education) that drive, or influence risk perceptions and consequent precautionary measures is important for risk management. Additionally, incorporating behaviors of private well users into a QMRA or socio-epidemiological model can highlight the role that well users' behaviors have in safe drinking-water management, as knowledge, attitudes, and perceptions have the potential to influence an individual's exposure to waterborne pathogens through infrequent health behaviors (i.e., well maintenance, treatment, and/or testing). Well user interview and focus group questions can be designed in the form of scenarios to determine the extent to which knowledge, attitudes, perceptions, and behaviors contribute to the likelihood of waterborne infections. For example, asking respondents whether or not they test after an extreme weather event. Results can inform both policymakers and well users on the influence that the social dimension of private well use can have on the environmental, and consequently human, health. Overall, acknowledging the human–water system as coupled by integrating social (knowledge, attitudes, perceptions, behaviors) and physical (e.g. hydrologic and microbiologic) aspects of water, will result in a more holistic, integrated approach toward risk communication, and ultimately private well management.

Discussion Questions

1 What are some effective approaches to enhancing private well management in a top-down and bottom-up fashion?
2 How can the integration of an individual's KAP improve and/or develop effective risk-communication strategies for private groundwater end-users?
3 What is needed to achieve sustainable management of private groundwater resources, from a social (awareness, perceptions, and behaviors) and physical (e.g. hydrological and microbiological) science perspective?

Assignments

1 Propose a behavior-change intervention or research study targeted for private groundwater end-users that is grounded in one of the seven socio-psychological models outlined in Figure 7.4.
2 Compose a short essay comparing behavioral barriers to private well risk mitigation identified in the current chapter to behavioral barriers to other risk-mitigation types (e.g., flood damage prevention, climate change adaptation).

References

Aldoory, L., Kim, J. N., & Tindall, N. (2010). The influence of perceived shared risk in crisis communication: Elaborating the situational theory of publics. *Public Relations Review, 36,* 134–140.

Alfarrah, N., & Walraevens, K. (2018). Groundwater overexploitation and seawater intrusion in coastal areas of arid and semi-arid regions. *Water, 10,* 143.

Andrade, L., O'Dwyer, J., O'Neill, E., & Hynds, P. (2018). Surface water flooding, groundwater contamination, and enteric disease in developed countries: A scoping review of connections and consequences. *Environmental Pollution, 236,* 540–549.

Andrade, L., O'Malley, K., Hynds, P., O'Neill, E., & O'Dwyer, J. (2019). Assessment of two behavioural models (HBM and RANAS) for predicting health behaviours in response to environmental threats: Surface water flooding as a source of groundwater contamination and subsequent waterborne infection in the Republic of Ireland. *Science of the Total Environment, 685,* 1019–1029.

Atkin, C. K., & Rice, R. E. (2012). Theory and principles of public communication campaigns. In R. E. Rice & C. K. Atkin (Eds.), *Public communication campaigns* (4th ed., pp. 2–19). Sage.

Balog-Way, D., McComas, K., & Besle, J. (2020). The evolving field of risk communication. *Risk Analysis, 40*(1), 2240–2262.

Barthel, R., & Seidl, R. (2017). Interdisciplinary collaboration between natural and social sciences – status and trends exemplified in groundwater research. *PLoS ONE, 12*(1), E0170754.

Barthel, R., Foster, S., & Villholth, K. G. (2017). Interdisciplinary and participatory approaches: The key to effective groundwater management. *Hydrogeology Journal, 25,* 1923–1926.

Boholm, Å. (2019). Risk communication as government agency organizational practice. *Risk Analysis, 39*(8), 1695–1707.

Bubeck, P., Botzen, W. J. W., Kreibich, H., & Aerts, J. C. J. H. (2013). Detailed insights into the influence of flood-coping appraisals on mitigation behaviour. *Global Environmental Change, 23*(5), 1327–1333.Chappells, H., Campbell, N., Drage, J., Fernandez, C. V., Parker, L., & Dummer, T. J. (2015). Understanding the translation of scientific knowledge about arsenic risk exposure among private well water users in Nova Scotia. *Science of the Total Environment, 505,* 1259–1273.

Chappells, H., Parker, L., Fernandez, C. V., Conrad, C., Drage, J., O'Toole, G., Campbell, N., & Dummer, T. J. B. (2014). Arsenic in private drinking water wells: An assessment of jurisdictional regulations and guidelines for risk remediation in North America. *Journal of Water & Health, 12*(3), 372–392.

Charrois, J. W. A. (2010). Private drinking water supplies: Challenges for public health. *Canadian Medical Association Journal, 182*(10), 1061–1064.

Chique, C., Hynds, P. D., Andrade, L., Burke, L., Morris, D., Ryan, M. P., & O'Dwyer, J. (2020). Cryptosporidium spp. in groundwater supplies intended for human consumption – A descriptive review of global prevalence, risk factors and knowledge gaps. *Water Research, 176,* 115726.

Cialdini, R. B., & Trost, M. R. (1998). Social influence: Social norms, conformity and compliance. In D. Gilbert, S. Fiske, & G. Lindzy (Eds.), *The handbook of social psychology* (Vol. 2, pp. 599–632). McGraw-Hill.

Cole, J. M., & Murphy, B. L. (2014). Rural hazard risk communication and public education: Strategic and tactical best practices. *International Journal of Disaster Risk Reduction, 10*(A), 292–304.

Colley, S. K., Kane, P. K., & MacDonald Gibson, J. (2019). Risk communication and factors influencing private well testing behavior: A systematic scoping review. *International Journal of Environmental Research and Public Health, 16*(22), 4333.

Demski, C., Capstick, S., Pidgeon, N., Sposato, R. G., & Spence, A. (2017). Experience of extreme weather affects climate change mitigation and adaptation responses. *Climatic Change, 140,* 149–164.

Di Pelino, S., Schuster-Wallace, C., Hynds, P. D., Dickson-Anderson, S. E., & Majury, A. (2019). A coupled-systems framework for reducing health risks associated with private drinking water wells. *Canadian Water Resources Journal/Revue Canadienne des Ressources Hydriques, 44*(3), 280–290.

Ferrer, R. A., & Klein, W. M. (2015). Risk perceptions and health behavior. *Current Opinion in Psychology, 5,* 85–89.

Tienen, M. N., & Arshad, M. (2016). The international scale of the groundwater issue. In A. J. Jakeman, O. Barreteau, R. J. Hunt, J. D. Rinaudo, & A. Ross (Eds.), *Integrated groundwater management: Concepts, approaches and challenges* (pp. 21–48). Springer Nature.

Figoli, A., Bundschuh, J., & Hoinkis, J. (2016). Fluoride, uranium and arsenic: Occurrence, mobility, chemistry, human health impacts and concerns. In A. Figoli, J. Hoinkis, & J. Bundschuh (Eds.), *Membrane technologies for water treatment: Removal of toxic trace elements with emphasis on arsenic, fluoride and uranium* (pp. 3–11). Taylor & Francis.

Flanagan, S. V., Marvinney, R. G., & Zheng, Y. (2015). Influences on domestic well water testing behavior in a Central Maine area with frequent groundwater arsenic occurrence. *Science of the Total Environment, 505,* 1274–1281.

Flanagan, S. V., Spayd, S. E., Procopio, N. A., Chillrud, S. N., Braman, S., & Zheng, Y. (2016). Arsenic in private well water part 1 of 3: Impact of the New Jersey Private Well Testing Act on household testing and mitigation behavior. *Science of the Total Environment, 562,* 999–1009.

Flanagan, S. V., Spayd, S. E., Procopio, N. A., Marvinney, R. G., Smith, A. E., Chillrud, S. N., Braman, S., & Zheng, Y. (2016). Arsenic in private well water part 3 of 3: Socioeconomic vulnerability to exposure in Maine and New Jersey. *Science of the Total Environment, 562,* 1019–1030.

Ford, L., Bharadwaj, L., McLeod, L., & Waldner, C. (2017). Human health risk assessment applied to rural populations dependent on unregulated drinking water sources: A scoping review. *International Journal for Environmental Research and Public Health, 14*(8), 846.

Foster, S., & Ait-Kadi, M. (2012). Integrated Water Resources Management (IWRM): How does groundwater fit in? *Hydrogeology Journal, 20,* 415–418.

Fox, M. A., Nachman, K. E., Anderson, B., Lam, J., & Resnick, B. (2016). Meeting the public health challenge of protecting private wells: Proceedings and recommendations from an expert panel workshop. *Science of the Total Environment, 554,* 113–118.

Fox-Rogers, L., Devitt, C., O'Neill, E., Brereton, F., & Clinch, J. P. (2016). Is there really "nothing you can do"? Pathways to enhanced flood-risk preparedness. *Journal of Hydrology, 543*(B), 330–343.

Gamhewage, G. (2014). *An introduction to risk communication.* World Health Organisation.

Gleeson, T., Befus, K. M., Jasechko, S., Luijendijk, E., & Bayani Cardenas, M. (2016). The global volume and distribution of modern groundwater. *Nature Geoscience, 9,* 161–167.

Gorelick, S. M., & Zheng, C. (2015). Global change and the groundwater management challenge. *Water Resources Research, 51,* 3031–3051.

Green, T. R. (2016). Linking climate change and groundwater. In A. J. Jakeman, O. Barreteau, R. J. Hunt, J. D. Rinaudo, & A. Ross (Eds.), *Integrated groundwater management: Concepts, approaches and challenges.* Springer Nature.Grönwall, J., & Danert, K. (2020). Regarding groundwater and drinking water access through a human rights lens: Self-supply as a norm. *Water, 12,* 419.

Grützmacher, G., Kumar, P. J. S., Rustler, M., Hannappel, S., & Sauer, U. (2013). Geogenic groundwater contamination – definition, occurrence and relevance for drinking water production. *Zentralblatt für Geologie und Paläontologie, Teil I, 1,* 69–75.

GWP (2012). *Groundwater resources and irrigated agriculture - making a beneficial relation more sustainable.* GWP Perspectives Paper. Global Water Partnership.

Heath, R. L. (2018). Risk management and communication. In R. L. Heath, & W. Johansen (Eds.), *The international encyclopedia of strategic communication.* Wiley-Blackwell.

Henry, H. F., & Suk, W. A. (2018). Public health and Karst groundwater contamination: From Multidisciplinary research to exposure prevention. In W. B. White, J. S. Herman, E. K. Herman, & M. Rutigliano (Eds.), *Karst groundwater contamination and public health* (pp. 7–14). Springer Nature.

Hexemer, A. M., Pintar, K., Bird, T. M., Zentner, S. E., Garcia, H. P., & Pollari, F. (2008). An investigation of bacteriological and chemical water quality and the barriers to private well water sampling in a Southwestern Ontario Community. *Journal of Water and Health, 6*(4), 521–525.

Honjo, K. (2004). Social epidemiology: Definition, history, and research examples. *Environmental Health and Preventive Medicine, 9*(5), 193–199.

Hooks, T., Schuitema, G., & McDermott, F. (2019). Risk perceptions toward drinking water quality among private well owners in Ireland: The illusion of control. *Risk Analysis, 39*(8), 1741–1754.

Hug, S. J., Winkel, L. H. E., Voegelin, A., Berg, M., & Johnson, A. C. (2020). Arsenic and other geogenic contaminants in groundwater – A global challenge. *CHIMIA International Journal for Chemistry, 74*(7–8), 524–537.

Hynds, P., Misstear, B., & Gill, L. (2013). Unregulated private wells in the Republic of Ireland: Consumer awareness, source susceptibility and protective actions. *Journal of Environmental Management, 127,* 278–288.

Hynds, P. D., Misstear, B. D., & Gill, L. W. (2012). Development of a microbial contamination susceptibility model for private domestic groundwater sources. *Water Resources Research, 48*(12). W12504. https://www.doi.org/10.1029/2012WR012492

Hynds, P., Regan, S., Andrade, L., Mooney, S., O'Malley, K., DiPelino, S., & O'Dwyer, J. (2018). Muddy waters: Refining the way forward for the "sustainability science" of socio hydrogeology. *Water, 10*(9), 1111.

Hynds, P., Regan, S., Mooney, E., & O'Dwyer, J. (2018). Development of a pocket technology groundwater risk application for local evaluation (GRAppLE) in highly groundwater reliant populations. *EGU General Assembly Conference Abstracts, 20,* 7685.

Imgrund, K., Kreutzwiser, R., & de Loë, R. (2011). Influences on the water testing behaviours of private well owners. *Journal of Water and Health, 9*(2), 241–252.

Jakeman, A. J., Barreteau, O., Hunt, R. J., Rinaudo, J. D., Ross, A., Arshad, M., & Hamilton, S. (2016). Integrated groundwater management: An overview of concepts and challenges. In A. J. Jakeman, O. Barreteau, R. J. Hunt, J. D. Rinaudo, & A. Ross (Eds.), *Integrated groundwater management: Concepts, approaches and challenges* (pp. 3–20). Springer Nature.

Jones, A. Q., Dewey, C. E., Doré, K., Majowicz, S. E., McEwen, S. A., David, W. T., Eric, M., Carr, D., & Henson, S. J. (2006). Public perceptions of drinking water: A postal survey of residents with private water supplies. *BMC Public Health, 6*(1), 94.

Jones, A. Q., Dewey, C. E., Doré, K., Majowicz, S. E., McEwen, S. A., Waltner-Toews, D., ... & Mathews, E. (2005). Public perception of drinking water from private water supplies: Focus group analyses. *BMC Public Health, 5*, 1–12.

Khan, S. J., Deere, D., Leusch, F. D. L., Humpage, A., Jenkins, M., & Cunliffe, D. (2015). Extreme weather events: Should drinking water quality management systems adapt to changing risk profiles? *Water Research, 85*, 124–136.

Kleindorfer, P. R., Kleindorfer, P. R., Kunreuther, H., Kunreuther, H. G., & Schoemaker, P. J. (1993). *Decision sciences: An integrative perspective.* Cambridge University Press.

Kløve, B., Ala-Aho, P., Bertrand, G., Gurdak, J. J., Kupfersberger, H., Kværner, J., Muotka, T., Mykrä, H., Preda, E. Rossi, P., Uvo, C. B., Velasco, E., & Pulido-Velazquez, M. (2014). Climate change impacts on groundwater and dependent ecosystems. *Journal of Hydrology, 518*(B), 250–266.

Knobeloch, L., Gorski, P., Christenson, M., & Anderson, H. (2013). Private drinking water quality in rural Wisconsin. *Journal of Environmental Health, 75*(7), 16–21.

Kreutzwiser, R., de Loë, R., & Imgrund, K. (2010). Out of sight, out of mind: Private water well stewardship in Ontario. Summary of the findings of the Ontario household water well owner survey 2008. Water Policy and Governance Group, University of Waterloo.

Kumar, M., Goswami, R., Patel, A. K., Srivastava, M., & Das, N. (2020). Scenario, perspectives and mechanism of arsenic and fluoride co-occurrence in the groundwater: A review. *Chemosphere, 249*, 126126.

Lall, U., Josset, L., & Russo, T. (2020). A snapshot of the world's groundwater challenges. *Annual Review of Environment and Resources, 45*, 171–194.

Langer, E. J., & Roth, J. (1975). Heads I win, tails it's chance: The illusion of control as a function of the sequence of outcomes in a purely chance task. *Journal of Personality and Social Psychology, 32*(6), 951.

Lapworth, D. J., Baran, N., Stuart, M. E., & Ward, R. S. (2012). Emerging organic contaminants in groundwater: A review of sources, fate and occurrence. *Environmental Pollution, 163*, 287–303.

Lavallee, S., Hynds, P. D., Brown, R. S., Schuster-Wallace, C., Dickson-Anderson, S., Di Pelino, S., Egan, R., & Majury, A. (2021). Examining influential drivers of private well users' perceptions in Ontario: A cross-sectional population study. *Science of the Total Environment, 763*, 142952.

Lee, D., & Murphy, H. M. (2020). Private wells and rural health: Groundwater contaminants of emerging concern. *Current Environmental Health Reports, 7*, 129–139.

Lefebvre, R. C. (2013). *Social marketing and social change: Strategies and tools to improve health, well-being, and the environment.* Jossey-Bass.

Li, P., Karunanidhi, D., Subramani, T., & Srinivasamoorthy, K. (2021). Sources and consequences of groundwater contamination. *Archives of Environmental Contamination and Toxicology, 80*, 1–10.

Lim, S., & Prakash, A. (2020). How the opposing pressures of industrialization and democratization influence clean water access in urban and rural areas: A panel study, 1991–2010. *Environmental Policy and Governance, 30*, 182–195.

Limaye, S. D. (2017). Socio-hydrogeology and low-income countries: Taking science to rural society. *Hydrogeology Journal, 25*(7), 1927–1930.

Little, K. E., Hayashi, M., & Liang, S. (2016). Community-based groundwater monitoring network using a citizen-science approach. *Groundwater, 54*(3), 317–324.

Lundgren, R. E., & McMakin, A. H. (2018). *Risk communication: A handbook for communicating environmental, safety, and health risks* (6th ed.). Wiley.

MacDonald Gibson, J., Fisher, M., Clonch, A., MacDonald, J. M., & Cook, P. J. (2020). Children drinking private well water have higher blood lead than those with city water. *Proceedings of the National Academy of Sciences, 117*(29), 16898–16907.

MacDonald Gibson, J., & Pieper, K. J. (2017). Strategies to improve private-well water quality: A North Carolina perspective. *Environmental Health Perspectives, 125*(7), 076001.

Maier, A., Krolik, J., Randhawa, K., & Majury, A. (2014). Bacteriological testing of private well water: A trends and guidelines assessment using five years of submissions data from southeastern Ontario. *Canadian Journal of Public Health, 105*(3), e203–e208.

Malecki, K. M., Schultz, A. A., Severtson, D. J., Anderson, H. A., & VanDerslice, J. A. (2017). Private well stewardship among a general population-based sample of private well owners. *Science of the Total Environment, 601*, 1533–1543.

Margat, J., & van der Gun, J. (2013). *Groundwater around the world: A geographic synopsis.* CRC Press.

McDowell, C. P., Andrade, L., O'Neill, E., O'Malley, K., O'Dwyer, J., & Hynds, P. D. (2020). Gender-related differences in flood risk perception and behaviours among private groundwater users in the Republic of Ireland. *International Journal of Environmental Research and Public Health, 17*(6), 2072.

McLeod, L., Bharadwaj, L., & Waldner, C. (2015). Risk factors associated with perceptions of drinking water quality in rural Saskatchewan. *Canadian Water Resources Journal/Revue canadienne des ressources hydriques, 40*(1), 36–46.

Minkler, M. (2004). Ethical challenges for the "outside" researcher in community-based participatory research. *Health Education & Behavior, 31*(6), 684–697.

Mitchell, M., Curtis, A., Sharp, E., & Mendham, E. (2012). Directions for social research to underpin improved groundwater management. *Journal of Hydrology, 448–449*, 223–231.

Molle, F., & Closas, A. (2020). Why is state-centered groundwater governance largely ineffective? A review. *WIREs Water, 7*, E1395.

Mooney, S., McDowell, C. P., O'Dwyer, J., & Hynds, P. D. (2020). Knowledge and behavioural interventions to reduce human health risk from private groundwater systems: A global review and pooled analysis based on development status. *Science of the Total Environment, 716*, 135338.

Mooney, S., O'Dwyer, J., & Hynds, P. D. (2020). Risk communication approaches for preventing private groundwater contamination in the Republic of Ireland: A mixed-methods study of multidisciplinary expert opinion. *Hydrogeology Journal, 28*, 1519–1538.

Morris, L., Wilson, S., & Kelly, W. (2016). Methods of conducting effective outreach to private well owners—a literature review and model approach. *Journal of Water and Health, 14*(2), 167–182.

Munene, A., & Hall, D. C. (2019). Factors influencing perceptions of private water quality in North America: A systematic review. *Systematic Reviews, 8*(1), 111.

Murphy, H. M., Prioleau, M. D., Borchardt, M. A., & Hynds, P. D. (2017). Epidemiological evidence of groundwater contribution to global enteric disease, 1948–2015. *Hydrogeology Journal, 25*(4), 981–1001.

Musacchio, A., Andrade, L., O'Neill, E., Re, V., O'Dwyer, J., & Hynds, P. D. (2021, March). Planning for the health impacts of climate change: Flooding, private groundwater contamination and waterborne infection–A cross-sectional study of risk perception, experience and behaviours in the Republic of Ireland. *Environmental Research*, 110707. https://doi.org/10.1016/j.envres.2021.110707

Niu, B., Loáiciga, H. A., Wang, Z., Zhan, F. B., & Hong, S. (2014). Twenty years of global groundwater research: A science citation index expanded-based bibliometric survey (1993–2012). *Journal of Hydrology, 519*(A), 966–975.

ÓhAiseadha, C., Hynds, P. D., Fallon, U. B., & O'Dwyer, J. (2017). A geostatistical investigation of agricultural and infrastructural risk factors associated with primary verotoxigenic E. coli (VTEC) infection in the Republic of Ireland, 2008–2013. *Epidemiology & Infection, 145*, 95–105.

Omarova, A., Tussupova, K., Berndtsson, R., Kalishev, M., & Sharapatova, K. (2018). Protozoan parasites in drinking water: A system approach for improved water, sanitation and hygiene in developing countries. *International Journal of Environmental Research and Public Health, 15*, 495.

Peterson, S. R., & Ashbolt, N. J. (2016). QMRA and water safety management: Review of application in drinking water systems. *Journal of Water and Health, 14*(4), 571–589.

Pittock, J., Hussey, K., & Stone, A. (2016). Groundwater management under global change: Sustaining biodiversity, energy and food supplies. In A. J. Jakeman, O. Barreteau, R. J. Hunt, J. D. Rinaudo, & A. Ross (Eds.), *Integrated groundwater management: Concepts, approaches and challenges* (pp. 75–96). Springer Nature.

Podgorski, J., & Berg, M. (2020). Global threat of arsenic in groundwater. *Science, 368*, 845–850.

Qayyum, S., Hynds, P., Richardson, H., McDermott, K., & Majury, A. (2020). A geostatistical study of socioeconomic status (SES), rurality, seasonality and index test results as drivers of free private groundwater testing in southern Ontario, 2012–2016. *Science of the Total Environment, 717*, 137188.

Re, V. (2015). Incorporating the social dimension into hydrogeochemical investigations for rural development: The Bir Al-Nas approach for socio-hydrogeology. *Hydrogeology Journal, 23*(7), 1293–1304.

Re, V., Sacchi, E., Kammoun, S., Tringali, C., Trabelsi, R., Zouari, K., & Daniele, S. (2017). Integrated socio-hydrogeological approach to tackle nitrate contamination in groundwater resources. The case of Grombalia Basin (Tunisia). *Science of the Total Environment, 593*, 664–676.

Renaud, J., Gagnon, F., Michaud, C., & Boivin, S. (2011). Evaluation of the effectiveness of arsenic screening promotion in private wells: A quasi-experimental study. *Health Promotion International, 26*(4), 465–475.

Richardson, S. D., & Kimura, S. Y. (2020). Water analysis: Emerging contaminants and current issues. *Analytical Chemistry, 92*(1), 473–505.

Roche, S. M., Jones-Bitton, A., Majowicz, S. E., Pintar, K. D. M., & Allison, D. (2013). Investigating public perceptions and knowledge translation priorities to improve water safety for residents with private water supplies: A cross-sectional study in Newfoundland and Labrador. *BMC Public Health, 13*(1), 1225.

Rowe, G., & Frewer, L. J. (2005). A typology of public engagement mechanisms. *Science, Technology, & Human Values, 30*(2), 251–290.

Schimpf, C., & Cude, C. (2020). A systematic literature review on water insecurity from an Oregon public health perspective. *International Journal of Environmental Research and Public Health, 17*(3), 1122.

Schuitema, G., Hooks, T., & McDermott, F. (2020). Water quality perceptions and private well management: The role of perceived risks, worry and control. *Journal of Environmental Management, 267*, 110654.

Severtson, D. J., Baumann, L. C., & Brown, R. L. (2006). Applying a health behavior theory to explore the influence of information and experience on arsenic risk representations, policy beliefs, and protective behavior. *Risk Analysis: An International Journal, 26*(2), 353–368.

Shaw, W. D., Walker, M., & Benson, M. (2005). Treating and drinking well water in the presence of health risks from arsenic contamination: Results from a U.S. hot spot. *Risk Analysis: An International Journal, 25*(6), 1531–1543.

Shreve, C., Begg, C., Fordham, M., & Müller, A. (2016). Operationalizing risk perception and preparedness behaviour research for a multi-hazard context. *Environmental Hazards, 15*(3), 227–245.

Simis, M. J., Madden, H. Cacciatore, M. A., & Yeo, S. K. (2016). The lure of rationality: Why does the deficit model persist in science communication? *Public Understanding of Science, 25*(4), 400–414.

Simpson, H. (2004). Promoting the management and protection of private water wells. *Journal of Toxicology and Environmental Health, Part A, 67*(20–22), 1679–1704.

Simpson, H. C., & de Loë, R. C. (2020). Challenges and opportunities from a paradigm shift in groundwater governance. *Hydrogeology Journal, 28*, 467–476.

Sivapalan, M., & Blöschl, G. (2015). Time scale interactions and the coevolution of humans and water. *Water Resources Research, 51*(9), 6988–7022.

Sivapalan, M., Savenije, H. H., & Blöschl, G. (2012). Socio-hydrology: A new science of people and water. *Hydrology Process, 26*(8), 1270–1276.

Smith, J. L. (2008). A critical appreciation of the "bottom-up" approach to sustainable water management: Embracing complexity rather than desirability. *Local Environment, 13*(4), 353–366.

Straub, C. L., & Leahy, J. E. (2014). Application of a modified health belief model to the pro- environmental behavior of private well water testing. *JAWRA Journal of the American Water Resources Association, 50*(6), 1515–1526.

Summers, R. J. (2010). *Alberta well water survey – A report prepared for Alberta Environment.* University of Alberta. http://aep.alberta.ca/water/programsandservices/groundwater/documents/AlbertaWater. WellSurvey-Report-Dec2010.pdf.

Talukder, M. R. R., Rutherford, S., Huang, C., Phung, D., & Zahirul Islam, M. Z. (2017). Drinking water salinity and risk of hypertension: A systematic review and meta-analysis. *Archives of Environmental & Occupational Health, 72*(3), 126–138.

Tandi, T. E., Kim, K., Cho, Y., & Choi, J. W. (2018). Public health concerns, risk perception and information sources in Cameroon. *Cogent Medicine, 5*(1), 1453005.

Thornton, T., & Leahy, J. (2012). Trust in citizen science research: A case study of the groundwater education through water evaluation & testing program. *The Journal of the American Water Resources Association, 48*(5), 1032–1040.

Tobias, R., & Berg, M. (2011). Sustainable use of arsenic-removing sand filter in Vietnam: Psychological and social factors. *Environmental Science and Technology, 45*(8), 3260–3267.

Ugas, M., Pearl, D. L., Zentner, S., Tschritter, D., Briggs, W., Manser, D., & Trotz-Williams, L. A. (2019). Examining the factors related to bacteriological testing of private wells in Southern Ontario. *Journal of Water and Health, 17*(6), 944–956.

United Nations General Assembly (2015). *Transforming our world: The 2030 agenda for sustainable development.* General Assembly 70 Session.

UN-Water (2018). *Groundwater overview: Making the invisible visible.* International Groundwater Resources Assessment Centre.

van der Linden, S. (2015). The social-psychological determinants of climate change risk perceptions: Towards a comprehensive model. *Journal of Environmental Psychology, 41,* 112–124.

Velis, M., Conti, K. I., & Biermann, F. (2017). Groundwater and human development: Synergies and trade-offs within the context of the sustainable development goals. *Sustainability Science, 12*(6), 1007–1017.

Villholth, K. (2016). Groundwater for food production and livelihoods - The nexus with climate change and transboundary groundwater management. In C. T. Hoanh, V. Smakhtin, & R. Johnston (Eds.), *Climate change and agricultural water management in developing countries* (pp. 154–175). CABI.

Villholth, K. G., & Conti, K. I. (2018). Groundwater governance: Rationale, definition, current state and heuristic framework. In K. G. Villholth, E. López-Gunn, K. I. Conti, A. Garrido, & J. van der Gun (Eds.), *Advances in groundwater governance* (pp. 3–31). CRC Press.

World Health Organization (WHO) (2016). Quantitative microbial risk assessment: Application for water safety management. https://apps.who.int/iris/handle/10665/246195

Zheng, Y., & Flanagan, S. V. (2017). The case for universal screening of private well water quality in the U.S. and testing requirements to achieve it: Evidence from arsenic. *Environmental Health Perspectives, 125*(8), 085002.

8

PUBLIC RESPONSES TO A PROPOSED WIND FARM AND THEIR APPLICATION TO TECHNICAL COMMUNICATION METHODS

Mary Le Rouge

The latest scientific reports published by the United Nations and the U.S. Global Change Research Program show evidence that environmental change is occurring (IPCC, 2021; U.S. Global Change Research Program, 2018). These reports argue that U.S. government policymakers must act to protect their constituents from the consequences of global warming. Although not all U.S. policymakers accept the science behind these reports, many do, including the sponsors of the Green New Deal (GND) resolution (U.S. Congress, 2019). Gaining political momentum to deal with the problem is one issue, but beyond this hurdle is the need to communicate complex scientific information to the public so that they may help make informed decisions about responses to environmental change through the public decision-making process.

Problem: Communicating this need to the public is challenged by the constraints of transcultural communication (Ding, 2014; Flower, 2008) and ideological conceptions of environmental collapse (Latour, 2018). In addition, people must be able to read and write in increasingly technical genres to make their voices heard regarding scientific issues surrounding environmental change (Brandt, 2015). The bar has been set higher for civic participation (Grabill, 2007; Simmons, 2008), which precipitates a need for advanced literacies that support transcultural empathy, cooperative decision-making, critical analysis of scientific data, and prototyping for future-oriented planning (Gross & Harmon, 2016; Potts, 2013; Sauer, 2003).

Solution: Improved communication of these issues through embodied metaphor will allow for a more just, flexible, and practical response to environmental change. This project studies public communication surrounding development of one of the first freshwater offshore wind farms in North America, on Lake Erie. This study provides a synthesis of information surrounding

DOI: 10.4324/9781003266549-10

contemporary environmental communication that will help push the next iteration of policy to better address environmental problems in ways that attend
to the concrete manifestations of environmental change. Global warming is
described as a slow-moving disaster by many, which makes addressing its consequences seem less urgent, but we are living today in a world that is currently
experiencing the effects of environmental change. How the problem is shared
and ideologically constructed makes a difference to varied populations, either
to their detriment or benefit.

Subject of Research

The subject of research for this chapter was a proposed wind farm to be situated
offshore on Lake Erie near Cleveland, Ohio. The organization that developed
this project, LEED Co. (Lake Erie Energy Development Co.), is a "public-
private nonprofit partnership devoted to catalyzing the offshore wind industry
in the Great Lakes Region" (LEEDCo, 2019). It partnered with Fred. Olsen
Renewables, headquartered in Oslo, Norway, to provide the wind turbines,
and with Case Western Reserve University's Great Lakes Energy Institute as
its research arm (Li & Yu, 2018). The company hoped to install 6 Vestas wind
turbines 8 miles offshore of Cleveland, Ohio, in a 20.7 MW pilot project called
Icebreaker Windpower Inc. that would be one of the first freshwater offshore
wind projects in North America. After proving the success of this pilot installation, the company could obtain permission to increase production by installing
more wind turbines on the lake. However, recent regulations, including energy
bill (H.B. 6), weakened Ohio's renewable energy standards, and approvals for
construction of new projects stalled (Tomich, 2019). Repeal of H.B. 6 was in
process in March 2021 because of the First Energy bailout scandal, which could
help remove some of the regulatory and financial barriers to this sustainable
energy project.

While the environmental assessment conducted by the U.S. Department of
Energy of the proposed LEEDCo pilot wind farm finds no significant impact
from the project, there are public concerns about adverse effects on fish because
of vibrations from the motors and changes in the lakebed topography, and
disruptions of bird migrations and bat populations because of the wide blade
circumference (U.S. Department of Energy, 2018). Fish are primarily impacted
by changes in behavior caused by avoidance of the noise from turbines, instead
of having adverse impacts as severe as mortality (Bailey et al., 2014). In terms
of adverse wind turbine effects on birds, research has shown that with proper
siting and turbine sizing, the effects on bird mortality are insignificant (Miao
et al., 2019). The public has also been concerned that water quality could be
compromised from the disruption of sediments on the lake bed, and there is the
possibility of oil spills leaking from the wind turbine generators and contam-
inating the water. However, probability estimates for a proposed Cape Wind

saltwater offshore wind farm in Nantucket show a low likelihood of significant spillage occurring from such installations (Etkin, 2006). There are also concerns from the local Cleveland population, especially those who use the lake for recreation and fishing and feel that navigating around the wind turbines would be difficult, and coastal inhabitants who worry that the visual impact of the wind turbines will make their property less appealing (Johnston, 2019). Although Cleveland is not known for the quality of its beaches, there are still many who travel to the lake during the summer for quick vacations, and having that view disrupted by wind turbines would not be ideal.

On the other hand, wind farms have been a proven source of reliable energy production for the past several decades that are sustainable and environmentally friendly. Moreover, the production of energy through wind turbines has become more cost-effective than energy production with fossil fuels (Brockway et al., 2019; IRENA, 2018). The construction and operation of wind turbines creates green jobs and reduces air pollution that would otherwise be created by using fossil fuels for electricity. In addition, such projects attract the relocation of large businesses that seek to use green energy in production processes. There are 738 megawatts of terrestrial installed wind capacity located in Ohio, mostly in the northwest where the land is very flat and rural (Tomich, 2019).

The Icebreaker Windbreaker Power Inc. project is a highly contentious example of an attempted move toward sustainable energy production that is enmeshed in issues of political controversy that were affected by misinformation campaigns, regulatory hurdles, environmental concerns, and local values. While siting the project offshore on the lake by several miles has the potential to reduce some of the local population's objections to the noise and altered scenery and might reduce the cost of the land leased for production, Icebreaker Wind also disrupts the community's relationship with the lake. Lake Erie has a very important role in the culture of Cleveland, a city that is sometimes called "the mistake on the lake."

In the cultural understanding of Clevelanders, the lake is north, and there is east and west of Cleveland, so if you know where the lake is, then you can find your way around town. There are local legends about the lake, including stories of sinking ships, fishing the native walleye, and old lighthouses that line the coast. The lake impacts the local weather in many ways, through lake effect rain and snow on the northeast, and reduced temperatures in the spring and warmer temperatures in the fall. This ecosystem makes the production of grapes and wine in the snow belt possible along the coast. If the lake has not frozen over, or has melted early, there is much more snow in the region. If the weather is bad, most likely the lake is to blame. All these factors play into the everyday lives of the local population.

Approval for construction of the wind farm was given in July 2020 by the Ohio Power Siting Board, with the stipulation that the turbines be turned off for eight months out of the year. This was meant to accommodate the migration of birds, called the "feathering" clause, even though LEEDCo followed all regulations for siting of the project to avoid bird migratory paths. This stipulation effectively killed the project for several months, but the feathering clause was lifted on October 8, 2020, after the board considered a movement to remove the extra language that had put the clause in place, text that had suddenly appeared in the document between the time of review and the vote that enacted the clause. Soon after this meeting, Chair of the Public Utilities Commission of Ohio and director of the Ohio Power Siting Board, Sam Randazzo resigned in the wake of the H.B. 6 scandal after the FBI searched his home in November 2020. He resigned because First Energy filed documents following the raid that indicated that he was a paid consultant for the company while serving on the Public Utilities Commission, which was a conflict of interest (Pelzer, 2020). With removal of the feathering clause, construction should be economically feasible, but it is unknown how the economic disruption of COVID-19 has affected funding for the project.

LEEDCo conducted numerous community "townhalls," meetings to address local concerns about the wind farm project, constructed documents meant to assuage unfounded fears about possible water contamination and bird mortality, and responded to media reports and public lawsuits that put their work in jeopardy of failing. All this community work was enough to finally make the Ohio Power Siting Board allow the project to commence construction under stipulations for continuous environmental monitoring. However, even during the meeting that questioned the "feathering" clause's mysterious appearance in the draft agreement that was approved and then revoked, board members questioned the public perception about wind turbines as dangerous to bird populations as the environmental assessment report denied significant problems with birds avoiding the rotating blades. The feathering clause was added by the legislature and was politically motivated by the oil and gas lobby, which played a role in constructing the public myth that bird mortality increases from wind turbines, much as the persistence of the idea that turbines cause cancer or are an expensive waste of time and money. But that the public would reinforce and amplify this misinformation despite scientific evidence to the contrary reflects a lack of trust in science, their inability to differentiate valid science from pseudo-science, and subscription to a more valued source of human understanding that subsumes the credibility of the technical documentation about the project.

Implications

This study has implications for the content and format of technical documents that are used to communicate with the public. For participants in this study, there was an overriding focus on how they understood environmental information through *embodied* means, or through the five senses and systems of the human body. From their huge drawings of the wind turbines in their imagined wind farms to the *anthropomorphism* (attribution of human qualities) of birds and agency given to Lake Erie, participants in the study repeatedly showed how limitations in the technical documentation served to box in and even contradict their personal conceptions of the wind farm. From the outcomes of this study, it is recommended that scientists and policymakers reevaluate how they frame the issues that they want to communicate to the public through technical documentation in embodied ways instead of prioritizing theoretical and scientific representations. Doing so will enable accommodations for a variety of groups from different ideological, socioeconomic, racial, and ethnic backgrounds to access technical documentation about environmental risk. By representing embodiment through metaphor in documentation, technical communicators can promote social justice initiatives by making information more accessible and actionable to audiences from diverse backgrounds.

This recommendation coincides with the findings of Beverly A. Sauer in the context of the mining industry and documentation of mining hazards for workers, but I have applied this finding to the context of the public's understanding of documentation about a proposed wind farm (Sauer, 2003). While Sauer focuses on gesture in the miners' expression of environmental risk, embodied cognition is investigated here in the theoretical representation of perception through metaphor and language usage.

The methodologies for field rhetoric that are supported by Candice Rai and Caroline Druschke bring a greater focus on materiality and embodiment to the study of a particular sustainable energy project to solve the *wicked* (overwhelmingly complex) problem of climate change (Rai & Druschke, 2018). Metaphorical construction of the public's understanding of this physical project in terms of the human body (instead of just experience of the physical environment through the senses of the body) was a useful way to describe points at which engagement with the public could be improved. Experience architecture is a means toward developing participation of the public in governmental policy making on energy issues (Salvo & Potts, 2017). However, UX is not only applicable in terms of technological systems architecture through software or hardware (which does point toward important applications for communication methods) but is also applicable in terms of the metaphorical construction of language architecture.

Writing systems involve the transference of energy, as in power, whether physical, ideological, or political. Climate change is the result of human extractive industries that have taken fossil fuels from the Earth and converted them into energy in processes that have created air pollutants and changed the Earth's atmosphere to induce global warming. It is literacy that made the transference of knowledge from one person to another more durable and allowed for the evolution of more complex systems of knowledge to the sciences and these advancements in technology. Therefore, literacy must play a role in fixing the problem of climate change—not that all these advancements in the lives of humans should be rolled back, as in some back-to-nature turning away from technology, but that new sources of energy can be used to replace the damaging industries from before. The problem with the fossil fuel industry is a disregard for the consequences of production, its byproducts, and a lack of sustainability.

In this system, environmental injustices increase risk for many people who do not even have a say in how their environment is managed. Finding practical solutions to the wicked problem of climate change is only as difficult as looking beyond the terministic screens of the scientific genre and promoting more citizen science. This should not only take place in the university classroom, but also in the community and with citizens of all ages. So, this project studied communication surrounding the first offshore freshwater wind farm project in North America, on Lake Erie near Cleveland. By focusing on the concrete manifestation of climate change through development of a local renewable energy project, it is possible to understand how information circulates and literacy functions in the community to promote or deny a practical response to the problem of climate change.

Technical Documentation

This research was conducted with the assumption that the technical documents created by LEEDCo, the Ohio Department of Natural Resources, the U.S. Department of Energy, and other experts were accurate scientific documents that attempt to truthfully show the reality of the proposed wind farm and its purported effects on the environment. While LEEDCo had a vested stake in presenting its project as viable, the state and national government and environmental organizations were ostensibly neutral parties. However, the documentation used by the Ohio Power Siting Board (OPSB) to decide on giving the greenlight to the project was tampered with at the state-government level to include a "feathering" clause to shut down operation of the wind turbines nine months out of the year, as was seen in one of the final meetings of the board regarding the project (OPSB, 2020). Public commentary on the project submitted to the OPSB showed a broad range of both support and opposition, including repeated negative commentary from two residents of Northeast Ohio, Robert Maloney and Susan Dempsey. These residents sued LEEDCo to stop construction of the Icebreaker Wind project with the funding of coal magnate Robert Murray and his lawyer John Stock (Tomich, 2019). Two groups that purport to exist for the protection of bird populations against wind turbines, the American Bird Conservatory (ABC) and the Black Swamp Bird Observatory (BSBO), also sued LEEDCo for supposedly not doing due diligence in the environmental assessment about bird migration and effects in terms of bird deaths (Renewz.biz, 2019). However, the Audubon Society backs the LEEDCo project—making the ABC and BSBO lawsuit suspect. So, opposition groups exist, and it is useful to study their rhetoric to understand how disinformation is built to hold back renewable energy projects from development.

The fossil fuel industry pays for and spreads disinformation about sustainable energy industries that serves to confuse the public. In the case of wind energy, this has even come from former President Trump, who has claimed that wind turbines cause cancer, and Republican Texas lawmakers, who blamed frozen wind turbines on the energy grid failure during the February 2021 winter storm that also froze gas pipelines and other energy sources (Douglas & Ramsey, 2021; King, 2019). Common misconceptions that have been spread via social media include questioning the existence of climate change, denying a human role in climate change, claims that wind turbines cannot possibly provide enough energy, and assertions that wind turbines kill a large number of birds. Much of the literature on this topic focuses on the spread of "fake news," often found on social media in the form of memes and doctored infographics that supposedly prove that the science is wrong. It is useful to take apart this type of disinformation and analyze how it operates to persuade through unspoken warrants, or *enthymemes*, provided by a mixture of visuals and text that are *collocated* (located in the same place). But it is not just studying the existence or spread of disinformation that is important, rather it is also imperative to study the reasons why the public find false information believable. Beyond the evaluation that audiences must make regarding the credibility of claims made about environmental information, which includes checking sources for proper citing, formatting, accreditation and *ethos* (character) of authors, and verification or cross-checking facts from multiple sources, there is the larger problem of belief systems that challenge science. If one subscribes to the science of climate change and believes in the accuracy of scientific research in general, memes that purport to show how these things are false seem absurd. However, if there is any doubt as to the validity of the science—and there are many reasons to feel this way—then there is a fault in the armor, and the disinformation will appear naturally correct.

Many of the arguments made by climate change deniers are based on what the public knows from their "natural" or immediate experience. It gets warm in the summer and cold in the winter, therefore temperature fluctuations on a global scale must be cyclical like the weather, not evidence of a dangerous warming trend in the climate. Birds hit the windows of houses all the time for no reason, so they must be harmed by wind turbines. Energy has traditionally come from the Earth from burning substances pulled from the ground, and so it cannot possibly come out of thin air. These kinds of arguments are used by anti-wind farm advocates because they are so powerful, drawing their strength from what the public understands of how the world works from personal experience. When these experiences are juxtaposed with scientific and technical documentation that contradicts what the public understands to be true, it forms cognitive dissonance. A decision is made—whether to subscribe to the ethos of the scientist or policymaker who created the report to the discredit of personal experience or remain suspicious of its content. Often, these documents are viewed with suspicion because of their complexity and lack of attention to human experience and belief systems. Participants in the study noted that the creators of the technical documents must have had some ulterior motive in making them, such as LEEDCo wanting to build its wind farm and make a profit, which then discredited the content. Those participants who took the content of the technical documents at face value, without concern for the motives of the creators, still tended to criticize the format and function of the documents themselves as overly "busy," illegible, or prohibitively technical. And most participants did not believe that the technical documents accurately reflected the way that they understood the wind farm would look and operate.

This is a problem not only for the field of technical communication, but for the representation of science, which came under constant attack by the Trump Administration from 2016 to 2020. Whether the anti-science movement is funded by the fossil fuel lobby or not, its power comes from the ability to sway public opinion based on appeals to an embodied understanding of the world that is usually lacking in technical documentation. Although common problems with stasis (determining "what is"), definition (of common terms), access (to relevant information), and aggregation (how information is packaged and distributed) trouble the public's reception of technical communication about environmental issues, it is the lack of focus on human embodiment in the content of technical documents that primarily reduces their functionality. Criticisms so often fall on the format of documents to the exclusion of the "messy" work of untangling the metaphors, symbolism, and basic ideologies that support the knowledge contained within. Technical communicators often work from the assumption that the public is (or should be) on the same page, prizing scientific objectivity over embodied knowledge, but the reverse is true. People understand the environment primarily through their physical bodies and through metaphors and symbols that reflect this embodiedness. By attending to these factors in scientific and technical documentation, communicators can regain the trust of the public. This is an opportunity to turn the weapons of miscommunication back on themselves, to make citizen science truly possible.

Much of the literature in the rhetorical communication about environmental issues has hinged on the shock and dismay of how little the public is engaged and their ideas valued in the decision-making process for large-scale projects that could harm their communities. This happens time and again, and the solutions we have so far provided mostly attend to creating virtual venues for the dissemination of information among the public through website creation and social media platforms. At the Symposium on Communicating Complex

Information (SCCI) conference at Old Dominion University in February 2021, presentations focused on user experience and UX in terms of accessibility of information in an online environment and improving the technical specifications of the human–computer interface. There is value in allowing the dialog of the public to an online platform when that information is listened to by project managers and policymakers. But as some noted in a discussion about COVID-19 dashboards provided to the public during the pandemic, having access to information (or an outlet for commentary) can act like a *panacea* (a cure-all). It is no different from inviting the public for commentary, logging it, and then forgetting about it in a bid to make them feel as if their input was valued, just as Jeffrey Grabill and Michele Simmons found in their studies of local informational meetings and websites about environmental risk (Grabill, 2007; Simmons, 2008). This is just the kind of activity that makes the public angry and increases distrust in those in power. The findings of my research will not provide a technological solution for this problem, such as an open-source GIS system or social media platform to increase public participation in environmental decision-making, which are solutions proposed by Grabill and Simmons and Liza Potts (Potts, 2013). Implementing such systems involves spending much time and money developing a tool that will quickly become obsolete, and their success really depends on how they are used—whether they are truly used to discover and integrate the public's commentary in decision-making, or to make the public feel *as if* they were heard.

The content of technical documentation needs to change so that it is acceptable for public consumption and engagement on a metaphorical basis. The government currently recommends "plain language" in technical documentation to improve the linguistic content of complex information, and this problem in language difficulty and education level of the public is well known. The problem is that technical documentation systematically erases the metaphorical and symbolic underpinnings of how the public understands the world. This style of documentation subjugates the embodied experience of humans to an objective, flattened rendition of what is deemed scientifically accurate and necessary toward the ends of efficient communication. However, text that is efficient is rarely accurate because it glosses over outliers and special situations that the most common, conservative, accepted functions in science cannot explain. However, it is exactly the unexplainable that often results in the most interesting and novel findings. We need a more open use of metaphor in technical documentation not only to regain the trust of the public but also to make new findings and explorations in science possible. When writers attempt to keep language static, as with technical communication, it loses its relevance to the socio-historical context. This happens at a disciplinary and industry level, putting pressure on the individual writer to follow conventions that may or may not make sense in the context of a particular text. Success in communicating that information becomes measured by its adherence to the traditional style guidelines previously set for that genre of text, not by how well that information is received by the intended audience or the public.

Therefore, experience and usability methods can be used as means toward improving technical communication, not in the sense that these terms usually refer to research in online engagement in computer programs or systems, but in terms of audience engagement with the content of technical documentation and its metaphorical constructions. These documents can very well exist online and have interactive functions, even using virtual reality to attempt greater connections with their audiences, but they all are constructed with underlying metaphors that affect their ability to engage with the public. But it is not as simple as listing commonly used metaphors for certain subjects and then implementing them in the technical documentation. There are nuances to language in actual use and context that make analysis of each unique situation important before attempting to implement a change in documentation. For technical communicators, evaluating audience affinity for metaphorical constructions of environmental information is a necessary step in communicating effectively with the public.

Embodiment and Metaphor

An *enactive conception of cognitive science* (how we understand things, in action) as described by Francisco J. Varela, Evan Thompson, and Eleanor Rosch in their book *The Embodied Mind* (1991, rev. 2016) helps to show how human experience is linked to environmental understanding through *structural coupling* (combining two disparate systems of knowledge). *Cognitive science* is loosely defined as "the study of the mind," but which has links to "linguistics, neuroscience, psychology, sometimes anthropology, and the philosophy of the mind" (Varela et al., 2016, p. 4). Cognition is *"mental representation*: the mind is thought to operate by manipulating symbols that represent features of the world or represent the world as being a certain way" (Varela et al., 2016, p. 8). These symbols are organized in a structure that can be *hierarchical* (by level of importance), composed of loosely affiliated *nodes* (points of importance or intersection), or categorized in such a way that *phenomenological* (direct physical) experiences have meaning. What Varela and colleagues posit is that "emergent" properties that arise out of the two-way interaction of human experience and the environment serve to give humans the experiences that they use in everyday survival. They refuse to subscribe to either an *objective* (not influenced by personal experience) or *subjective* (relying on personal experience) understanding of the world, and instead see a "middle way" through embodiment that is informed by the meditational teachings of Buddhism.

Human scale and orientation are the primary features of embodied experience as it relates to how we understand the world. Francisco Varela and colleagues describes this notion well:

> By using the term *embodied* we mean to highlight two points: first that cognition depends upon the kinds of experience that come from having a body with various sensorimotor capacities, and second, that these individual sensorimotor capacities are themselves embedded in a more encompassing biological, psychological, and cultural context.
>
> *(Varela, Thompson, & Rosch, 2016, p. 173)*

Taking the cultural context into account, George Lakoff and Mark Johnson, in their book *Metaphors We Live By* (1980), argue that "Metaphors may create realities for us, especially social realities. A metaphor may thus be a guide for future action … metaphors can be self-fulfilling prophecies" (Lakoff & Johnson, 1980, p. 156). They cite Amory Lovins in Chapter 23, who discusses the metaphorical difference between following a HARD ENERGY PATH and a SOFT ENERGY PATH. Hard energy is viewed as "inflexible, nonrenewable, needing military defense and geopolitical control, irreversibly destructive of the environment," whereas soft energy is characterized as "flexible, renewable, not needing military defense or geopolitical control, and not destructive of the environment" (Lakoff & Johnson, 1980, p. 157). Thinking of energy sources this way creates an artificial *dichotomy* (of two opposing means of thought), however, that is not always true because renewable energy may very well need military defense or geopolitical control in the future, and in some ways the process of obtaining renewable energy sources *is* destructive of the environment. A more nuanced understanding than this of the metaphors used in practice regarding energy sources and the environment is necessary to comprehend how language about energy policy can be pragmatically shaped.

This chapter is theoretically based in the *embodied cognition* (knowledge gained from our physical bodies in context) posited by Varela and colleagues and evidenced through the linguistic metaphor systems outlined by George Lakoff and Mark Johnson (Lakoff & Johnson, 1980). It is supported by primary research conducted in 2019–2020 on public responses in Northeast Ohio to a proposed offshore wind farm on Lake Erie, the pilot Icebreaker Wind project. Evidence that participants preferred to explore relationships with the wind farm through embodied experience is shown through their use of *anthropomorphism* (the attribution of human characteristics to the nonhuman) and *personification* (giving the characteristics of a person to the nonhuman) with metaphor when describing nonhuman aspects of the project such as the Earth, mechanical objects of the wind farm, Lake Erie, and birds through surveys, oral interviews, and drawings. Embodied cognitivism allows us to evaluate the types of memories and second-hand knowledge that participants shared in their surveys and interviews that lead to some common structural formulas that they used to understand and communicate about technical environmental documentation. These findings point to ways that technical documentation can be improved to better communicate with the public through embodied representations of environmental issues that are constructed with prototypical (base or elemental forms that are used to build more complex) metaphors.

For example, participants in the wind farm study frequently noted the importance of birds and the risk that wind turbines posed to their lives during migration. It is easy to state that there is much misinformation circulated about the number of bird deaths attributable to wind turbines, and that these participants are simply misguided or simple minded. The technical document that participants reviewed in the study showed a bar graph with an astronomical number of bird deaths attributed to cats versus a comparatively miniscule amount attributable to wind turbines. Most participants laughed at this comparison and noted that the document was circumspect because of its poor formatting. But this very visual, statistical, and textual evidence of the safety of wind turbines to birds did not change participant's worries that birds would be unduly harmed by the structures. It is the conceptual framework itself of this document that seeds doubt in the public's understanding of wind turbine effects on birds, not its statistical content. Why show the number of bird deaths attributed to wind turbines if it is not a problem? What needs to change here is the conceptual mapping of the way this information is shown, not WIND TURBINES AS CAUSE OF BIRD DEATHS but WIND TURBINES AS PROTECTOR OF BIRD HABITAT. The reduction in fossil fuel use brought about by the wind power installation will result in an overall slowing down of temperature increase and habitat loss for birds. This idea could be physically reinforced by wind farm developers by providing nesting areas or habitats for birds close to their wind farms or even on the wind turbines. The improved technical documentation would counter public misunderstanding by providing a more accurate picture of how birds are holistically affected by the environmental factors surrounding their habitat and migrations, and the language used in the document would prioritize this metaphor.

In another example, participants consistently drew their vision of the wind farm to huge scale and talked about wind turbines as being huge, even after seeing the photographic simulations showing the turbines from the shore as it was almost impossible to see the horizon. There is much concern from the local government about the potential visual impact of siting wind farms near where the local population lives. But to the contrary, many participants in the study were upset with how small the turbines looked in the simulation and instead described how they thought they were beautiful structures. In the news, articles focus on the population against such a project for visual reasons because there is a struggle and tension there. It becomes a Not in My Back Yard (NIMBY) debate. The creators of the technical document that participants viewed in the study assumed that the public would prefer the wind farm to be almost invisible in the photos, with the minimum visual impact possible.

But participants instead viewed the smallness of the turbines in the photos as deceptive, an indication of the unimportance of the project or an understatement of its effects. This information subscribed to the metaphor of WIND TURBINES AS A NUISANCE, even if it was trying to show the opposite effect. What is needed instead is to engage with the metaphor of WIND TURBINES AS A VALUABLE RESOURCE by showing the increased energy available to the local population and decreased environmental pollution. These things are hard to show in photographs but can be described in terms of how much energy would be created for an individual to use and how their health (and the health of the surrounding environment) would be improved by the amount of CO_2 that would not be created by the fossil fuel use that is avoided. This idea could be reinforced by wind farm developers by making these facts clearly available from the outset, something that participants called for repeatedly in my study. Developers might also consider doing away with the idea that the public does not want to see wind turbines in their backyard because this does not seem to be the case—they have been accepted by the public much as electric towers or lighthouses and can be billed as positive symbols of the societal shift to renewable energies. Wind farm developers could physically manifest this shift in the public's conceptualization of wind turbines by building them even more aesthetically obvious on the skyline, by making a statement about their right to take up space and reflecting their importance to our future goal of energy sustainability. The technical documentation would bring this viewpoint to the fore, not in some bid to hide how the turbines would disrupt the landscape, but to reflect how their visual presence is now acceptable and valuable to the public.

Finally, there is the example of public understanding of Lake Erie and analysis of the first technical document to show its limitations and make recommendations. There are common conceptions about the natural environment in public knowledge that are frequently dismissed as "folk" or too personal to count as real knowledge in science. These include the Northeast Ohio public's understanding of Lake Erie, which is known to be a changeable and often malevolent force. In the first technical document, the dangers of the lake environment for the operation of the wind turbines are shown in multiple photographs, one of waves crashing against a lighthouse, and another, of sea ice in Europe.

A graphic of average wave heights on Lake Erie shows anticipated wave activity on the wind turbine base and the text discusses the turbines' ability to withstand these forces. But participants were not convinced of the accuracy of this document because of their innate knowledge of the lake. The document assumed the idea of LAKE ERIE AS A MEASURABLE FORCE and that the risk to the turbines was therefore controllable. But participants viewed LAKE ERIE AS AN UNCONTROLLABLE FORCE and as posing a considerable risk to the wind farm. That the technical documentation did not acknowledge this made participants question the practical feasibility of the project and the honesty of the developer when assessing the risk. By using statistics to prove the measurability of the lake's dangers to the wind farm, the technical documentation had the opposite effect on participant response. Developers should consider acknowledging that assessments of risk, especially regarding forces of nature, are limited and that such projects are very much subject to environmental dangers. These assessments can be listed in the technical documentation as "what if" scenarios with possible solutions for situations such as extreme ice flow, wind, and temperatures in technical documentation. The physical manifestation of attending to these risks could be shown by incorporating design features in the turbines that provide secondary and tertiary fail safes for wind, wave, and ice risk mitigation.

Formula for Metaphor

The change of *domain* (area of knowledge) that is necessary to implement these three adjustments to metaphorical construction can be shown in a formula. Ferdinand de Saussure was a Swiss linguist who first captured the idea of the signified (large "S" on top) and the signifier (small "s" on bottom) in his formula for the *algorithm* (a set of rules to be followed in a calculation) of linguistics. The "sound-image" stands for a mentally produced sound-image in the brain tracked to the "concept" that it represents. The symbol system of *semiology*, or *semiotics*, gains its meaning from social usage, and is therefore humanistic, looking at language as it is used in the human context (Saussure, 2011 [1916]) (Figure 8.1).

Psychoanalyst Jacques Lacan reformulated this *schema*, or plan, in two ways in his paper "The Instance of the Letter in the Unconscious," first as seen in Figure 8.2.

$$linguistic\ algorithm = \left(\frac{S}{s} \right) = \frac{signified}{signified} = \frac{Concept}{sound - image}$$

FIGURE 8.1 Saussure's formula for the algorithm of linguistics.

$$\frac{s}{S} = \frac{TREE}{\text{(tree image)}} = \frac{signified}{signified} = \frac{sound - image}{concept}$$

FIGURE 8.2 Lacan's reformulation of the algorithm of linguistics.

In other words, Lacan believed that the oval surrounding the sign and signifier should be eliminated because the relationship was not set into a unified function but rather was a fluid, ever-changing concept (Fink, 2004). He also reverses the operation, where the signifier (small "s") takes prominence over the signified (big "S"), signaling that the sign has a life of its own that is not dependent on signifying anything at all. He exchanges the image of a tree for the signified and the word *TREE* for the signifier, denoting the linguistic function of the word in relationship with the concept. Jacques Lacan's second reformulation of Saussure's linguistic algorithm uses an *anecdote*, or story. In the second variation, there are two competing signifiers (a gentlemen's and a ladies' bathroom) that both stand in for the concept of the train station. Lacan's story about two children arriving at a train station describes how one says they have arrived at "gentlemen" and the other says they have arrived at "ladies" when they have both arrived at the train station. This anecdotal description of metonymy is then transformed into the formula (Figure 8.3).

Where in *metonymy* (a figure of speech where a word is substituted to mean another thing) the function of the linguistic construction is approximately equal to (E) maintaining the bar (−) between the signifier in its many iterations $\left(S...S^{1}\right)$ and the signified (*s*) because the domain is the same, part to whole, and nothing is added. It has a referential function to meaning that already exists in this domain. However, in the case of metaphor, there is an additive function (Figure 8.4).

Where the plus sign indicates both crossing the bar between signifier and signified and adding new meaning(s) to the original signifier. These two formulas I take to be accurate enough renditions of one instance of metonymy or metaphor, although I would question why, if Lacan initially reversed the signifier to the top of the equation, he did not continue with this in his formula with the small "s" appearing above a large "S," rather stating that the large "S" now stands for the signifier instead of the signified. But it also seems to be splitting hairs to argue about which side of the equation should have prominence in the formula because they are as two sides of a coin. Both Saussure and Lacan were describing the relationship between signifier and signified in the field of linguistics, exploring how language acts as a stand-in for signification.

$$\frac{S}{s} = f(S)\frac{1}{s} \text{ therefore } f(S...S^{1})S \cong f(S-)s$$

FIGURE 8.3 Lacan's formula for metonymy.

$$f\left(\frac{S^{1}}{S}\right)S \cong S(+)s$$

FIGURE 8.4 Lacan's formula for metaphor.

$$f\left(\frac{M...M^1}{M}\right) M \cong M(+)m$$

FIGURE 8.5 Le Rouge's formula for metaphor.

Here, we are focusing instead on conceptual metaphor systems. If the target-source system of conceptual metaphors of Lakoff and Johnson is applied to these formulas, one can see that a metaphor encompasses both the target (signified) and the source (signifier) of a conceptual domain (Lakoff & Johnson, 1999). As in the train station anecdote, competing or opposite arguments exist in the same domain and can even evoke the same meaning with metonymy. Both *ladies* and *gentlemen* signs refer to arrival at the train station. But when the conceptual metaphor is limited to a part–whole understanding, there is no knowledge added or gained: it still means we have arrived at the train station. However, changing the metaphors used to refer to the meaning also results in creation of a new conceptual domain that pulls the argument out of a *dialogical* (two-sided), oppositional *morass* (a place where you get bogged down). It is not just that the function of metaphor is additive and crosses the bar between signified and signifier to add meaning to the original conceptual domain, but it also creates opportunities for entirely new conceptual domains to exist, built from previous metaphors. The train station is not just a ladies' or gentlemen's bathroom but a location that gives access to multiple destinations. This formula shows how the new conceptual domain operates in relation to the parent formula. As with metonymy, two or more competing metaphors exist, but instead of just acting as reference, they cross the bar to add meaning. Here, I change the formula to refer to metaphor (M) and meaning (m) (Figure 8.5).

The new domain of the meaning includes multiple metaphors, each adding their own meanings as in layers. It is not enough to identify the main metaphor used in an argument; it is necessary to investigate all the tangential meanings that attach to it, and then figure out the construction that will enroll multiple metaphors that provide the highest degree of congruence with the cognitive domain of the public and the domain of science. Enrollment of these metaphors is indicated by the use of linguistic and visual cues.

Argument and Metaphor

For example, we can return to the example of WIND TURBINES AS CAUSE OF BIRD DEATHS modified by WIND TURBINES AS PROTECTOR OF BIRD HABITAT. When listening to an argument from a participant who was against the construction of the wind farm, one main metaphor brought up was that of the EAGLE AS THE SYMBOL OF THE UNITED STATES. She argued that eagles have special eyesight and a head/eyebrow construction that allows them to only see directly below them when in flight. She reasoned that although eagles are known to have very good eyesight, they cannot see straight ahead of them while in flight and are therefore more likely to run into wind turbine blades. Her argument is strong not because of the logic about eagle vision (which is not true—eagles can see very well in front during flight), but because the eagle is a powerful symbol of national pride. Her statement makes building the wind farm an affront to the nation, even if the logic behind it is questionable. In her statement, WIND TURBINES AS CAUSE OF BIRD DEATHS is supported by a completely unrelated metaphor from a different conceptual domain that has been added to the formula in an illogical, but effective statement that wind turbines are anti-American. This type of logic, which Stephen Toulmin describes as deriving from an unstated *warrant* (or guarantee), is the basis for most communication, including on social media in the form of memes where an image is linked to text that makes an unstated, but immediate argument (Toulmin, 2003). This is also called *enthymeme*. The Data is supported by the Warrant (whether stated or unstated) and Backing, leading logically to the Qualifier and Claim, unless there is a Rebuttal (counterargument) (Figure 8.6).

In *spurious* (false) arguments about environmental issues, such as the one described above, the argument that wind turbines should not be constructed on Lake Erie takes the following form (Figure 8.7).

But this kind of argument works despite its faulty logic because it appeals to national pride in a physical way, which all the data showing the safety of wind turbines in relation to birds will not be able to compete with. With a new conceptual mapping, the following argument could take its place (Figure 8.8).

Making this argument in the technical documentation would satisfy both the public's need to equate a large-scale sustainable energy project with national

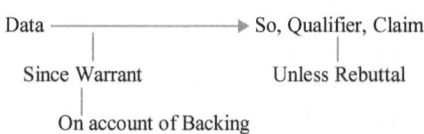

FIGURE 8.6 Toulmin's formula for argument.

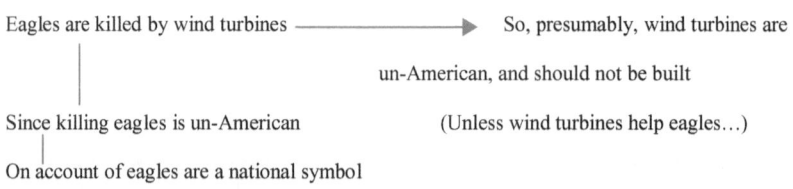

FIGURE 8.7 Spurious eagle argument.

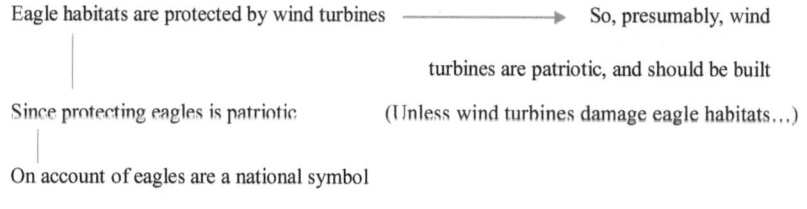

FIGURE 8.8   Modified eagle argument.

pride, and the scientist's need to prove that this project will not negatively impact local bird populations during migration and will instead improve bird habitats.

Representation of Metaphor Aptness

Metaphor aptness can be measured and recorded using survey data to scale up the number of responses possible in future research. An example of such a measurement tool can take a format as simple as a customer satisfaction survey that asks questions of the public, scientists, and policy makers about the perceived fitness of certain metaphors in relation to an environmental project. These metaphors would need to be discovered and listed from a preliminary ethnographic investigation of the local population's communicative practices (because this information is usually not documented) and also outlined from technical documentation about the project. Participants in the survey would then choose how closely they subscribed to the metaphor.

This data can be represented on a scatterplot graph so that *aptness* (the fit) of the metaphor to reflect meaning in terms of the public domain (on the x axis) and the scientific or technical domain (on the y axis) is compared. However, the question of aptness cannot be quantitatively deduced from source data—it can only be qualitatively assigned through ethnographic research and then represented quantitatively. Data point M^3 could represent the positive aptness of the eagle metaphor for the public and its negative scientific aptness. What technical communicators should be looking for are metaphors, such as M^1 and M^2, that succeed in both the public and scientific domains, the sweet spot for effective communication with the public that also supports scientific truth (Figure 8.9).

Such graphs have been used to show terminological correlations at the word or phrasal level of language using Antconc or similar programs to count instances of collocation, where words tend to appear together (Herndl et al., 2018). However, examining instances of metaphorical construction and usage cannot be fully done by a computer yet. Natural Language Processing programs have come a long way toward being able to do this, but in fact, identifying metaphors and their meanings is one of the biggest barriers to the accuracy of artificial intelligence. A computer will not realize that reference to

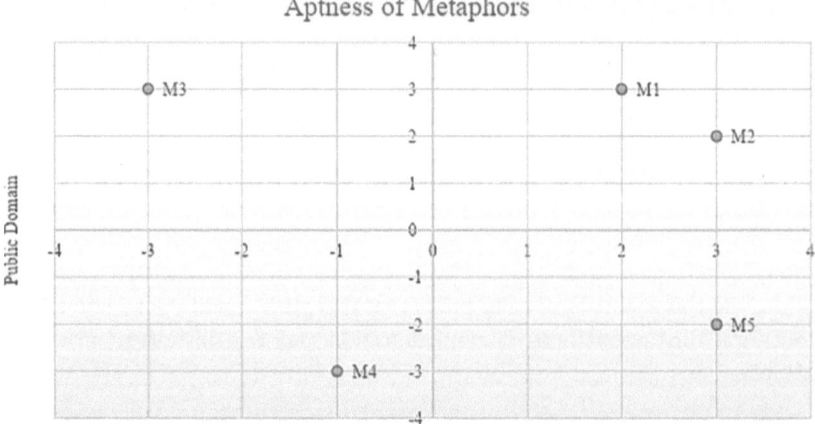

FIGURE 8.9 Aptness of metaphors.

an eagle in the context of wind turbines and bird deaths has nothing to do with the perception of eagles or environmental hazards of turbine construction but is a metaphorical cue that wind turbines are anti-American. This reference is unexpected because it is working at the level of metaphor, not word or phrase. The public is extremely efficient at creating, processing, and understanding complex metaphors, and technical communicators would do better by attending to this fact.

Discussion Questions

1 How is the concept of embodiment defined in this chapter? What is the relationship between embodiment and metaphor? Are there ways that we see the world that can only be understood through the lens of our bodies?

2 How would you react if a large wind farm were proposed for construction in your backyard? What concerns might you have about this project in your community? How would other people in your neighborhood react?

3 There are times when words are purposefully meant to mean more than one thing at a time, such as in poetry, and in some jokes. Can you think of examples of this happening in your everyday life?

Assignments

1 Choose a local environmental issue that is important where you live or go to school. Make a list of all the ways that metaphors are used to describe the issue. Are these metaphors used more in the public or scientific domains? Are there some that would be useful to both the public and experts? Assign them levels of aptness from 1 to 5 and graph them as shown in Table 8.1.

2 (a) Use Toulmin's formula for argument to describe a common argument in a local environmental debate. How does including different warrants and their backing change the success or failure of the argument? (b) Find a meme on social media that comments on an environmental topic. Fully deconstruct its meaning using Toulmin's formula.

References

Bailey, H., Brookes, K. L., & Tompson, P. M. (2014). Assessing environmental impacts of offshore wind farms: Lessons learned and recommendations for the future. *Aquatic Biosystems, 10*(8). https://aquaticbiosystems.biomedcentral.com/track/pdf/10.1186/2046-9063-10-8.pdf

Brandt, D. (2015). *The rise of writing: Redefining mass literacy.* Cambridge: Cambridge University Press.

Brockway, P. E., Owen, A., Brand-Correa, L. I., & Hardt, L. (2019). Estimation of global final-state energy-return-on-investment for fossil fuels with comparison to renewable energy sources. *Nature Energy, 4*, 612–621.

Ding, H. (2014). *Rhetoric of a global epidemic: Transcultural communication about SARS.* Carbondale: Southern Illinois University Press.

Douglas, E., & Ramsey, R. (2021, February 16). *No, frozen wind turbines aren't the main culprit for Texas' power outages.* Retrieved from The Texas Tribune: https://www.texastribune.org/2021/02/16/texas-wind-turbines-frozen/.

Etkin, D. S. (2006). *Oil spill probability analysis for the Cape Wind energy project in Nantucket Sound.* Cortland Manor, New York: Environmental Research Consulting.

Fink, B. (2004). *Lacan to the letter: Reading Ecrits closely.* University of Minnesota Press.

Flower, L. (2008). *Community literacy and the rhetoric of public engagement.* Carbondale: Southern Illinois University Press.

Grabill, J. (2007). *Writing community change: Designing technologies for citizen action.* Hampton Press.

Gross, A., & Harmon, J. (2016). *The internet revolution in the sciences and humanities.* Oxford: Oxford University Press.

Heartland Institute (2021). *Ivar Giaever.* Retrieved from Heartland Institute: https://www.heartland.org/about-us/who-we-are/ivar-giaever.

Herndl, C., Hopton, S., Cutlip, L., Polush, E., Cruse, R., & Shelley, M. (2018). What's a farm? The languages of space and place. In C. Rai & C. Druscke (Eds.), *Field rhetoric: Ethnography, ecology, and engagement in the places of persuasion* (pp. 61–94). University of Alabama Press.

International Renewable Energy Agency (2019, May). *Renewable power generation costs in 2018.* Retrieved from IRENA: https://www.irena.org/publications/2019/May/Renewable-power-generation-costs-in-2018.

IPCC (2021, August). *Climate change 2021: The physical science basis.* Retrieved from UN Intergovernmental Panel on Climate Change: https://www.ipcc.ch/report/ar6/wg1/downloads/report/IPCC_AR6_WGI_Full_Report.pdf.

Johnston, L. (2019, April). *Lake Erie groups rev up opposition to Cleveland wind turbine project, as developers negotiate with state.* Retrieved from Cleveland.com: https://www.cleveland.com/news/2019/04/lake-erie-groups-rev-up-opposition-to-cleveland-wind-turbine-project-as-developers-negotiate-with-state.html.

King, L. (2019, April 3). *Do wind farms cause cancer? Some claims Trump made about the industry are just hot air.* Retrieved from USA Today: https://www.usatoday.com/

story/news/politics/2019/04/03/cancer-causing-wind-turbines-president-donald-trump-claim-blown-away/3352175002/.

Lakoff, G., & Johnson, M. (1980). *Metaphors we live by.* University of Chicago Press.

Latour, B. (2018). *Down to earth: Politics in the new climatic regime.* Cambridge: Polity.

LEEDCo (2019, October 4). *Icebreaker wind.* Retrieved from LEEDCo: http://www.leedco.org.

Li, J., & Yu, X. (2018). Onshore and offshore wind energy potential assessment near Lake Erie shoreline: A spatial and temporal analysis. *Energy, 147,* 1092–1107.

Miao, R., Ghosh, P. N., Khanna, M., Wang, W., & Rong, J. (2019). Effect of wind turbines on bird abundance: A national scale analysis based on fixed effect models. *Energy Policy, 132,* 357–366.

Ohio Power Siting Board (2020, October 8). *Ohio power siting board meeting - Oct. 8.* Retrieved from YouTube: https://www.youtube.com/watch?v=AJDbNB-rPjk.

Pelzer, J. (2020, November 20). *Sam Randazzo resigns as Public Utilities Commission of Ohio chair.* Retrieved from Cleveland.com: https://www.cleveland.com/open/2020/11/sam-randazzo-resigns-as-public-utilities-commission-of-ohio-chair.html.

Potts, L. (2013). *Social media in disaster response: How experience architects can build for participation.* Routledge.

Rai, C., & Druschke, C. (Eds.). (2018). *Field rhetoric: Ethnography, ecology, and engagement in the places of persuasion.* University of Alabama Press.

Renewz.biz. (2019, December 12). *Bird groups file lawsuit over Icebreaker.* Retrieved from Renewz.biz: https://renews.biz/56960/bird-groups-file-lawsuit-over-icebreaker/.

Salvo, M. J., & Potts, L. (Eds.). (2017). *Rhetoric and experience architecture.* Parlor Press.

Sauer, B. A. (2003). *The rhetoric of risk: Technical documentation in hazardous environments.* Lawrence Erlbaum.

Saussure, F. d. (2011 [1916]). *Course in general linguistics* (C. Bally, & A. Sechehaye, Eds.). Columbia University Press.

Simmons, M. (2008). *Participation and power: Civic discourse in environmental policy decisions.* SUNY Press.

Tomich, J. (2019, July 12). *Strangled Ohio wind industry: 'We don't want to give up'.* Retrieved from E&E News Renewable Energy: https://www.eenews.net/stories/1060727747.

Toulmin, S. (2003). *The uses of argument.* Cambridge University Press.

U.S. Congress. (2019, February 7). *H.Res. 109- Recognizing the duty of the federal government to create a green new deal.* Retrieved from Congress.gov: https://www.congress.gov/bill/116th-congress/house-resolution/109.

U.S. Department of Energy (2018). *Final environmental assessment LEEDCo project Icebreaker Lake Erie, City of Cleveland, Cuyahoga County, Ohio.* Retrieved from Energy.gov: https://www.energy.gov/sites/prod/files/2018/09/f55/EA-2045-LEEDCo-Final%20EA-2018.pdf.

U.S. Global Change Research Program (2018, November). *Fourth national climate assessment.* Retrieved from GlobalChange.gov: U.S. Global Change Research Program: https://nca2018.globalchange.gov.

Varela, F. J., Thompson, E., & Rosch, E. (2016). *The embodied mind: Cognitive science and human experience* (Rev. ed.). MIT Press.

9

EVALUATING ECOLOGICAL PERCEPTIONS AND APPROACHES IN THE FOURTH NATIONAL CLIMATE ASSESSMENT REPORT

Diane Martinez

The authors of the Fourth National Climate Assessment (NCA4) were clear that Earth's climate is changing "primarily as a result of human activities" and future impacts and risks of climate change depend on actions taken today (USGCRP, 2018, p. 34). These impacts and risks, to the degree that scientists know now, make up the content of NCA4. The message in NCA4 is vital for readers to understand, and consequently act on, if humans are going to alter life on Earth through life-saving measures or continue toward possible extinction. How effective, though, is NCA4 in delivering that message? One way to answer that question is to examine the document for effective strategies in environmental communication.

The field of environmental communication has grown considerably since the 1960s when the connection between human activities and the environment became part of mainstream consciousness. This social awakening is most often attributed to Rachel Carson's book *Silent Spring* that warned of the environmental impact from the indiscriminate use of dichlorodiphenyltrichloroethane (DDT) and other pesticides. In the 1970s, further awareness about human influence on the environment was brought to the public's attention via reports about climate change, referred to as global warming at that time. It was well known, even then, that increased greenhouse gas emissions would subsequently amplify the amount of carbon dioxide in the atmosphere and create warming trends that would affect global climate.

One consequence of this environmental enlightenment was that environmental communication burgeoned as a field as climate change captured the attention of scientists, the government, and the public. Genres in environmental communication grew alongside technology over the decades and now include scholarly articles, scientific and government reports, white papers, books

DOI: 10.4324/9781003266549-11

and textbooks, blogs, environmental literature, narratives, essays, video blogs, documentaries, and video games. Content-wise, initial environmental communication took on "differing subjects, approaches, and even conceptions of communication" (Cox & Depoe, 2015, p. 13); however, during the first decade of this century, scholars began deliberating how environmental communication is shaped by social, cultural, and ideological factors (Cox & Depoe, 2015). One such relationship under close examination then, and now, is how climate-change risk is communicated to the public.[1]

According to NASA, an organization that has been involved in climate studies since the 1970s, *climate change* is "a long-term change in the average weather patterns that have come to define Earth's local, regional and global climates" (NASA, "Overview," 2021, para. 5). Changes in these patterns have mild to dire consequences for all life on this planet. Since the 1960s, scientists have known that human activity is altering climate patterns not on a long-term, planetary, timeline, but on a generational timeline, and some effects are noticeable today. Human activity that creates these altered climate patterns can be mitigated to prevent further impact. It is important, then, that the public is made aware of human activities that affect natural planetary systems and cause climate change, as well as risks to human health and welfare that result from these changes in climate. This type of information is referred to as *climate-change risk communication*.

In terms of studying climate-change risk communication and the public's reaction, some interesting findings have surfaced. One group of researchers found that "citizens are exposed to many messages about climate change on a daily basis, yet studies show a declining trend in public understanding of human-caused climate change" (Schweizer et al., 2013, p. 43). Similarly, having more knowledge about climate change does not always translate into "greater action or policy change" (Cox, 2015, p. 372). How, then, can climate-change risk communication, such as NCA4, generate much-needed action?

There is no one answer to this question, but some scholars suggest an ecological approach to environmental communication will help (e.g., DiCaglio et al., 2018; Druschke & McGreavy, 2016; Schweizer et al., 2013). *Ecology* is the study of relationships between organisms and their physical environment. An ecological approach to environmental communication would focus on the relationship between humans and the environment and encourage people to make essential changes (DiCaglio et al., 2018). Underlying difficulties of an ecological approach, however, are scale and interrelationships. *Scale* is the idea of how well readers relate to information. Beverly A. Sauer (2003) argued that "embodied knowledge," or knowledge that readers can easily connect to

---

1 I acknowledge that there are many kinds of publics and the exposure to and understanding of environmental communication in its multiple forms and genres will vary among these publics.

their own experiences, is missing from institutional communication practices that "may inadvertently silence or render invisible the kinds of information that decision-makers need to assess and manage risk" (p. 5). Thus, effective environmental communication scales information so that readers make a connection between themselves and the issue being discussed. Likewise, the idea of *interrelationships* is an ecological principle that focuses on the inseparable relationship and impact between organisms and their environment (DiCaglio et al., 2018). According to DiCaglio et al., "to experience ecology is to see interconnection in relation to the scale at which you live, learning to notice and feel the relationship between what is around you and the rest of the planet" (p 443). In terms of climate-change risk communication, it is crucial that readers see humans, individually and collectively, as an element that impacts the natural environment.

It is also important that climate-change risk communication conveys the uneven distribution of the effects of climate change on human health and welfare, often referred to as "climate justice." *Climate justice* "is a term, and more than that a movement, that acknowledges climate change can have differing social, economic, public health, and other adverse impacts on underprivileged populations" (Simmons, 2020, para. 2). Climate change disproportionately affects people in low-income countries and the poor in high-income countries, women, people of color, and indigenous communities (Levy & Patz, 2015). Inequities associated with climate change have historical roots in race, class, gender, economics, and other factors. Effective environmental communication emphasizes the connection between humans and the environment, but it also stresses how environmental impact is experienced across the globe—how actions in one part of the world affect the environment and people in other parts of the world.

In this chapter, I pull from studies that combine ecology and environmental communication to provide guidance on my evaluation of NCA4 in terms of its approach to communicating climate-change impacts and risks. I first provide background information on NCA4 and the ecological theories of Earth Systems Science (ESS) and Gaia theory, which are grounded in the concept of a human–environment connection. A literature review on the challenges of environmental communication is also provided as background. Finally, I explain my methodology and discuss my results.

Problem: There are missed opportunities in NCA4 to present readers with information on a scale they can comprehend and see themselves as part of the climate-change process. Consequently, necessary changes to mitigate further human influences on climate and the environment are not likely to occur as a result of this report.

Solution: I recommend an ecological and ethical approach to writing the next reports, NCA5 and NCA6, that focus on scale, interrelationships, and climate justice as ways to improve their efficacy in transforming human activities that continue to exacerbate climate change. Likewise, the insights gained from this study can be used for future environmental communication in multiple genres.

Background

The effort behind NCA4 was coordinated by the United States Global Change Research Program (USGCRP), established in 1989 and composed of 13 federal agencies. USGCRP is mandated by Congress to coordinate federal research and report on the "forces shaping the global environment, both human and natural, and their impacts on society" (USGCRP, "About USGCRP," n.d., para. 1). The aim of NCA4 is to provide findings from the USGCRP that analyze the effects of climate change, as well as current trends, both human-made and natural, regarding global changes. NCA4 was released in two volumes, the first volume intended for a scientific audience. Volume II: Impact, Risks, and Adaptation in the United States was written for Congress and the President of the United States, but its audience also includes "decision-makers, utility and natural resource managers, public health officials, emergency planners, and other stakeholders" (USGCRP, 2018, p. 1). It could be argued that the public is one of those stakeholders since it is stated that the report aims to present "technical information in a manner more accessible to a broad audience" (USGCRP, 2018, p. 4), which is evidenced in the attention given to design and relatively non-technical language. Likewise, legislators might use or reference this report when speaking to their constituents about issues of climate change and their political decisions and actions. NCA4 also has been the focus of many mainstream news articles that reported on the scientific findings in the report and emphasized the warnings, predictions, and cost of climate change (e.g., American Institute of Physics, Global News, Union of Concerned Scientists, The Weather Channel, Public Broadcasting Service, Science Daily (as reported by NOAA), Vox, and CBS News). Volume II is the report under study in this research because the audience is much broader than Volume I.

There is a tendency to think about climate change in terms of linear and immediate cause and effect, but planetary systems (e.g., weather, ecosystems, biosphere, atmosphere) respond to outside forces in degrees of variability that are not on a one-to-one scale with immediate or noticeable results. Therefore, it is important to explore ecological theories that provide a foundational understanding of the interconnectedness of planetary systems keeping in mind that humans are part of the whole Earth System. ESS and Gaia theory are two such theories.

Earth System Science (ESS)

The ESS is a scientific theory that "Earth operates as a single, complex, adaptive system driven by the diverse interactions between energy, matter and organisms"

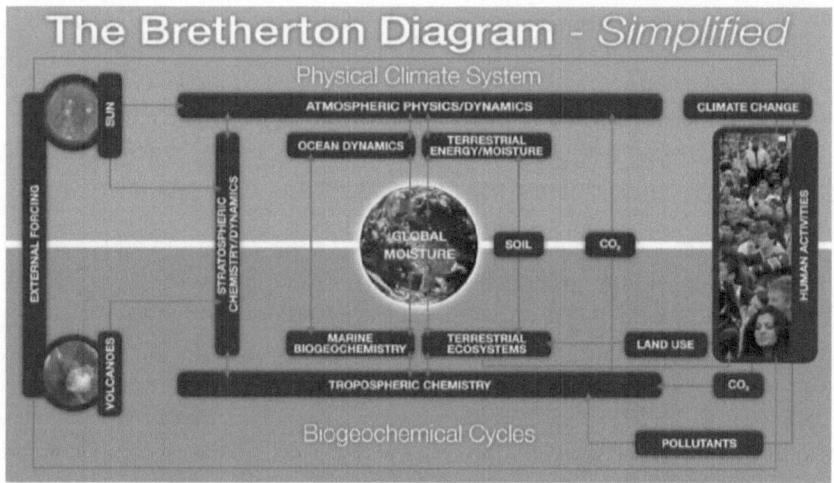

FIGURE 9.1 Bretherton Diagram.

(Steffen et al., 2020, p. 54), of which human activity is part (Fellows, 2019; Steffen et al., 2020). Conceptualizations of ESS were heralded by scientists such as James Hutton, an 18th-century geologist who originated the idea of geologic uniformitarianism; Alexander von Humboldt, an 18th- and 19th-century geographer, naturalist, and explorer; and Vladimir Vernadsky, originator of the theory of the biosphere in the early 20th century (Steffen et al., 2020). In the 1980s, NASA brought ESS into mainstream science and developed a visual representation of the Earth System called the Bretherton diagram (see Figure 9.1). The Bretherton diagram "was the first systems-dynamics representation of the Earth System" that showed a connection between the physical climate system and a wide array of "forcings and feedbacks. Humans constituted a single box of their own connected to the rest of the Earth System through three forcings (carbon dioxide, pollutant emissions and land-use change)" (Steffen et al., 2020, p. 56). This diagram depicts the inseparable connection and interaction between humans and the environment, which is a top priority for ESS in terms of climate (Fellows, 2019).

The direct impact that humans have on the environment is an important concept that must be part of climate-change risk communication if readers are to realize the impacts and risks they are essentially bringing upon themselves. This idea was taken even further by James Lovelock and Lynn Margulis with Gaia theory.

Gaia Theory

Gaia is a term that refers to Earth as a complex living system of which human beings are a part. Gaia is an alternative perspective to the more popular view that "sees nature as a primitive force to be subdued or conquered" (Lovelock, 1979, p. 12). Instead, "Gaia offers a perspective of Earth as a living subject rather than an inanimate object" (Crist & Rinker, 2010, p. 11). *Gaia theory* is a unique perspective that asserts a connection between living and non-living elements of Earth that have evolved and self-regulate. Planetary systems including "global temperature, atmospheric content, ocean salinity, and other factors" self-regulate to continually support life despite external forces that might otherwise throw the system out of balance (Environment and Ecology, n.d., para. 4). For example, prior to Gaia theory, serendipity and distance from the sun (the "Golidlocks view") were reasons why the Earth was the only habitable planet in the solar system (Crist & Rinker, 2010). Gaia theory offers an alternative view. According to Gaia theory, a force, such as the brightness of the sun, which has increased by 25–30% over 3.8 billion years, would make Earth uninhabitable. But as a self-regulating system that reacts to such forces, Earth remains habitable and hospitable to the living and non-living elements that compose the planet (Crist & Rinker, 2010). According to Lovelock, the originator of Gaia theory, the goal of Gaia is *homeostasis*, or balance, that provides "an optimal physical and chemical environment for life on this planet" (Lovelock, 1979, p. 11).

The idea of the Earth being a self-regulating entity has met with extreme skepticism because self-regulation implies foresight and intention (Moody, 2012), as well as agency (Crist & Rinker, 2010). Although ESS relies on the same principles as Gaia theory regarding a single Earth System, ESS has not been debated as much as Gaia theory because of one main difference: ESS is used to make predictions based on the mechanistic interactions among planetary elements, including humans, not because the Earth has an internal goal (Fellows, 2019; Moody, 2012).

No matter if the Earth is or is not self-regulating, it is not immune to changes, and the changes imposed by humans (who are self-regulating) are beyond the natural factors that affected it before the Industrial Revolution. There are environmental consequences for human activity on this planet, and ESS and Gaia theory provide fundamental ecological principles for understanding this interdependence, which could be included in climate-change risk communication. Communicating intricacies about this interdependence is only half the equation in creating environmental communication that encourages change. Effective communication about the impact of human activity on a scale that presses readers to comprehend their own embodied experiences in relation to larger planetary systems is also essential. How these two goals can be achieved is a point of research among scholars of environmental communication.

Issues in Environmental Communication

Environmental communication is a broad interdisciplinary field that brings together expertise in life science, social science, politics, economics, history, psychology, humanities, communication, and many other disciplines (DiCaglio et al., 2018; Hansen & Cox, 2015; Moser, 2016). It is also an evolving field due to numerous complex challenges associated with engaging readers to the point of changing behaviors that negatively affect the environment. I cannot cover all challenges associated with environmental communication (of which climate-change risk communication is part of), but I will touch on three commonly discussed issues: engaging readers, explaining cause and effect (i.e., interrelationships), and scaling information.

Engaging readers to make significant and necessary lifestyle and societal changes is one key challenge of climate-change risk communication. Early on, climate-change communication was mostly focused on raising awareness and explaining the science behind climate change (Moser, 2016). This attention on awareness and understanding led to what researchers call the "deficit model" of communication (Coppola, 1997; DiCaglio et al., 2018; Druschke & McGreavy, 2016; Hanson-Easey et al., 2015; Sterman, 2011). The *deficit model* is the idea that the public is skeptical, apathetic, or even hostile, to scientific claims about climate change simply because they do not understand the science. This model subscribes to a binary view of communication and promotes the idea that communication flows only one way, from experts to "a passive and trusting public" (Druschke & McGreavey, 2016, p. 47). With environmental communication, the overarching idea is that if the public was given more information and understood the science behind climate change, then they would make necessary changes. Some environmental communication experts have criticized this notion (Coppola, 1997; Cox, 2015; Druschke & McGreavy, 2016; Grabill & Simmons, 1998; Hanson-Easey et al., 2015; Schweizer et al., 2013; Sterman, 2011), and studies show that even though the public has been made aware of climate matters, they have become more passive and even "redefine their obligations to solve environmental problems as simply being informed about the issues" (Coppola, 1997, p. 10). Being informed, however, does not equate to buy-in, and reader engagement, much less change, is low for the deficit model of communication.

To increase engagement, some environmental communication scholars emphasize tapping into readers' values and attitudes. Schweizer et al. (2013) proposed a place-based framework built on audience cultural values and beliefs. A *place-based framework*, they claimed, increased public engagement on environmental issues because it is built on the idea that people form attachments with landscapes and value those places. When environmental communication is based in local and familiar places, people are motivated to do what is necessary to preserve that place (also noted in Akerlof et al., 2013). Grabill and Simmons (1998), too, stated that "people's risk perceptions are determined by real and localized situations," not hypothetical situations and places that readers cannot relate to (p. 419). Respectful communication practices also affect public engagement on environmental matters. Druschke and McGreavy (2016) asserted that effective communication is two-way communication that "starts with recognizing and valuing where audiences are coming from, and then working to incorporate those perspectives" (Druschke & McGreavy, 2016, p. 47). It is possible that "much of the frustration and confusion around environmental degradation" stems from forcing people to think on a scale that has no real meaning to them (DiCaglio et al., 2018, p. 440). Addressing personal relevance through values and place, then, becomes a key factor in effective, change-inducing, environmental communication.

Cause and effect are additional challenges in environmental communication. ESS and Gaia theory point to the non-linearity of changes within Earth's System where influences are not seen or experienced immediately, or, in some instances, within one's lifetime. Gardner (2011) echoes this point that causes and effects of climate change are "spread across space, time, and species" making the problem one that is "profoundly intergenerational" (p. 8). Furthermore, when people do not immediately experience or see how their actions affect the Earth on a planetary scale, they may "inadvertently 'trigger tipping'" points (Sterman, 2011, p. 819) leaving a more dire problem for future generations. How, then, do environmental communicators entice readers to make needed changes now?

One way that immediate changes may come about is by using *ecological rhetoric*, a combination of ecological principles and rhetoric (DiCaglio et al., 2018; Druschke & McGreavy, 2016). A collaboration between ecology and rhetoric presents opportunities to rethink environmental communication so that it does not simply inform readers using the deficit model of communication but instead focuses on relationships and incorporates two-way communication that addresses the values of readers on a scale they comprehend. Just as ESS and Gaia theory point to the undeniable relationship between humans and their impact on the environment, environmental communicators can use this same principle to analyze how rhetoric impacts readers, which is the basis of my analysis of NCA4.

Methodology

NCA4 is an example of technical communication that communicates causes of climate change, the impact of climate change on Earth's natural systems, and risks to human health and welfare caused by climate change, all within the context of what these impacts and risks mean to the United States specifically. To evaluate the effectiveness of NCA4 in inducing needed changes, individually and collectively, I analyzed the report for ecological rhetoric, most especially in terms of how well scale and interrelationships were used or addressed.

As part of my analysis, I pull from several scholarly works (e.g., Cox, 2015; DiCaglio et al., 2018; Druschke & McGreavy, 2016) to examine NCA4 in terms of scale when communicating impacts, risks, and change. Specifically, I aimed to answer the question: On what scale(s) are impacts, risks, and change communicated in NCA4? I then examined how likely information in the report resonates with readers on a level that will bring about change. According to DiCaglio et al. (2018), "society" is not where experience and change occur, and "generic communications ... may not help an individual to see interconnection"; instead, effective rhetoric "begins with individuals working toward a new way of understanding and relating to their lived contexts" (p. 444). Thus, I considered if the information in NCA4 is scaled to levels that intended audiences could relate to.

ESS and Gaia theory are models of Earth's System in terms of relationships; these theories also provide opportunities for environmental communicators to consider connections between environmental texts and reader response. Using the foundational principles of interrelationships as expressed in ESS and Gaia theory (Earth as a single, complex, adaptive and interconnected system of living and non-living matter that includes humans), I examined the report to answer two additional questions: How is the principle of interrelationships represented in NCA4? What role do humans play, if any, in these representations within the report?

To answer these questions, I reviewed the document for instances where climate change, Earth, or Earth's System are presented wholly or through division. In instances where interconnectedness is mentioned or represented, I examined what elements were included and what elements are shown to impact other elements. I then considered the implications of these representations regarding the audiences and purpose of NCA4, especially in terms of opportunities for readers to understand and relate their actions to Earth's System and in terms of climate justice—how their actions affect others. The following results and discussion focus on the representation of scale and interrelationships in NCA4. Given that the document is over 1,500 pages, my findings in that section are illustrative and not comprehensive.

Communicating Impacts and Risks

It is acknowledged throughout NCA4 that the "world we live in is a web of natural, built, and social systems" all which affect one another and are affected by climate change (p. 640). How climate change influences these systems depends on how well scientists can predict "all the ways in which climate-related stressors might lead to severe or widespread consequences" for these systems (p. 640). NCA4 is dedicated to explaining those stressors and consequences to the extent known now; thus, Chapters 2 through 27 are devoted to impacts and risks of climate change in relation to 16 national topics and 10 regions throughout the United States and U.S.-affiliated Pacific islands. One outstanding characterization of how impacts and risks are communicated in these chapters is seen in how the relationship between nature and humans is depicted.

In the report, the relationship between nature and humans is dichotomous where humans are victims of climate change and the natural processes that occur due to climate change. In much of the report, climate change and changes in natural systems that result from climate change (e.g., increased or decreased rain, rising temperatures, rising sea levels) are characterized as separate entities that threaten and victimize the United States, its citizens, national infrastructure, and way of life as shown in the following examples, which are representative of the writing throughout the report.

> Climate change is expected to cause growing losses to American infrastructure and property and impede the rate of economic growth over this century.
>
> *(p. 25)*

> Future changes in precipitation and the potential for more extreme rainfall events will exacerbate water-related challenges in the Northern Great Plains.
>
> *(p. 149)*

> The ability of U.S. forests to continue to provide goods and services is threatened by climate change and associated increases in extreme events and disturbances. For example, severe drought and insect outbreaks have killed hundreds of millions of trees across the United States over the past 20 years, and wildfires have burned at least 3.7 million acres annually in all but 3 years from 2000 to 2016.
>
> *(p. 234)*

When read over and over, readers may disassociate themselves with their role in climate change because climate change and natural processes act independently from humans and always in threatening ways. Thus, readers do not see their own actions as being responsible for climate change or any of the resulting impacts of climate change because they are victims of these natural forces.

It is acknowledged many times throughout NCA4 that climate change is caused by human activities, but it may be impossible for readers to internalize that information when their accountability is conveyed only in vague terms and on a scale that is too large for most readers to personally relate to as shown in the following excerpts.

> Global action to significantly cut greenhouse gas emissions can substantially reduce climate-related risks and increase opportunities for these populations in the longer term.
>
> *(p. 25)*

With substantial and sustained reductions in greenhouse gas emissions ...
the increase in global annual average temperature relative to preindustrial
times could be limited to less than 3.6°F (2°C).

(p. 42)

In addition, without substantial and sustained mitigation efforts to reduce
global greenhouse gas emissions, the need for adaptation and resilience
investments to address the impacts of climate change on the energy sector
is expected to increase if the most severe consequences are to be avoided
in the long term.

(p. 189)

In these examples, which are illustrative of the text throughout the report,
specific global actions that will reduce greenhouse gas emissions are not mentioned, which leaves readers with a void in knowing what should be done and
by whom. Readers are unable to see themselves as part of climate change because the only mention of human involvement is "human activities," which are
most commonly referred to as increasing greenhouse gas emissions, deforestation, and land-use change, not something that is relatable to everyday activities
of readers. If the goal of climate-change risk communication is to encourage
change, then the rhetoric must be scaled to levels that readers can relate to. It
must answer the question: How do I, and anything I do, connect with the entire planet? This sort of relatability is missing in NCA4.

Another characteristic is how the relationship between nature and humans
is always negative. By focusing on this negative connection, "we are *unlikely
to experience the world ecologically*" but rather become confused, bitter, or unable
"to fully internalize the connections underlying the negative environmental
effects" [italics in original] (DiCaglio et al., 2018, p. 443). In other words,
humans will not see themselves as part of the problem—or the ones actually
causing the problem—but only as victims. This negativity leads to the idea that
we need to "rid ourselves of these connections" (DiCaglio et al., 2018, p. 443)
and furthers the disparity between individual actions and the environment.

Effective climate-change risk communication brings readers to a place of
embodiment where they experience a direct connection between their actions
and Earth. Whitmarsh (2015), however, found that this is not something readers do on their own. She explained that there is a:

...general tendency to underestimate one's own contribution to causing
climate change. People tend to identify causes of climate change with
other people or countries, such as SUV drivers, firms, the US or China
... and with more 'distant' activities, namely industry and deforestation,
rather than their own actions.

(p. 340)

A great deal of *NCA4* is written on a scale that has no direct relevance to intended audiences because the language is vague or beyond individual or collective relatability or accountability. If the report is intended to simply inform readers about the current state of climate change, then this approach fulfills that purpose, and it is an example of the deficit model of communication. If, however, the intent is to change behaviors, as indicated in the Overview, where it is stated that "This report characterizes specific risks across regions and sectors in an effort to help people assess the risks they face, create and implement a response plan, and monitor and evaluate the efficacy of a given action" (USGCRP, 2018, p. 44), then information in the report has to reach audiences at the level of "a given action." For instance, even though it might be assumed that most readers know what "greenhouse gas emissions" means, there is an even greater assumption that readers will interpret that term and the data provided in NCA4 at a level where they exist and operate on a day-to-day basis, which is unlikely. A further problem is how effects of climate change are separated through the illusion of boundaries.

Nationalism, Not Interconnectedness

The United States is not the only country that assesses climate impact, risks, and adaptations. Appendix 4 in NCA4 lists several countries and their approaches to assessing climate change for their own purposes. This appendix is intended to place NCA4 "within the broader international landscape of assessment activities and to compare it with other approaches" (USGCRP, 2018, p. 1431). Within the global context, Earth is divided into countries as seen in the report's Figure A4.1 (USGCRP, 2018, p. 1431) (Figure 9.2), which presents a flat map of selected regions of the globe, highlighting the countries used as a comparison in Appendix 4. Climate change is a global phenomenon that is not restricted by borders, and some parts of the world experience the effects of climate change more drastically than other parts. Figure A4.1 depicts the globe through the eyes of nationalism, a world where each country assesses climate change in isolation.[2] Although there is acknowledgment in *NCA4* that climate change is global, the report's focus is on the "changes that affect Americans' lives, communities, livelihood, now and in the future" (p. 36). Other countries' reports mentioned in Appendix 4 appear to have taken a similar, nationalistic, view of climate change.

A nationalistic assessment of climate change is problematic for many reasons, the main ones being that it is unrealistic and socially irresponsible. In NCA4, maps of the United States depict several issues related to climate change

2 With the exception of the European Union, which assesses climate change for 32 member countries but that report was not referenced in Appendix 4.

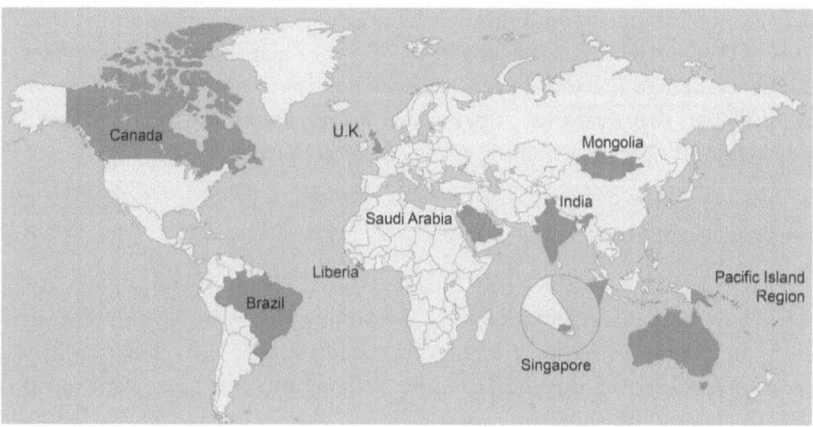

FIGURE 9.2 World Map.

including current and projected impacts on temperature, precipitation, sea level, and hurricane tracks, for instance. Effects of climate change are not evenly distributed throughout the world, and the constant reminder of how the nation and world are divided into discrete sections and what areas can anticipate the most or least extreme effects may create the illusion that certain areas are relatively safe or not large climate-change contributors since they experience fewer effects. But that is a false conception. China and the United States, for instance, are the two largest emitters of carbon dioxide (Frohlich & Blossom, 2019), and even though these countries are affected by climate change, their economies are more likely to continue to prosper. Low-income countries, on the other hand, produce a fraction of the carbon dioxide emitted by high-income countries but experience disproportionately higher health and economic risk and consequences due to climate change (Levy & Patz, 2015). Consequently, change may not be seen as a priority in high-income areas since the future does not look as dire as it does in other areas. Illustrating incongruent effects of climate change is necessary for reporting accurate climate-change science; however, there is still the issue of how people connect their individual actions in North Carolina, for instance, to how climate change manifests in Washington state, much less in Malaysia. One issue with the maps and individual regions in NCA4 is that they reinforce our habituated way of "thinking of the world as made up of discrete objects" (DiCaglio et al., 2018, p. 440). Breaking up the world into distinct borders creates distance between the report content and readers' embodied experiences, as well as their cognitive ability to see the Earth System as a single entity, an entity of which they are a part. If environmental communication is to move readers beyond an intellectual comprehension of ecology, then we need to find ways to connect with readers on a personal scale and to help them see themselves connected to larger ecological processes and their humanitarian responsibilities no matter where they reside on the shaded maps of climate change. If the goal of NCA4 is to help readers weigh evidence of climate change impacts during decision-making processes, then there are a few strategies that could be employed in upcoming reports that may increase the likelihood of readers making actual, needed changes.

Recommendations

The contributors to NCA4 had a tall order to fill given the report's scope and diverse audiences. My intention in studying this report is to consider its effects on readers in terms of inducing needed change. I conclude with three recommendations for upcoming reports, which also can be applied to other climate-change risk communication.

First, Whitmarsh (2015) reported that the public tends to understand climate change in a generic sense with weather being the most common identifier. One possible reason that publics connect climate change and weather is that most people get their information about climate change from the news where it is often coupled with images of weather catastrophes (Schweizer et al., 2013). The significance of this association is that "the public understands weather and natural disasters as *acts of god* and fails to see that human actions and lifestyle choices are capable of influencing the pace of climate change" [italics in original] (Schweizer et al., 2013, p. 44). The common strategy in NCA4 of using climate change and climate-related processes as separate entities that threaten humans contributes to the dissociation of humans seeing themselves as influencers of climate-change causes or mitigation. Instead, they see themselves only as victims. Including the principles of ESS and Gaia theory would provide an ecological foundation to the report, a relationship that is not emphasized enough, at least in direct ways. Additionally, sentence structure that presents humans as the initial actors who affect planetary systems would show the direct connection between human decisions and the environment, which, in turn, circle back to their own health and well-being.

Second, it would be more effective for readers to realize their embodied participation in climate change cause or mitigation if explicit activities were shown to increase or decrease greenhouse gas emissions. There are instances where examples are given, such as efforts toward adaptation like building roads, seawalls, developing flood maps, and setting new building codes and land-use ordinances, but there are two problems with this level of reporting: (1) It creates the appearance that others (unnamed and unknown) will do the work that needs to be done, and (2) There is no inroad for readers to see their personal choices and activities as having any impact, negatively or positively, on Earth or climate change. By showing the impact of specific activities, readers can see who and what most affect the environment. They can then make informed decisions about what to do and who or what to support and hold accountable. Activity ranking would give readers a more realistic, concrete, image of human-caused climate-change cause and mitigation, especially if the first recommendation were implemented. Environmental impact, however, is only one facet of interrelationships, and more discussion about social impact from a global perspective should be integrated in all chapters. Climate change must be represented as a collective, global, issue, not a nationalistic one.

Last, fear-inducing language takes over NCA4. Given the critical situation of climate change as it stands right now, it seems appropriate that this report would indicate the urgent need to address climate change before things get worse. However, approaches like simply informing readers (via the deficit model) or framing information in "'apocalyptic' rhetoric" (Hansen & Cox, 2015, p. 17) have very little appeal or effect on audiences (Coppola, 1997; Hansen & Cox, 2015). Instead of framing the text in terms of disastrous climate events that victimize humans, it would be more effective to help readers "shift to thinking ecologically" (DiCaglio et al., 2018, p. 441) through more promising frames that include "'energy security,' 'morality and ethics,' and 'public health'" (Hansen & Cox, 2015, p. 18) or place (Nielsen, 2017; Schweitzer et al., 2013). In other words, framing nature as something worth having now and in the future can illuminate the interconnection between humans and the environment, and ultimately, the welfare of each other and the planet.

Conclusion

NCA4 is an important text in the conversation on climate change cause and mitigation because it originates from a government agency established for the purpose of monitoring and reporting climate change. The American president, Congress, and people expect this report to be accurate in its science and an influential resource as well. In seeking answers to the report's effectiveness in communicating climate-change risk, we can speculate on how readers react to the report's content. Thus, continued work in this area would be to assess and analyze changes in environmental policies and activities from the time this report was released to the next iterations of it, although many other factors must be part of that analysis. In upcoming NCA reports and other environmental communication, incorporating approaches to communicating climate-change risk using ecological and embodied perspectives is more likely to result in readers taking necessary and much-needed action.

Discussion Questions

1 Search for three news articles that reference the Fourth National Climate Assessment (NCA4). Discuss how NCA4 was referenced and whether information about climate change causes, effects, and mitigation factors was scaled in the news stories themselves. In what terms was interconnectedness addressed in the news stories?

2 Conduct preliminary research on what climate justice is, what communities experience the most negative aspects of climate change, and why. Trace the multiple dimensions of climate change and associated social justice issues found in your research through some sort of graphic. Discuss how prevalent these issues are in news stories, class discussions, and other conversations.

3 Read the "Overview" of NCA4 and discuss your reaction to the information in that part of the report, especially in terms of scale and interconnectedness. Discuss how effective the message is for you and your peers in inducing changes in your own behaviors.

Assignments

1 Personal reflection essay. Choose one chapter from *NCA4*, and read it in its entirety. While reading, evaluate how you see yourself affected by the problem presented in the chapter. Consider your role in creating the problem discussed. Do you see yourself as part of the problem? If so, discuss how. If not, explain why not and who you see as being responsible. Consider your role in solving the problem. Is there anything you can do to help solve the problem? If so, explain. If not, explain why you cannot contribute to the solution.

2 Environmental article. Choose a single topic associated with climate change or environmentalism and write a persuasive article that directly shows and explains human impacts on the environment and scales information about the problem and solutions.

References

Akerlof, K., Maibach, E. W., Fitzgerald, D., Cedeno, A.Y., & Neuman, A. (2013). Do people "personally experience" global warming, and if so how, and does it matter? *Global Environmental Change, 23*, 81–91.

American Institute of Physics (2018, December 6). Trump Administration takes aim at national climate assessment. https://www.aip.org/fyi/2018/trump-administration-takes-aim-national-climate-assessment.

Associated Press. (2018, November 24). U.S. should expect worsening weather disasters, new government climate report warns. *Weather Channel*. https://weather.com/news/climate/news/2018-11-23-fourth-national-climate-assessment.

Coppola, N. W. (1997). Rhetorical analysis of stakeholders in environmental communication: A model. *Technical Communication Quarterly, 6*(1), 9–24.

Cox, R. (2015). Scale, complexity, and communicative systems. *Environmental Communication, 9*(3), 370–378.

Cox, R., & Depoe, S. (2015). Emergence and growth of the "field" of environmental communication. In A. Hansen and R. Cox (Eds.), *The Routledge handbook of environment and communication* (pp. 13–25). Routledge.

Crist, E., & Rinker, H. B. (2010). *Gaia in turmoil*. MIT Press.

DiCaglio, J., Barlow, K. M., & Johnson, J. S. (2018). Rhetorical recommendations built on ecological experience: A reassessment of the challenge of environmental communication. *Environmental Communication, 12*(4), 438–450.

Druschke, C. G., & McGreavy, B. (2016). Why rhetoric matters for ecology. *Frontiers in Ecology and the Environment, 14*(1), 46–52.

Environment and Ecology (n.d.). Gaia theory: Model and metaphor for the 21st century. http://www.environment-ecology.com/gaia/711-gaia-theory-model-and-metaphor-for-the-21st-century.html.

Fellows, A. (2019). *Gaia, psyche and deep ecology*. Routledge.

Frohlich, T. C., & Blossom, L. (2019 July 14). These countries produce the most CO_2 emissions. *USA Today*. https://www.usatoday.com/story/money/2019/07/14/china-us-countries-that-produce-the-most-co-2-emissions/39548763.

Grabill, J. T., & Simmons, W. M. (1998). Toward a critical rhetoric of risk communication: Producing citizens and the role of technical communicators. *Technical Communication Quarterly, 7*(4), 415–441.

Hansen, A., & Cox, R. (2015). Introduction: Environment and communication. In A. Hansen & R. Cox (Eds.), *The Routledge handbook of environment and communication* (pp. 1–10). Routledge.

Hanson-Easey, S., Williams, S., Hansen, A., Forgarty, K., & Bi, P. (2015). Speaking of climate change: A discursive analysis of lay understandings. *Science Communication, 37*(2), 217 239.

Irfan, U. (2018, November 24). 3 big takeaways from the major U.S. climate report. *Vox*. https://www.vox.com/2018/11/24/18109883/climate-report-2018-national-assessment.

Levy, B. S., & Patz, J. A. (2015). Climate change, human rights, and social justice. *Annals of Global Health, 81*(3), 310–322.

Lovelock, J. E. (1979). *Gaia: A new look at life on earth*. Oxford University Press.

Moody, D. E. (2012). Seven misconceptions regarding the Gaia hypothesis. *Climate Change, 113*, 277–284.

Moser, S. C. (2016). Reflections on climate change communication research and practice in the second decade of the 21st century: What more is there to say? *WIREs Clim Change 2016, 7*, 345–369. https://doi.org/10.1002/wcc.403.

NASA (2021). NASA resources. Overview: Weather, global warming and climate change. https://climate.nasa.gov/resources/global-warming-vs-climate-change.

Nielsen, E. B. (2017). Climate crisis made manifest: The shift from a topos of *time* to a topos of *place*. In D. G. Ross (Ed.), *Topic-driven environmental rhetoric* (pp. 87–105). Routledge.

NOAA (2018, November 25). New federal climate assessment for U.S. released. *Science Daily*. https://www.sciencedaily.com/releases/2018/11/181125113728.htm.

Public Broadcasting Service (2018, November 28). Climate update: The Fourth National Climate Assessment. *PBS Peril and Promise: The Challenge of Climate Change*. pbs.org/wnet/peril-and-promise/2018/11/climate-update-the-fourth-national-climate-assessment.

Sauer, B. A. (2003). *The rhetoric of risk: Technical documentation in hazardous environments*. Routledge.

Schweizer, S., Davis, S., & Thompson, J. L. (2013). Changing the conversation about climate change: A theoretical framework for place-based climate change engagement. *Environmental Communication, 7*(1), 42–62.

Silverstein, J. (2018, November 24). Mass deaths and mayhem: National Climate Assessment's most shocking warnings. *CBSNews*. https://www.cbsnews.com/news/national-climate-assessments-most-shocking-climate-change-warnings.

Simmons, D. (2020, July 29). What is "climate justice"? *Climate Connections*. https://yaleclimateconnections.org/2020/07/what-is-climate-justice.

Steffen, W., Richardson, K., Rockström, Schellnhuber, H. J., Dube, O. P., Dutreuil, S., Lenton, T. M., & Lubchenco, J. (2020). The emergence and evolution of earth system science. *Nature Reviews Earth and the Environment, 1*, 54–63.

Sterman, J. D. (2011). Communicating climate change risks in a skeptical world. *Climate Change, 108*, 811–826.

Stober, E. (2018, November 23). Here's what you need to know about the devastating U.S. climate change report. *Global News.* https://globalnews.ca/news/4694012/us-climate-change-report.

Union of Concerned Scientists (2018, November 22). Latest National Climate Assessment shows US already suffering damages from climate change: Statement by Brenda Ekwurzel, NCA4 report author, Senior UCS Climate Scientist. Union of Concerned Scientists/About/News. https://www.ucsusa.org/about/news/latest-national-climate-assessment.

U.S. Global Change Research Program (2017). *Climate science special report: Fourth National Climate Assessment, Volume I.* Wuebbles, D. J., D. W. Fahey, K. A. Hibbard, D. J. Dokken, B. C. Stewart, & T. K. Maycock (Eds.), U.S. Global Change Research Program. doi: 10.7930/J0J964J6.

U.S. Global Change Research Program (2018). *Impacts, risks, and adaptation in the United States, Fourth National Climate Assessment, Volume II.* Reidmiller, D. R., C. W. Avery, D. R. Easterling, K. E. Kunkel, K. L. M. Lewis, T. K. Maycock, & B. C. Stewart (Eds.), U.S. Global Change Research Program. doi: 10.7930/NCA4.2018.

U.S. Global Change Research Program (n.d.). GCRA mandate. GlobalChange.gov. https://www.globalchange.gov/content/whats-new-nca4.

Whitmarsh, L. (2015). Analysing public perceptions, understanding and images of environmental change. In A. Hansen & R. Cox (Eds.), *The Routledge handbook of environment and communication* (pp. 339–367). Routledge.

III

Representations of Human Beings and Earth Together

10

RECONCILING GESTURES

Overcoming Obstacles to Transcultural Risk Communication in South African Coal Mines

Beverly A. Sauer

Problem: Video-conference calls, isolation, social media, and television have been the primary means of communicating risk since March 2020, when the United States went into lockdown to control the pandemic. This difficult year also revealed inequities in the distribution of broadband and computer-mediated instruction, the unequal distribution of risk in largely Black and Brown communities, the burden of risk among frontline workers, and the need to remediate systemic economic and social injustice that culminated in the death of George Floyd and the rise of the Black Lives Matter movement.

As we move into a post-pandemic workplace, technical communicators must develop new skills to communicate risk effectively across racial and economic barriers with diverse audiences who may not share the same education, language, culture, or experience. To balance the tension between risk and control in uncertain environments, communicators must persuade audiences that risk is serious enough to warrant concern, but they must not frighten workers who may refuse to work in dangerous or unpredictable situations. They must keep their messages concise so that audiences will not tune out critical safety messages, but they must provide sufficient information so that audiences believe that experts are not concealing or holding back critical information. Most importantly, communicators must ensure that their explanations align with workers' knowledge and actual experience. In short, effective risk-communication messages must build upon and address the knowledge and beliefs of diverse audiences who may not share common languages, education, culture, or experience.

Solution: My research in South Africa provided an opportunity to investigate how analysis of speech and gesture might help risk communicators overcome differences in language, education, culture, and experience. Difficult

DOI: 10.4324/9781003266549-13

cross-cultural contexts as in South Africa provide extreme cases that highlight aspects of risk communication less visible in ordinary American workplace communication. The lingering effects of apartheid on Black education, the deep distrust between races, and the linguistic complexity of South Africa's 11 official languages complicate the most well-intentioned risk messages. Translation remains the best practical solution for creating effective risk messages for audiences who do not share a common language or culture, but local translators have difficulty articulating technical concepts in the nine so-called African languages.

In 1997, 2005, and 2008, I lived and worked in South Africa, where I collected video-taped interviews of South Africans to discover how we determine what individuals know about risk in difficult and complex social and cultural environments where participants do not share a common language, culture, education, or experience. Funded by grants from the National Science Foundation,[1] I interviewed working coal miners, managers, trainers, researchers, apprentices, and ordinary citizens with no experience in mining and observed training onsite in South African mines to locate sites of misunderstanding in risk communication designed to reduce methane explosions.

All interview data was translated onsite at the mines and re-translated by a second interpreter with knowledge of the 11 official South African languages as well as Fanakalo—a pidgin language spoken in the mines meant to overcome problems of interpretation. This process was reflexive (bi-directional) to the extent that it did not simply examine unidirectional presentations from speaker to audience. Instead, this research was designed to create an inclusive risk-communication environment that elicits the understandings and cultural values of all participants.

It is perhaps not surprising that a French-speaking Congolese PhD student in Mining Engineering at Witwatersrand University, an Afrikaans-speaking engineering consultant at the Council and Scientific research, a novice coal miner, and an educated Sowetan with no experience of mining provide very different accounts of risk. As the examples in this chapter reveal, however, highly educated individuals often created highly confident but misleading or confusing explanations for the most basic concepts in mine safety, while miners denied an education in science under apartheid pointed to apparent errors in scientific explanations from master's-level trainers. The analysis of gesture thus enables us to reconcile sites of apparent misunderstanding and error without privileging the credentials of those with advanced degrees and facility in English.

1   National Science Foundation grant #SES-0428345. The Multimodal Dimensions of Risk Communication in Difficult Cross-Cultural Contexts; NSF grant #9812059. Translating Risk in Speech and Gesture.

Why South African Coal Mines?

My research in South Africa was an outgrowth of several projects studying risk communication and embodied knowledge in American and British coal mines. In *The Rhetoric of Risk* (2003), I described how experienced miners relied upon embodied sensory information—information they sensed (heat and vibrations), smelled (smoke and fumes), heard (creaks), and even tasted (the acidity of coal). Miners not only sensed information in their bodies, they also represented that information in their bodies—in speech and gesture.[2] Miners developed complex gestural representations of the risks they perceived and the environments in which they worked. Too often, however, information represented by gestures was not captured in writing or speech alone.

Research in U.S. mines raised questions about how culture might affect communication about risk and safety. Funded by a second grant from the U.S. National Science Foundation, I interviewed British coal miners during a period of extreme economic turmoil in British mines. Following a strike that devastated the economies of British mining communities, British Coal had privatized the formerly government-owned mines and shifted to the more "economically feasible" and less labor-intensive American methods of roof-bolting that failed to give warning prior to a massive mine collapse (see Sauer, 1996, 2003a). Almost immediately, British mines experienced their first fatal mine collapse in many years. Miners blamed management for compromising safety to reduce costs of production.

In a global economy, the newly privatized British mines also faced economic competition from countries such as South Africa, where cheap coal produced by cheap labor undercut the price of coal in Britain and led to cost-cutting measures that would ultimately affect risk and safety. I thus turned my attention to the problem of risk communication in South Africa (Sauer, 2003c, 2004, 2006).

Language Policy and Education in Post-Apartheid South Africa

As South Africa struggles to remediate gaps in the education of Black students more generally, policymakers disagree about the quality and appropriate character of education for victims of apartheid: How and where should training be delivered? Is science education necessary for workers in labor-intensive industries like mining? How does onsite practical skills training limit or foster economic advancement in post-apartheid SA? And most important: What are the components of a socially just education that can apply to the conditions of work in uncertain and hazardous industries? (Ndimande, 2013).

---

2 Women in the mining community also measured risk in ordinary events like the increasing dust on miners' clothes—sign of an imminent coal dust explosion (Sauer, 1993).

Improving cross-cultural risk communication is particularly important in SA where historically segregated institutions seek to advance previously disadvantaged populations into positions of authority where they can affect risk outcomes. While U.S. agencies assume that miners have a right to understand the rationale for workplace safety (Sauer, 2003a), Black Zulu-speaking miners in South Africa have received little or no formal training prior to the 1994 elections and little formal education in science.

As South Africans struggle to create a more democratic distribution of wealth and power, government agencies are struggling to renegotiate socially just notions of expertise and authority without sacrificing safety (Kraak, 2002; SA Language-in-Education Policy, 1997). While the 1994 elections promised democratic reform, Black miners need more than basic skills and competency in English to move into supervisory positions where they can achieve economic parity and develop more effective means of protecting the health and safety of colleagues. South Africa's Language-in-Education Policy (1997) guarantees South Africans the right to be educated in their language of choice where practicable, but the government lacks funding and must depend upon private industry to educate workers (Policy Document on Adult Basic Education and Training [ABET], 2001).

Following the 1994 elections, the new government recognized the need to provide assistance to "victims in respect of higher education and training" under apartheid. The National Qualifications Framework (NQF) Act 67 of 2008 (National Qualifications Act, 2008) provides for a single integrated system to oversee and manage the development of a single integrated national framework for learners to "facilitate access to, and mobility and progression within, education, training and career paths" and "accelerate the redress of past unfair discrimination in education, training and employment opportunities" for all South Africans. The NQF defines 10 levels of qualification, from NQF levels 1 and 2 (=Grades 9 and 10) through NQF levels 9 and 10 (=master's degrees and Doctoral Degrees).

Within the NQF, the Occupational Qualifications Framework (OQF) provides for the certification of skills related to specific occupations, identifies the needs of industries and the labor market in general, clarifies the relationship between workplace education and traditional educational career paths, clarifies issues related to mobility within education and training, and provides consistency in the development of learning standards.

In theory, both the NQF and the OQF Qualifications Framework are grounded in three equal standards: knowledge and theory, practical skills, and work experience. In practice, however, the process of technical training and vocational/occupational certification is largely unidirectional and retains much of the hierarchical economic and social structures that privileged scientific and abstract knowledge under apartheid (Mnguni, 1999).

To implement the NQF, the Chamber of Mine's Department of Education needed to answer questions about the relationship between experience and education as they related to the skill certification and promotion of Black workers: (1) What specific practical skills demonstrate competence in mining at all levels? (2) How can industry overcome gaps in the knowledge and theory component required to demonstrate competence?

Given the presumed historic deficits in Black workers' education and the difficulty of quantifying theoretical and practical knowledge, it is not surprising that the structures of mine safety training organized under the OQF continue to replicate the hierarchical institutional structures that had disenfranchised Black workers under apartheid. Thus, the framework presumes that master's-level trainers (Level 9) would have the correct knowledge to provide the theoretical training required by the OQF. At the same time, Level 1 would presume that workers would lack the knowledge necessary to understand mine practices.

As the following examples demonstrate, the NQF framework counts years of service in a particular job but may not count the embodied experience of miners, trainers, foremen, apprentices, and managers—what we might call actual experience underground. Moreover, assessment methods rely upon oral and written responses—often in English—and thus may fail to take into account the knowledge miners express in gesture or in a language that does not easily translate into English.

The Rhetoric of Risk: Representing Embodied Knowledge in Hazardous Worksites

Although it is easy to dichotomize practical experience and academic knowledge, U.S. agencies value practical knowledge and draw upon miners' experience to develop formal scientific standards so that the information can be used as evidence for future risk management legislation and regulation. Because spatial relationships are critical in defining and locating hazards, the process of managing risk underground also involves what Levelt (1996) calls perspective taking—the process of reconstructing what we observe, experience, or hear—from a position as character or observer in the event (cf. McNeill, 1992).

Risk communication in coal mines illustrates this process. Since the earliest Roman mines, miners have developed complex vocabularies to describe events and conditions underground (Hoover & Hoover, 1950). These vocabularies articulate the spatial and temporal experience of individuals within highly structured institutions that situate individuals differently in relation to risk. Miners, for example, locate themselves in relation to physical objects, events, and conditions. Management views events from a position above and outside the underground world of miners. Miners describe themselves in relation to the "bleeder line" (Sauer, 2003a). Managers call the section "the number one entry of the first left section." The "bleeder" describes a critical component of the mine's ventilation system underground. The "first left section" describes the section in relation to the mine's plan of development, viewed from a systems perspective. The number of the section tells us little about the potential hazard or benefit of a particular location in a crisis. In the 1984 Wilberg Mine fire, the sole surviving miner used the direction of air flow to help him locate an exit (Huntley et al., 1992, cf. Sauer, 2003a).

Unfortunately, embodied experience is sometimes so site-specific that even experienced miners cannot transfer knowledge from one site to another in the same mine (Mark & DeMarco, 1993). After an accident, management and investigators must (1) capture and evaluate the testimony of many individuals who experience the disaster differently because they are located in different sites inside and outside the mines and (2) reconcile these narratives with data from scientific tests and measurement. In a coal mine explosion, for example, the force of the explosion at any point in the mine can be calculated by comparing how each individual felt a "pop" in their ears at different locations in the mine tunnel (Sauer, 2003). (The process is familiar to airplane passengers who sense changes in cabin pressure.)

But how can we translate what "pop" means in transcultural risk-communication messages? As the following discussion demonstrates, the analysis of speech and gesture is fundamental to making visible how individuals represent their experiences of risk.

The Role of Gesture in Speakers' Representations of Risk

Within complex three-dimensional environments such as mines, individuals use gesture to represent dynamic processes and three-dimensional spatial relationships critical to effective risk decision-making. Individuals use gesture to represent objects and actions. Counterintuitively, the miners we interviewed for *The Rhetoric of Risk* (2003a) also represented abstract scientific concepts in gesture—like the dynamics of rock strata (layers) inside the mine roof (ceiling of the mine tunnel). Gesture thus provides an additional channel of information that enables individuals to integrate theory and practice, find patterns in practices and procedures, and move flexibly between theoretical analysis and embodied (local) experience (Sauer, 2003a; cf. McNeill, 1992). The analysis of gesture thus helps make visible both experiential knowledge and the abstract theoretical frameworks critical to understanding risk.

Because speech alone cannot represent the spatial and temporal complexity of hazardous environments, risk-communication messages are necessarily *multi-modal* (McNeill & Duncan, 2000; Sauer, 1999). That is, risk messages combine multiple modalities (forms) of language, including written and oral speech, tone, gesture, movement, and visual information, to create more complex meanings than are possible in speech or its transcription in writing alone.

Some gestures have very specific meaning within a culture—like the peace sign in America. Linguistic anthropologists like Heather Brookes have developed extensive dictionaries of gestural meanings within specific cultures like South African townships. These codified gestures function as signals or directions and are generally pre-planned. In this chapter, however, we focus on naturalistic gestures—unplanned gestures that arise naturally as an accompaniment to speech (McNeill, 1992).

Gestural Viewpoint as Situated Knowledge

To understand events in a disaster, investigators must evaluate many different narratives that reflect each individual's physical and institutional location. Previously, I argued that this notion of risk goes beyond a simple relativism (each person has his or her own viewpoint or interpretation) to argue that each representation is "situated" (located or placed) at specific points of time and space in organizations and environments that influence how individuals observe and experience risk (Sauer, 1998a, 1998b, 1999; cf. Emmory & Reilly, 1995; Liddell & Metzger, 1998; McNeill, 1992). A speaker's viewpoint may literally be situated inside the hazardous environment or above and outside in the mine offices. Speakers can also assume viewpoints above and outside their current location—as when mine foremen describe how an open door in a mine passageway can affect the flow of ventilation throughout the entire system. In this case, speakers construct an internal—imagined—representation of mine space based upon a mine map, theoretical knowledge, or direct experience traveling mine passageways.

In hazardous environments, having access to many different viewpoints may provide risk decision-makers with more information than any single viewpoint alone. As psycholinguistic research in gesture suggests, moreover, speakers are not limited to a single viewpoint when they talk about risk. Instead, they can represent different viewpoints as character or observer in their narrative. Even incorrect representations may help learners process embodied and theoretical knowledge. As Alibali and Goldin-Meadow (1993) demonstrate, students learning a mathematical concept create many different gestural representations of mathematical concepts during a period of transition that predicts their readiness to learn. This research suggests that adults—continually learning about risk through experience—might also create multiple and sometimes conflicting representations as they transition to a more nuanced understanding of risk.

Psycholinguistic and cultural studies of gesture thus support the notion that speech and gesture together provide a more complete representation of what individuals know than either speech or gesture alone (McNeill & Duncan, 2000; Sauer, 1999). In simple terms, gesture provides an additional channel of information that enables individuals to find patterns in practices and procedures and move flexibly between different viewpoints—acting out their own and others' reactions, for example, in an accident (Sauer, 2003a; cf. McNeill, 1992). Unfortunately, these different gestural representations may also create confusion in difficult cross-cultural contexts where speakers do not share a common language or cultural referent (cf. Church & Goldin-Meadow, 1986). In complex transcultural environments, researchers need consistent and culturally sensitive methodologies to capture and interpret the full body of temporal, spatial, and material knowledge that speakers represent in gesture.

While it may seem naive to argue that systemic racism is at work in this difficult process of capturing cultural meanings embodied in gesture (Gumperz, 1992, p. 303; cf. Gumperz & Cook-Gumperz, 1982), the examples in this chapter illustrate how uniformed processes of translation and a history of mistrust create confusion and misunderstanding that exacerbate underlying racial tensions and undermine the structures of knowledge assessment at the heart of SA's attempt to remediate the effects of apartheid on Black workers.

Making Competence Visible

The methods employed in this study were designed to assess practical and theoretical competence across individuals who differed in education, language, culture, and experience. Unlike many studies of Black workers' knowledge and understanding produced under the racist apartheid regime (Pheta, 1982), this study is reflexive to the extent that it looks at both trainer and worker misunderstandings. More important, the analysis looks at the ways that gesture makes visible both competence and misunderstandings for workers at all levels.

While it is easy to blame the lack of education under apartheid for miners' questions regarding subjects such as methane, their confusion did not ultimately reflect a lack of knowledge but rather questions regarding incorrect and ambiguous information presented in the training session. Similarly, we did not expect that mine researchers at the Council of Scientific and Industrial Research (CSIR) and a PhD student in the Department of Mining Engineering at Witwatersrand University would have difficulty answering the most basic questions about why methane explodes.

This chapter uses two data sets. In the following discussion, we summarize the results of Part 1, which laid the foundation for the research we focus on in Part 2.

Part 1: Identifying Sources of Misunderstanding in Coal Mine Explosives Training Programs

In July 1997, we observed and analyzed safety training practices conducted by CSIR trainers at the Kloppersbos training center near Pretoria, SA, to identify sites of misunderstanding. The training included three parallel sessions explaining the science of methane explosions as well as small experiment-level demonstrations and a massive, staged explosion.[3] At the time of this research, the CSIR was developing outcomes-based training schemes to support the NQF. We also interviewed an officer at the SA Chamber of Mines responsible for formulating NQF mine competency standards prior to training.

Because we were making judgments about subjects' competence, accuracy of translation was important. All video- and audio-taped data was translated onsite and reinterpreted (back-translated) by an interpreter with fluency in the nine so-called "African" languages, English, Afrikaans, and Fanakalo—a pidgin language used in mining. The interpreter also created an ongoing protocol in the transcript that indicated problems in interpretation and sites of potential ambiguity or changes in tone that affected the interpretation. This elaborate transcription process allowed us to check the accuracy of the onsite translation and, more important, to identify terms and concepts that might be open to more than one interpretation (Tschabalala, 1997).

Not surprisingly, problems of translation and interpretation created sites of misunderstanding and confusion in training. Because translators lacked equivalent Zulu or Ndebele terms for terms or concepts in English, they invented terms, Africanized English terms, or borrowed terms with similar cultural meanings. Africanized terms generally use the English or Afrikaans term but add a Zulu prefix, as in iMethane or le-Methane or the Africanized Afrikaans term iVentilasi (Ventilation)—in much the same way that French language speakers create le Hot Dog or le Laptop (Kashula & Anthoniessen, 1995).

In the Kloppersbos training session, the translator borrowed the Zulu term *imbawula* to translate the English term methane. In Black townships, *Imbawula* refers to a brazier or small stove. South African miners know that braziers produce a toxic and lethal gas (carbon monoxide) that silently kills inhabitants when they take the braziers inside their houses. In theory, the notion of "danger" embodied in the term *imbawula* enables miners to comprehend the danger inherent in an invisible gas as methane, but the term is confusing because the meanings embodied in the terms *imbawula* and methane are not technically equivalent. Ultimately, the term *imbawula* produces scientifically incorrect risk messages because *imbawula* (a burning coal stove) produces carbon monoxide (CO), not methane (CH_4), though both are dangerous gases. More important, these misunderstandings undercut trust—especially given the antagonism between the trainer and miners prior to the 1994 elections.

---

3   Altogether we collected data from four 30 minute training sessions inside the Kloppersbos training facility; approximately 30 minutes of video-taped interviews with mine management and trainers before and after training sessions and interviews; and an additional three hours of audio-taped interviews with individual miners following the sessions. We also videotaped trainers, safety managers, and researchers in safety at the CSIR in Johannesburg, SA (approximately 16 hours).

Given deficits in miners' education under apartheid, it was not surprising that trainers blamed miners' second-language limitations and gaps in science education when miners seemed to challenge trainers' explanations of how methane explodes. Trainers were partially correct.

Under apartheid, the 1953 Bantu Education Act introduced a system of racially segregated, underfunded mother-tongue language education for students of color and secured a cheap unskilled labor force for the white electorate (Moore, 2017). Following the 1976 student uprisings in Soweto, violence promulgated against Blacks disrupted their education and created a lost generation of Black workers who received little or no science education following the 1976 Soweto uprisings.

At the same time, however, the extreme racial segregation under apartheid produced an unexpected outcome. Under apartheid, Black workers developed highly structured work teams that operated independently within traditional mine management structures. As a result, Black miners developed their own vocabulary for identifying risk. These terms were frequently inaccessible to Afrikaans- and English-speaking management who managed mine operations in offices at a safe distance from the dangers of the underground mine (Moodie & Ndatshe, 1994).

As this analysis suggests, experienced Zulu-speaking miners expressed legitimate concerns about the science presented in training sessions. Thus, miners who understood the dangers of methane (*imbawula*) could not transfer that information to the situation in mines because the situations are not scientifically the same (Tschabalala, 1997).

Part 2: Identifying Meanings in Gesture

Part 1 provided a long list of misunderstandings resulting from what we believed to be problems of translation and confusing gestures, but it was not clear whether these problems reflected cultural differences or particular features of the training session. In 2005, we returned to South Africa to investigate how differences in gesture and semantic meaning might reflect more general patterns of language, education, and/or experience within specific language and cultural groups. We also considered the null hypothesis: that differences in gestural viewpoint are idiosyncratic and reflect individual stylistic differences not accounted for by culture, race, education, or experience.

To account for differences across various languages, all subjects were asked to explain terms in their mother tongue (L1) and other (L2) languages if applicable. Older experienced Black miners with limited formal education did not generally speak English or Afrikaans. Shift Bosses, on the other hand, spoke Afrikaans as their mother tongue (L1) but conducted training sessions in English (their L2 language). These training sessions were then translated into Zulu by a second (Black) trainer onsite at the mines. Younger Black miners—and union representatives in particular—spoke English in school as a second language. Miners and shift bosses also communicated in Fanakalo, a highly simplified pidgin language used as a language of instruction in South African coal mines, although recent NQF training schemes have argued for the elimination of Fanakalo as a language of instruction because it oversimplifies risk decision-making.

As the following examples demonstrate, the NQF standards reveal little about what individuals at different levels actually know when they talk about risks in their environment. If we quantified the level of each individual based upon practical knowledge displayed in their interviews, miners might easily outscore our PhD student despite his degree—especially since he also has no experience in a South African coal mine. Measured consistently across culture and language, subjects' responses did not differ as much as we might imagine as we moved up the scale of education.

What Gesture Reveals

The following examples give a small sample of the entire data collected during this study. However, the finely grained analysis of speech and gesture is important because it points to the problem at the heart of the NQF framework: How do we make visible the embodied and institutional knowledge of individuals? What assumptions govern the link between education and expertise in managing hazardous workplaces? Our answers to those questions are critical to the extent that the levels in the NQF framework implicitly define the economic value of particular types of knowledge in a hierarchy that heavily privileges advanced education.

The following examples are presented in ascending order of mine experience and education. As these examples suggest, language influences the types of information and the viewpoint represented in subjects' responses, especially for Afrikaans-speaking subjects for whom Afrikaans has remained the language of the mine workplace.

When subjects respond in English—frequently, the language of education for the group represented here, their English and Afrikaans responses are not equivalent in the sense of a word-for-word translation. Gestures also differ within subject responses across different languages.

The chart in Figure 10.1 summarizes the results of this analysis across language, education, experience, and culture. As this chart suggests, subjects with no mining experience draw upon the local meaning of *imbawula* in South African townships to answer the question "What is *imbawula*?" Learners in a mine-training program describe the function and manufacture of the *imbawula*/brazier in townships. Experienced miners, shift bosses, and mine safety researchers share much the same understanding of *imbawula* as a gas produced during the production of coal, and express it in much the same terms. The PhD

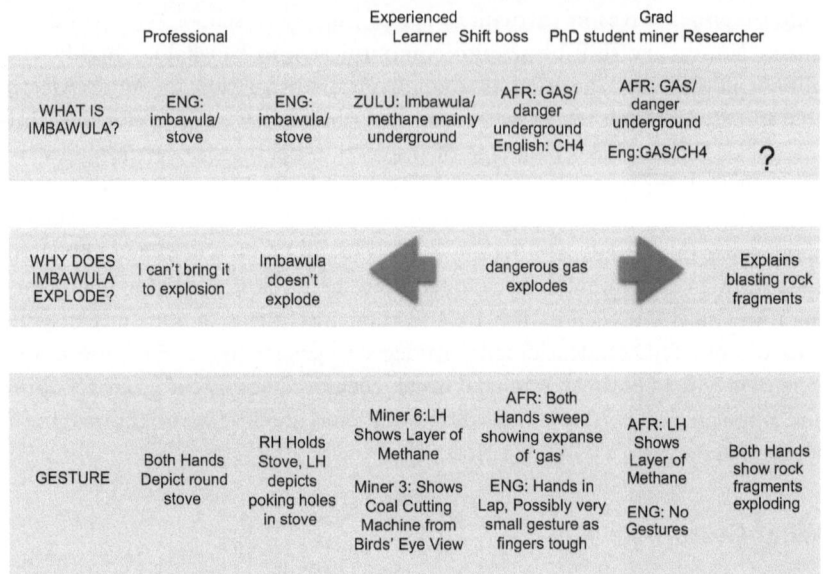

FIGURE 10.1 Comparison of responses across education, experience, and culture.

student is wildly off mark to the extent that he draws from his experience in Congolese diamond mines.

As these examples suggest, moreover, the embodied experience of a common work environment may create a culture of its own that influences how individuals communicate across language and culture.

Working Professionals in Soweto Township
Subjects 1 and 2–Soweto
English–Speaking, Non–Mining, Educated to Fifth Form, Self–Identify as Zulu

Educated English-speaking subjects interviewed in Soweto township with no mining experience represent *imbawula* in gesture as a small round stove (Figure 10.2). It is thus not surprising that they vigorously deny that *imbawula* explodes. Subject 1 Soweto explains carefully that *imbawula* is a stove made from a "a big tin, this size (showing with his hands), about 25 liters." According to Subject 1, the process of burning is slow—not the sudden eruption of an explosion:

> You put wood. First you newspapers, you just fold them, put them and then you put wood and then from there you put coal, amalahle [coal], and then you light with the matches. Then it will burn. But it will take time, about 20 to 30 minutes before getting hot.

Similarly, Subject 2 Soweto knows *imbawula* but cannot explain "explosion *imbawula*": "Well, I don't know how to bring it to explosion."

FIGURE 10.2   Non-mining subject represents Imbawula as a small stove.

Apprentice Miner in Mine-Training Program
Learner 2-Sasol
English-Speaking, Less than 1 month Mine Safety Training, Educated to Fifth Form

English-speaking subjects interviewed in a new miner training program at the Sasol Mine demonstrate how learners draw upon previous cultural understandings as they attempt to explain technical processes. Thus, Learner 2 Sasol explains that *imbawula* "is when they take the drum, and they poke holes in it" to make a stove. He is confused by the term "explosion *imbawula*" (~methane explosion), however, since "methane [e.g. Imbawula] doesn't explosion."[4]

Experienced Miners
Miner 6 and Miner 3-Middleburg
Non-English-Speaking, Self-Identify as Zulu, No Formal Education

Because most of the older experienced miners we interviewed had little knowledge of English, we might have expected them to carry over the cultural meaning of *imbawula* as a brazier. Instead, experienced miners at the Middleburg mine defined *imbawula* as a dangerous gas ("poison") produced in the areas in which they work—without reference to the brazier or stove.

Miner 6 Middleburg depicts methane as a layer of gas at the top ("roof") of the mine space underground (Figure 10.3). His explanation demonstrates his understanding of the imminent danger posed by methane if there is no proper ventilation (umoya) to evacuate the methane from the mine as well as a deeper understanding of the potential ambiguity in the term. His explanation is not entirely correct, however, since methane is not "smoke." More importantly, this example shows why miners confuse the term "layer" (iRoof layer) in the context of methane with the layer of rock (strata) at the roof of the mine.

4   As a thought experiment, try thinking about why methane explodes in a coal mine but methane/gas stoves don't explode.

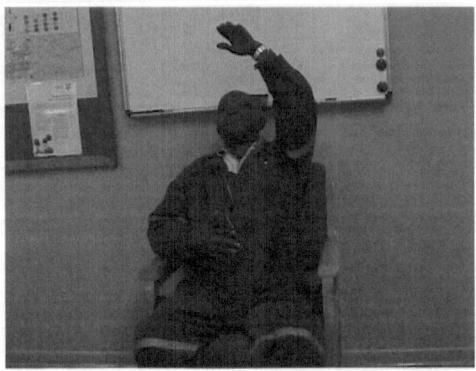

FIGURE 10.3 Experienced miners represent the layer of methane at the roof (top) of the mine workspace.

Often, terms in Zulu or Xhosa may have two distinct meanings in English. The term *Ululasa* in Xhosa, for example, encompasses the spectrum of green and blue in English. To avoid confusion, translators distinguish "ululasa the color of the sky ('blue') vs. ululasa the color of the grass ('green')" (Kashula & Anthoniesson, 1995). Miner 6 Middleburg similarly couples two terms, *imbawula* and le-Methane—the Africanized form of the term "methane." This doubling up of the terms may indicate his attempt to distinguish the meaning *imbawula*/le-Methane as a "dangerous gas" from its local meaning (*imbawula/* stove). Miner 6 also employs this doubling when he describes "*umoya*/wind" (~ventilation) in the passage below:

> MINER 6 Middleburg:
> Oh. Imbawula ngumsi ofumaneka kwiindawo esisebenza kuzo, em-igodini. Ene futhi, le-methane, imbawula, yinto efumaneka phezulu, etafuleni. And then iyiphoyizeni le nto. Ukuba ngaba ithi idubule, it is, ibulala abantu, in the work place. And then ifuna ukuba, ukuyikhusela, kufuneka ukuba kube khona umoya, wind. Ya.
> [Oh. Imbawula is smoke [gas] that is found in the areas in which we work, underground. And also, this methane, imbawula, is something that is found above, in the roof. And then it is a poison, this thing. If it should explode, it is, it kills people. And then it needs, in order to prevent it, there must be ventilation, wind. Yeah.]

Miner 3 Middleburg also describes *imbawula* in terms of its location under-ground, but he assumes a different perspective—as if he is looking down from an imaginary point in space above the working mine. In his gestures, his hands show the cutting process which produces *imbawula*. The following passage reveals his hesitancy and lack of a vocabulary to explain what methane "is" though he knows we find it in a mine when we work with coal. In this passage, he uses both hands to show the coal-cutting process (Figure 10.4), although he does not describe the process in speech:

MINER 3 Middleburg:
Imbawula kush'ukutsi yi … Imbawula siyithola la, ikakhulukazi la emgodini, la sisebenza khona emalahleni.

[Imbawula, I can say, is … We find imbawula right here, mainly here underground, where we work with coal.]

Trainer 2–Middleburg
Afrikaans-Speaking/ L2 English, Educated Shift Boss with Experience Underground, Self-Identifies as Afrikaans, Completed Secondary Education

Miners and their Afrikaans-speaking shift bosses share a common workspace and common experience underground despite the separation implied by apartheid regulation. But shift bosses are also more educated in terms of mine theory and gas mechanics.

In Afrikaans (his mother tongue), he explains that *imbawula* means "gas." But he also adds "It means danger." Like the experienced miners above, he describes *imbawula* from a practical viewpoint as a product of the workspace of the mine:

SHIFT BOSS [AFRIKAANS]: *Imbawula? Imbawula, eh, dit beteken vir my gas. Dit beteken gevaar. Dis gas wat ons ondergrond kry in die steenkool myne. Dit kan ontploffings oorsaak. Dit word gemeet met spesiale instrumente. Die word opgelei. Ja.*

[Imbawula, eh, to me it means gas. It means danger. It is gas that we find underground in coal mines. It can cause explosions. It is measured using special instruments. They are trained. Yeah]

FIGURE 10.4 Experienced miner represents a position outside of and above the mine space as he depicts the working machinery that cuts the rock and releases methane.

In English, the shift boss translates *imbawula* by its chemical formula CH4. While the shift boss's explanations may thus seem more scientific, his explanations in English and Afrikaans are nearly identical to experienced miners' explanation that *imbawula*/methane is a dangerous gas. His response demonstrates how merely using scientific terms like "CH4" might count as greater theoretical competence in an NQF ranking:

> SHIFT BOSS: In English? Ah well, imbawula, that's methane, CH4. That's a gas we get underground. People get trained to search for that, to measure it, to detect it and all the seriousness of methane down underground, what it can do. It can cause explosions. It can cause coal dust explosions that come from methane explosion. The people get trained to detect it with special instruments. Yeah.

Subjects were also asked to represent concepts in gesture without speaking. Shift Boss explains "that would be very difficult to do that, very difficult. If you haven't got a clue what the guy's talking about … it's very difficult to explain by hand what a *imbawula* is." Yet freed from the need to explain in speech, both hands sweep outward, he depicts the expansion of the gas in the mine. This large gesture contrasts the absence of gesture accompanying speech in his two responses above and suggests that gestures—though perhaps difficult to imagine—provide a more accurate representation of his embodied understanding (Figure 10.5).

Post Graduate Researcher
CSIR Researcher 2
Afrikaans-speaking Graduate-Level Mine Safety Researcher, No Underground Mining Experience, Self-Identifies as Afrikaans, L2 English

FIGURE 10.5 Without speech, Trainer 2 uses both hands to represent the expansion of methane within the mine. Both hands sweep in a large circle as if filling the entire mine space.

Levelt (1998) and Levinson (1996), among others, argue that speech and gesture express different aspects of spatial and embodied experience that differ across different languages. These differences may explain in part the different gestures that subjects employ as they themselves use different languages—in this case Afrikaans and English.

CSIR Researcher 2, a graduate-level researcher at the CSIR, demonstrates how language might affect what we perceive as competency in an individual respondent. When CSIR Researcher CSIR answers in his mother tongue Afrikaans, he assumes the same viewpoint as Miner 6 above—holding his hand above his head as he looks up toward the invisible "layer" of methane at the top (roof) of the mine tunnel. In this gesture, CSIR Researcher also flattens his hand to represent the "layer" itself. Like Trainer Middleburg and Miner 6-Middleburg, his gesture situates *imbawula* inside the mine workspace (Figure 10.6). Given his lack of experience underground, it is difficult to assess whether his gesture reflects actual embodied experience or what Church and Goldin-Meadow (1986) call gestural uptake—when speakers imitate the gestures of others in the same way that one might repeat a statement to confirm or acknowledge the statement.

When the CSIR Researcher responds in English, by contrast, he employs no gestures, and his hesitant response might classify his theoretical competency at the level of lower-level working miners.

> CSIR-Researcher: Methane is a gas that can ... that can (um) ... (pause.) A consequence of methane is a big explosion in coal mines.

CSIR Researcher's lack of gestures in English might also reflect his lack of embodied experience with the subject he is describing. Though individuals differ in the degree to which they employ gestures idiosyncratically, new learners in a

FIGURE 10.6   Afrikaans-speaking researcher also locates Methane as a layer at the roof (top) of the mine.

British mine employed elaborate gestures to describe objects they had recently worked with. However, these same learners employed no gestures when they were relying on book-learning and lacked embodied experience underground (Sauer, 2003a).

PhD student, Witwatersrand University
Wits Grad 3
French-speaking Congolese Diamond Mine Experience

Wits Grad 3, a PhD student in Mining Engineering at Witwatersrand University, confesses: "*Imbawula* I don't know about" and then, with a very high degree of confidence, proceeds to depict an explosion in speech and gesture as he incorrectly applies his institutional experience in Congolese diamond mines to explain the meaning of explosion *imbawula*. In his gestures, he flails both arms wildly to represent the blasting fragments of rock.

What We Learn from the Study of Gesture

Too often, knowledge displayed in gestures is ignored or cut out of the frame when talking heads gather to solve critical problems in management and science or assess worker competencies. More insidiously, workers scanning email during video conference calls do not attend to gestural clues in online communication. As the technology for capturing and analyzing gesture improves, we look forward to technical communicators who can communicate, interpret, and assess knowledge represented in gesture.

For practitioners of technical communication, this project demonstrates how the analysis of speech and gesture can help individuals and institutions reconcile apparent differences in meaning that derail the best-intentioned risk communication (Sauer, 2013). This project is not solely a critique of elite discourses, however. As we argue in this chapter, the process of reconciliation requires that both sides participate in and understand how diverse audiences interpret and act upon information embodied in speech and gesture.

This study also has important implications for risk decision science across many different industries and disciplines. In the years since the publication of *The Rhetoric of Risk*, social and behavioral science researchers and practitioners have recognized that risk messages must take into account the lived experience and cultural contexts that affect risk outcomes (Brest & Krieger, 2010; Kahnemann & Tversky, 1990; Morgan et al., 2002; Sauer, 2004). Influenced by the death of George Floyd and the Black Lives Matter movement, technical communication has also taken a more inclusive social justice turn.

For South Africans specifically, this analysis reveals the degree to which the NQF's competency standards rely upon and continue to reinforce inequities of education, language, and culture. While embodied knowledge is difficult and expensive to capture and assess, this analysis reveals the need to interrogate the accuracy of educated subjects' claims in speech and gesture in any assessment of subjects' theoretical and practical competence. It is thus not surprising that educational degrees and years of work experience provide more easily measured competencies than difficult to unpack translations across the 11 official South African languages.

On a practical level, the methodology employed in this chapter can be applied to improve institutional training and certification, hiring, and problem resolution in human resource departments in an increasingly multicultural and transnational workforce (Sauer, 2003b, 2003c). Changes in the work environment resulting from shifting demographics and generational change also argue for changes in current training practices to re-skill existing workforces in increasingly automated environments.

Globally, job losses and changes in the work environment resulting from the recent COVID pandemic have created a more urgent need to re-skill workers. In South Africa, the government has charged the OQF to provide a greater emphasis on advanced cognitive skills—including communication and teamwork— for workers and management. Employers need more broadly-defined skill sets to assess what workers know and to allow previously-disenfranchised workers to transport their skills and experience to new industries (Beneke, 2020). This process is particularly important in coal-dependent economies as climate change and automation reduce the number of well-paying coal mine jobs.

As *The Rhetoric of Risk* suggests, expert systems are constructed on a foundation of frequently unarticulated tacit knowledge (Sauer, 2003a). Although the move offshore has relocated many of the problems of industrial labor, as the COVID pandemic demonstrated, eCommerce is still highly dependent on the labor of frontline workers who sort, dig, carry, manufacture, organize, and deliver products within a digital economy. High-level engineers and management will also be working differently in video conferences and online platforms where talking hands might ultimately make a critical difference in how we manage risk.

Discussion Questions

1 This chapter suggests that embodied knowledge provides warnings of imminent or impending disasters. Based upon press reports or online discussions of previous disasters, what embodied signals did individuals miss before the disaster? Why did they dismiss embodied warnings? How did subjects express their embodied concerns? Why can't we evacuate a building every time a resident hears creaking noises in the night? How can we maintain healthy skepticism without dismissing important signals prior to a disaster?

2 This chapter is partly intended to make you aware of the role of gesture in technical communication across different cultural and social contexts. To what extent have you begun to notice gesture in online, classroom, an in-person conversations? How do speakers' gestures (or the absence of gesture) affect how you evaluate communication messages? To what extent do you notice your own gestures and the gestures of others in ordinary conversation? How does this experience enable you to attend more carefully to the messages communicated in gesture?

Assignments

1   Video record several friends or colleagues describing the same task or instructions. If possible, ask individuals who speak more than one language to explain the task in their primary (L1) and secondary (L2) languages. As you watch the videos, describe the shape and movement of gestures that speakers employ as they talk. Are there similarities or differences across language or culture? Does education and experience make a difference? Do speakers who share the same language use the same gestures to describe the common task? Use the examples in this chapter to help you explain how you might reconcile differences.

2   The COVID pandemic has provided the opportunity to observe many individuals as they present risk messages to the public. Review at least two different risk-communication videos online. Identify issues or terms that might confuse or upset audiences who do not share the same education, experience, language, or social background. Describe the gestures (or absence of gestures) that individuals use in their presentation. What gestures are more effective in conveying the risk message? What gestures detract from the risk message? If speakers do not gesture or if gestures distract from the risk message, explain how you might advise the speaker to revise or improve the risk message.

3   Review video of individuals presenting a complex technical concept or risk messages in a TED talk or other online forum. Be sure that speakers' hands are visible.

   a   Transcribe a portion of the presentation to identify possible sources of misunderstanding or miscommunication in speech alone.

   b   Re-watch the same portion of the presentation twice: first without sound and then without gestures.

Describe how speech, tone, and gesture work together to convey a complex technical message—or create confusion depending on the message and context.

References

Agricola, G. (1950). *De Re Metallica* (H. Hoover, & L. H. Hoover, Trans.). Dover. (Original work published in 1556).

Alibali, M., & Goldin-Meadow, S. (1993). Gesture-speech mismatch and mechanisms of learning: What the hands reveal about a child's state of mind. *Cognitive Psychology, 25*, 468–523.

Beneke, S. (2020, July 29). Re-skilling of mine workers to be considered. Skills portal. https://www.skillsportal.co.za/content/re-skilling-mine-workers-be-considered.

Brest, P., & Krieger, L. H. (2010). *Problem solving, decision making, and professional judgment: A guide for lawyers and policy makers*. Oxford University Press.

Brookes, H. (2004). A repertoire of South African quotable gestures. *Journal of Linguistic Anthropology, 14*(2), 186–224. https://www.jstor.org/stable/43102646.

Church, R. B., & Goldin-Meadow, S. (1986). The mismatch between gesture and speech as an index of transitional knowledge. *Cognition, 23*, 43–71.

Emmory, K., & Reilly, J. S. (Eds.) (1995). *Language, gesture, and space*. Erlbaum.

Gumperz, J. J. (1992). Interviewing in intercultural situations. In P. Drew & J. Heritage (Eds.), *Talk at work: Interaction in institutional settings* (pp. 302–327). Cambridge University Press.

Gumperz, J. J., & Cook-Gumperz, J. (1982). Language and the communication of social identity. In J. J. Gumperz (Ed.), *Language and Social Identity* (pp. 1–21). Cambridge University Press.

Huntley, D. W., Painter, R. J., Oakes, J. K., Cavanaugh, D. R., & Denning, W. G. (1992). *Report of investigation: Underground coal mine fire. Wilberg Mine. I.D. No. 42–00080. Emery Mining Corporation, Orangeville, Emery County, Utah. December 19, 1984.* MSHA.

Justice and Constitutional Development, Department of Government Notices. R. 852. Promotion of National Unity and Reconciliation Act (34/1995): Regulations relating to assistance to victims in respect of Higher Education and Training. Page 3, gazette number 38157.

Kahneman, D., & Tversky, P. (1982). Subjective probability. In D. D. Kahneman, A. Tversky, & P. Slovic (Eds.), *Judgment under uncertainty: Heuristics and biases* (p. 46). American Association for the Advancement of Science.

Kaschula, R., & Anthonissen, C. (1995). *Communicating across cultures in South Africa: Toward a critical language awareness.* Witwatersrand University Press.

Kraak, A. (2002). *Competing education and training policy discourses: A 'systematic' versus 'unit standards' approach.* IISRC.

Policy Document on Adult Basic Education and Training (ABET) (2001). South Africa: Department of Education. http://education.pwv.gov.za/DoE Sites/ABET/ABET Policy.htm.

Levelt, W. J. M. (1996). Perspective taking and ellipsis in spatial descriptions. In P. Bloom, M. A. Peterson, & M. F. Garrett (Eds.), *Language and space* (pp. 76–107). MIT Press.

Levinson, S. C. (1996). Language and space. *Annual Review of Anthropology, 25,* 353–382.

Liddell, S. K., & Metzger, M. (1998). Gesture in sign language discourse. *Journal of Pragmatics, 30*(6), 657–698.

Mark, C., & DeMarco, M. J. (1993). Longwalling under difficult conditions in U.S. coal mines. *CIM Bulletin,* April, 31–38.

McNeill, D. (1992). *Hand and mind: What gestures reveal about thought.* Chicago University Press.

McNeill, D., & Duncan, S. D. (2000). Growth points in thinking-for-speaking. In D. McNeill (Ed.), *Language and gesture* (pp. 141–161). Cambridge University Press.

Mnguni, M. H. (1999). *Education as a social institution and ideological process: From the Negritude Education in Senegal to Bantu Education in South Africa (Vol. 9).* Waxmann Munster.

Moodie, T. D., & Ndatshe, V. (1994). *Men, mines, and migration.* University of California Press.

Moore, N. (2017). Segregated schools of thought: The Bantu Education Act (1953) revisited.

Morgan, M. G., Fischhoff, B., Bostrom, A., & Atman, C. (2002). *Risk communication: A mental models approach.* Cambridge University Press.

National Qualifications Framework (2008). South African Government. https://www.gov.za/documents/language-education-policy-0.

Ndimande, B. (2013). From Bantu education to the fight for socially just education. *Equity & Excellence in Education, 46*(1), 20–35.

Pheta, R. T. (1982). Claimed proficiency of black mine employees in various languages. (Research Report #5/83. Project #GHQ01, Nov. 1982). Johannesburg, South Africa, Chamber of Mines.

Policy Document on Adult Basic Education and Training (ABET) (2001). South Africa Department of Education.

Promotion of National Unity and Reconciliation Act (34/1995): Regulations relating to assistance to victims in respect of Higher Education and Training. (Justice and Constitutional Development, Department of Government Notices, R. 8952). Regulation R 852, 3 November 2014).

Sauer, B. A. (1996). Communicating risk in a cross-cultural environment: A cross-cultural comparison of rhetorical and social understandings in US and British mine safety training programs. *Journal of Business and Technical Communication, 10*(3), 306–329.

Sauer, B. A. (1998). Embodied knowledge: The textual representation of embodied sensory information in a dynamic and uncertain material environment. *Written Communication, 15*(2), 131–169.

Sauer, B. A. (1999). Embodied experience: Representing risk in speech and gesture. *Discourse studies, 1*(3), 321–354.

Sauer, B. A. (2003a). *The rhetoric of risk: Technical documentation in hazardous environments.* Routledge.

Sauer, B. A. (2003b). Language acquisition and the discourses of risk. Chicago Linguistics Society Proceedings. Proceedings of the Annual Meeting of the Chicago Linguistic Society, 39/2. http://chicagolinguisticsociety.org/public/cls39-2-toc.pdf.

Sauer, B. A. (2003c). Gesture and the (workplace) imagination: What gesture reveals about management's attitudes in post-apartheid South Africa. In A. Lorenzo, F. Ramallo, & X. Rodriguez-Yáñez (Eds.), *Discourse, communication and enterprise* (pp. 272–291). Palgrave-Macmillan.

Sauer, B. A. (2004). What gestural space reveals about trainers' assumptions in South African coal mine safety training. SAM Proceedings (Society for the Advancement of Management).

Sauer, B. A. (2006). Linguistic scholarship, management education, and science literacy in post-Apartheid South Africa: What analysis of gesture can contribute to the processes of democratic transformation. Eastern Academy of Management Proceedings. May, 2006.

Shrader-Frechette, K. S. (1990). Scientific method, anti-foundationalism and public decisionmaking. *Risk: Issues in Health and Safety, 1*, 23–41. http://www.fplc.edu/risk/vol1/winter/Shrader.htm.

South African Department of Education (1997, January). *General policy for technikon instructional programmes* (REPORT 150). Department of Education.

Tshabalala, H. (1997–2008). Translation of CSIR training documents. Cape Town, South Africa.

11

REANIMATING RISKS

Forest Giants and their Role in Technical Communication

Cooper Day and Christopher Scheidler

In this chapter, we suggest an approach to climate change risk communication through a vital materialist reanimation, an extension of calls for environmental justice in technical communication (Sackey, 2018). Our idea of reanimation draws from Jane Bennett's (2010) vital materialism and Donna Haraway's (1985) rendering of the interconnected nonhuman elements of the world to recognize risk that is oftentimes latent and emergent. In our rendering, *reanimation* is an impending state that is reached through continued interactions and *sedimentations* (into layers); this is to say, reanimation is a risk, like the centuries of burning fossil fuel and accumulated refuse that lies in wait. In other words, in each object is a potential new form of the object that threatens some change to the environment. Like the wooden pallet that has an abundance of potentiality, our everyday actions similarly work to reanimate new risks of climate change. By attending to the reanimation potentials of matter rather than framing environmental risk primarily as risk that is forthcoming based on future actions, we argue against a future of risk communication that gestures toward technological one-shot "fixes," and instead argue for risk communication that works toward systemic change (Thackara, 2017).

Problem: The scale and scope of environmental risks is difficult to communicate because they are immersive and *latent* complex risks that have not yet been revealed. In other words, because we are part of the environment and because we do not immediately feel the changes of the environment, it is easy to miss the forest for the trees. A new landfill does not immediately poison the water supply. A new power plant does not immediately raise the temperature. And tossing your old computer in the recycling box puts it out of sight and out of mind. Such complexity obscures the long life of materials and objects and, ultimately, gets in the way of more sustainable relationships with the environment.

DOI: 10.4324/9781003266549-14

Solution: Thinking about complex problems by imagining them as an emerging *animus*, a thing that is coming to life, helps us track the disparate ecologies and systems from which the risk emerges. Understanding that no objects or materials are inherently static allows us to trace their histories and force us to be accountable for their futures. And considering how our interactions with materials and objects leave them, and the environment, changed, points to how we reanimate materials like petroleum and lithium into new arrangements that leave a lasting impact on the planet.

This chapter is an attempt to pave a way for reconfiguring how we think about systemic change. In the following pages, we examine a global art installation of sculpted forest giants. These giants embody the very exigence of the risk stemming from climate change. Made of recycled materials, forest deadfall, and visitor contributions, these giants extend the notion of risk communication through their reanimated nonhuman, cyborg-netic lives (van Dooren et al., 2016). By treating the giants as reanimated hybrids of human, nature, and industrial materials, we complicate the discursive subject/object of technical communication. Such an approach is necessary for conceptualizing the distributed *emergence* (coming into being) that is especially inevitable with contemporary technical communication.

The Difficulty of Communicating Climate Change

The giants provide ample proving ground for testing the limits of technical communication—especially as it relates to non-discursive elements. There is no lack of communication between art installation and art appreciator. Given Dambo's activist persuasion—that is, creating purposeful communication to diverse audiences—we frame this work as risk communication (Bivens et al., 2020). Technical communication researchers have often examined risk communication at both (1) the textual level (i.e., the discursive practices that assist technical writers in the invention process) and (2) the sociocultural level (i.e., the discursive practices across workgroups) (Lancaster, 2018). The first approach, however, runs the risk of oversimplifying risk communication. Thus, Mannings's "persuasion matrix," which includes "destination, source, message channel, and receiver," was intended to help tech writers during the invention process in the early drafting stages (1982). Manning constructed the ACTIR (Action, Credibility, Target, Identification, and Readability) formula to help technical writers assess their risk communication. Research on the latter, sociocultural level, has found that much failed communication can be linked to social differences between groups (Coogan, 2002). Such failures lead technical communication scholars to question understandings of the complexities behind language at the textual level and the sociocultural (Andersson et al., 2020; Lancaster, 2018; Sauer, 2003).

We do not aim to complicate the definition of technical communication. We find that technical communication scholars, however, have a unique perspective to understand risk communication, which we argue can be extended to more artistic forms. We must begin merging our approaches to account for the more complicated ecological system(s) from which communication emerges (Lancaster, 2018; Spinuzzi, 2003)—especially considering that mass communication emerges from multiple and distributed sources (Edbauer-Rice, 2005). In a time of anti-intellectualism and science denial, environmental communication scholars must further explore what modes of communication might be more effective.

Considering the giants, we are specifically concerned with new definitions of risk—or some version of risk that attends toward the inchoate, indeterminate, and newly latent risk of changed climates and emerging crises. Given the inadequacies of the *deficit model* of communicating science, where the public is seen as intellectually unable to understand complex scientific concepts, we are especially interested in the affectual capacities of art and non-discursive communication (Bucchi, 2008). Rather than framing risk as "objective" risk or not (AdomBent & Godemann, 2011), we argue that the environmental crisis poses a world-altering threat on the planet that exceeds our capacities for measure. In this way, we avoid framing risk as a perception of experts or publics and then framing risk communication as the two-way channel between each. Instead, we suggest an approach of risk communication that begins with the indeterminate space of the nonhuman to transcend issues of self-interest when dealing with a planetary crisis. In other words, paying attention to the nonhuman elements provides us with new understandings of risk and new understandings of technical communication. For instance, how might a petition to end mountaintop removal mining be changed if we started from the inherent right of the mountain to exist relatively unbothered, rather than starting at the cruel economies of resource extraction?

Nonhuman and Communication

We approach this project in the interest of making a technical and environmental rhetoric that holds up for the 21st century, especially as technical communicators take a more visible and intentional role in environmental justice. Killingsworth and Palmer acknowledge that it would be "acting irresponsibly to prescribe definite rhetorical solutions to current discourse problems" (2012, p. 271); therefore, there is much room for communications scholars in the realm of environmental concerns. By recognizing the potential failure of rhetoric, as these authors do, we must question our very understanding of "rhetorician" or technical communicator in these environmental efforts (Cryer, 2018; Rickert, 2013; Rivers, 2017). Nathanial Rivers argues, "environmentalism specifically needs a more intense rhetoric—one engaged not simply in human discourse, but in the nonhuman, in the object" (2015, p. 422). Killingsworth and Palmer note an issue of "fading interest" in environmental concerns due to "the gap between science and general human experience" (2012, p. 270).

Some scholars have attempted to theorize a way forward in thinking about the relationship between humans and nonhumans, or as Donna Haraway (2016) likes to situate it: nonhuman relationships. Haraway describes what she calls the *chthulucene*, as opposed to the *anthropocene* or human era, as a new, and necessary, name for the complex and ongoing earthly forces of which humans are a part. She argues that the most important part of living in the chthulucene is making kin, which, according to her, must extend beyond the human because it is a necessary effort for all survival on earth. The *chthulucene*, as Haraway defines it, accounts for the complicated nature of existence that includes nonhuman actants in timespaces that are of past times, the present, and even times to come. Haraway's chthulucene is intended to be open to any and all knowledge-making, new and old, human and nonhuman.

Looking beyond human discourse remains a difficult task. We often must make choices on which "voices" (despite the human framing of "voices") to privilege in this conversation and how to make these contributions count. Expanding the rhetorical situation to nonhumans can help reveal the complexity of our knowledge-making practices but also points to existing human voices that remain ignored or silenced. Yet, "understanding" the rhetorics of nonhumans remains a challenging, if not contradictory, endeavor. One way that we have found helpful for thinking about this is to recognize that *ontology*, or the way things exist, is dynamic rather than static. When the ice in your iced coffee melts, is it no longer iced coffee? The wooden pallet breaks down or is turned into other things. Less obvious, however, is that the relationships things have with each other change even as they seem to remain the same. Scholars have often thought through this relationship. Nathaniel Rivers (2015), for example, explores how object-oriented ontology (OOO) complicates human/nonhuman division. He aligns with Scot Barnett's definition of OOO:

> The project of the object-oriented philosophy involves two key moves: first, the recognition of the ontology of individual objects or tool-beings and their perpetual withdrawal from other objects in the world; and second the attunement to the reality and implications of these objects coming into relation with one another and how those relations in turn produce new objects whose depths, like any other object, can never be fully known or expressed in language.
>
> *(Barnett, 2011)*

The move toward OOO or new materialism is common within current conversations about the environment. However, difficulties arise as "rhetorical studies is expanding its horizons toward non-discursive, material, affective, ecological, and energetic aspects, as it simultaneously considers what the discipline might offer to questions with worldly environmental stakes" (Druschke, 2019). Such expansions are worth exploring. Despite scholars such as Graham

Harman (2005) arguing that objects have complete autonomy, the implications of such a premise have yet to be fully realized. These theoretical orientations are useful in that they attempt to flatten any sort of hierarchy or taxonomy, but they do not prescribe a singular way of interpreting or extending meaning-making to nonhumans. As technical communicators, part of our skillset is the ability to legibly make meaning from our careful observations.

Thinking on nonhumans has been taken up in multiple fields and, if nothing else, "emphasizes that we humans are not in this alone and, machinations of modernity and man to the contrary, never have been" (Jones B., 2019, p. 5). Yet, the difficult question remains: How do we read, interpret, or challenge these human/nonhuman relationships to better understand and foster more ethical relationships with nonhuman kin? Technical communication embracing "wicked problems" gives us hope that our field can begin to address these issues (Buchanan, 1992; Wickman, 2014). For us, we start by treating things in the environment with the same care and respect that we might give if we were studying people in public. The work that follows is a summary and analysis of our ethnographic field report based on the forest giants at the Bernheim Arboretum. An *ethnographic field report* generally describes the actions, practices, and culture of people (actors) in a particular environment. The purpose of an ethnographic field report is to better understand the complex interactions of lived and everyday experiences of the people we study to arrive at the answer to a deeper understanding of the cultural norms and practices. Our report is a bit different in that it elevates art sculptures to a human-adjacent status. Conceptualizing these sculptures as actors allows us to probe their origins for deeper understanding. In showing this work, we hope to inspire a template for future scholars to track their own environmental hybrids of industry, logistics, nature, and even art. Afterward, we explore how this interaction of arboretum and art reanimates the boundaries between nature, industrial and consumer processes, and humans, and allows for a reconfiguring of relationships between each and might allow for new solutions to old problems.

Bernheim Forest

Located 20 miles outside Louisville, Kentucky, Bernheim Arboretum and Research Forest is one of the largest research forests in the United States. Its significance within the community is much more than just a place to spend time in nature. The forest serves as a place for researchers to further their knowledge of the natural world. For instance, the forest houses an education center that teaches visitors about the natural environment. As a 600-acre arboretum, Bernheim is home to more than 8,000 varieties of trees and plants. The many trails and paths introduce Bernheim's 500,000+ annual visitors to plant and tree varieties by way of plaques that provide the name, Latin genus, and common location. Throughout the year, Bernheim hosts a variety of events ranging from tree planting, to bird watching, to "Bernheim at night" hiking. Some of these events focus on education whereas others provide people volunteer opportunities to help Bernheim with conservation efforts. Bernheim has a large following in the surrounding communities and thus shapes perceptions of the environment through its many forms of outreach.

Bernheim also has several art programs/initiatives in an attempt to promote multidisciplinary relationships with the natural environment. Art is one of the more popular forms of outreach stemming from Bernheim. Every year, Bernheim hires up to four artists-in-residence who are given a stipend, as well as room and board, while they work on art installations "for the purpose of encouraging visitors' deep connections to nature" (Bernheim, 2021). Art pieces ranging from sculptures to prints can be seen throughout the forest and ultimately display a form of communication that fulfills Bernheim's goal, encouraging a connection with nature. Likely the most popular work is the Forest Giants.

Becoming Alive

Three forest giants—Mama Loumari, Little Nis, and Little Elina—made of local, "natural" materials were created by Danish artist Thomas Dambo. These giants are the remembered story and folklore of Dambo's childhood—each with their own relationship and part in the larger story of the global installation. Across the globe, he has installed similar giants and Trolls and he maintains a website where interested visitors can learn more (Dambo, 2021). Though the giants "arrived," or were completed, in Kentucky on March 15, 2019, Dambo supports the idea of visitors contributing by adding to the giants using upcycled materials (items that are brought back to use from the trash). This public contribution was made even more relevant in the summer of 2019 when visitors stole jewelry from Mama Loumari and the public responded by bringing her new upcycled jewelry. Over their inaugural spring and summer, the giants were a popular addition to the forest. Visitors from surrounding areas looking to get outside, are welcomed by the giants into the natural space of the Bernheim forest.

Little Nis, one of the forest giants, sits on the bank of a pond and stares down. With both hands flat on the ground, Nis leans over the pool looking down at his reflection. Little Nis's recognition of himself tells us that we are dealing with a seemingly sentient form of art. From a distance, one of the first things visitors notice is the top of Nis's smooth bald head. This feature is quite striking when remembering he is made of recycled wood. The curves and shapes of the giants are impressively life-like. Upon closer inspection, we see each piece of wood is differentiated. Despite the smooth appearance from a distance, the matter that makes up Nis is sourced from different materials. On one part of his body, an ink-stained stamp, indecipherable now, is there hinting at the former life of the materials. Approaching Nis and examining his body leaves us wondering—where did these pieces come from? Perhaps the planks of his arms are made from an abandoned bookshelf or upcycled pallet on its last leg.

Further along the trail, comes Mama Loumari. Her area is larger and includes more artifacts than the other giants. While the other two giants are in more open areas, Mama Loumari is found sitting in the shade with her back against a tree. Behind her, in the middle of a ring of small trees, sits a wide range of artifacts and goods that "belong" to her. Some of these appear to be recycled materials used as tools and others are made into things like a crib for her unborn child. There is even a dragon skull that, according to the lore posted on a nearby placard, was found in a mystical forest 200,000 years ago by Mama Loumari's partner, Isak Heartstone.

All of these items, as well as Mama Loumari herself, are made of recycled wood materials. Mama Loumari's skin is made of deliberately cut pieces of wood to get the shape of her body. For example, longer, smooth cuts of wood are used for her fingers; rounded pieces are used to make the shape of her ears. Her clothes and hair are made of more rough, "natural" looking cuts. Her clothes are made of what also appear to be cuts of plywood or similar industrial-grade materials, only these cuts are darker and are layered in such a way that looks like clothing. Her hair is a collection of smaller sticks and twigs, giving her long pulled back hair. These twigs and branches were not taken from living trees but were salvaged from the deadfall of the forest. As she lives in the forest, exposed to the elements, the branches that make her hair will, in time, decay and fall out. Like many of us, Mama Loumari will age and her appearance will change. Some change, however, may be more imminent. For instance, right now she is lounging with her hand resting on her large belly, portraying to viewers that she is pregnant—a fact confirmed by a sign that reads: "Please do not climb on Mama Loumari: We don't want to harm the baby giant." What, we wonder, is the gestation period for a forest giant?

Questions about the living needs of the giants are common among visitors. We are invited, as visitors to the forest, to respect the physical needs of Mama Loumari and her unborn child. Just as we are invited to respect their emotional needs—which was, of course, an issue when visitors stole jewelry from the giants. In this way, the forest positions the giants as adjacent to the human tourists that pay them a visit.

The final giant, Little Elina, is located in a more open space near the man-made Lake Nevin. Elina is positioned in a relaxed pose laying on her side with her legs crossed and her right arm crossing her body to pick up a large rock—a "rock" that humans would refer to as a boulder given its large mass. This boulder gives visitors a sense of scale—if this Giant child can play with a boulder with such ease, what will she grow into? Like her mother, Loumari, Elina's hair is made of long sticks and twigs, but she has much longer hair that is not kept as neatly as that of her mother (Figure 11.1).

FIGURE 11.1 Little Elina laying on her side with her legs crossed and her right arm crossing her body to pick up a large rock.

All of the giants appear to be relaxed in their Bernheim home. Despite being giants, they are welcoming to the many guests of Bernheim. Through recycled materials, Dambo has brought to life items that might otherwise be seen as waste; he has blurred the lines of what came first: the giant or the materials used to make them. In some ways, the giants exist outside of their body—as part of the story Dambo tells and, more aptly, when giants move or are relocated. For instance, will Mama Loumari ever give birth? Will the child be made of recycled pieces of the mother? The official plan from Bernheim is that the giants will remain in place until they deteriorate. Of course, deterioration is a material change and a conceptual change. At some point, that is, park rangers and forest managers will make the call to disassemble the giants. Will their wood be turned to mulch? Will parts of them be preserved? When Mama Loumari, Little Elina, and Little Nis cease to be physically present, will their influence on the forest end?

We know that matter and the influences of matter do not simply appear out of nowhere. Nothing is truly made of new materials. The wooden pallets and shelves that make up the giants previously had other lives. Perhaps the shelves were handed down from college student to college student until they eventually reached the trash. And the pallets were perhaps traded, but rarely tracked, from one large-scale shipper to the next. They may have served time on factory and warehouse floors, grocery store back rooms, and small farms—mirroring a work trajectory shared by both of the authors of this piece. These materials are made of wood and lived a life as a tree before they were cut, processed, and built into their consumer shape. The possibility exists that some of the wood was timbered and milled from the Colorado Pine forests, the current home to Isak Hearthstone, husband to Loumari and father to Nis and Elina. However, we cannot know for sure because the materials were not treated as potential for life or art until they had been discarded and deemed not useful as a commodity.

By treating Dambo's giants as reanimated hybrids of human, nature, and industrial materials, we are reminded of both the artificiality of these boundaries and the vitality and permanence of matter. We cannot, for certain, track each piece of the giants to some imaginary starting point; we cannot find the forest that grew the tree or the mill that milled the soft pine that became shipping pallets, nor can we track the warehouses and shipping destinations of the pallets. This imprecision is a feature of what Dambo communicates. The value and potentials for things only seemingly matter when those potentials square with human interests.

At the most immediate and local level, the giants can be visited and interacted with physically. The boundaries between living, industrial materials, and nature are blurred. We see Mama Loumari made from deadfall branches and reused wooden refuse, while we are also reminded to not climb on her stomach out of fear of her unborn child. As we tread off the concrete and on the dirt path to see Little Elina, we leave our impact on the ground, each step further packing down the ground and solidifying the path. Globally, the giants' stories inspire similar artistic pursuits that support environmental efforts either by Dambo or other artists (see similar eco-art such as that created by Andy Goldsworthy). Additionally, the popularity of Dambo's art in Bernheim may play a part in new installations. Either way, the giants embody the message of looking ahead in a time of crisis; they put a face to the name so to speak. Art can be a productive way to influence public perception, and, in this case, improve the lived relationship with the natural environment to where humans see themselves as part of the environment—as opposed to maintaining the false nature/culture binary that promotes the othering of the natural world and furthers an almost lost-cause or fatalist narrative when it comes to climate crisis.

Risk communication surrounding the climate crisis often leads to this crippling fear of this lost-cause narrative. Dambo's compositions, however, look ahead and suggest that production can be sustainable in a time of crisis, though we must acknowledge he is not saving the world through his giants or solving the climate crisis per se. Yet, people interacting with the giants are confronted with the fact that the materials used to create the giants are up/recycled, displaying how recycled materials can be productive in many ways. On the other hand, the fact that the giants are made of recycled materials is only novel because of the fact that environmental risk exists. Dambo not only displays how we can reuse materials but also creates nonhumans that highlight the impending environmental doom to which humans contribute.

Even the recent pandemic has influenced the giants. The arbertorum was closed to non-members during the most significant periods of Covid-19 lockdown. However, in recent months, the forest has reopened and now a sign has been placed that says, "Forest Giants are Social Distancing: Please Enjoy Them From Afar." In this way, Bernheim purposefully includes the giants in a process of risk communication, ultimately reanimating risk such as COVID-19.

Reanimation

We see the metaphor of reanimation present throughout the giants in two meaningful ways: first, we conceptualize the ontologies of vital materialism via reanimation. Strategically, despite how well this vital materialism squares with our own theoretical intuition, we recognize that pointing toward seemingly inert materials and granting them an always-already vitality is a large leap for some technical writing professionals and academics. Toward this end, we feel that reanimation serves as a bridge between these two ontological starting points. In other words, we forgo the issue of a priori origins of vitality (founded on logical deduction) and focus, instead, on vitality as emergent from interactions. In our case, the giants have become alive through art. When Dambo "rescued" wooden materials destined for landfills and other end-of-life resting spots, he granted these materials a new life in the forest. Of course, this art is additionally enlivened by the discursive power of "reduce, reuse, recycle" as an environmental slogan.

Second, if we can get our readers to accept the possibility of reanimation, we also believe that we can harness reanimation as a discursive tool. Toward this end, we position technical writers as having the potential to reanimate risks to audiences that may otherwise find the risks too distant, inchoate, and complex. In this sense, we want to recognize the inherent danger of defining life in proximity to human, human-like forms, or human self-interest. We have picked the giants as a case because they are not human; nevertheless, they are decidedly anthropomorphic. We want to recognize the limits of anthropomorphism—especially to the degree that such techniques are based on the perceived privileged status of the human. As Zakiyyah Iman Jackson (2020) reminds us, "it is crucial to critically engage with what it means to *be* in a biopolitical context that is characterized by entanglements of humans both historically recent and distant, nonhumans both big and small, and environments both near and far" (2020). In other words, the privileges and rights of people are not and have not been universally applied across time and space.

To our first point, the forest giants illustrate an ontology of multiplicity. Partially made of leftover materials of consumer systems, designed in the image of a children's story, co-created by artifacts left behind from visitors, and residing within an educational arboretum, the giants are the product of multiple boundaries interacting and intersecting. Materials like discarded shipping pallets, boughs, and twigs begin to come alive and take new forms alongside Dambo's globalized story of forest giants. In some installments, these intersections are even more pronounced with giants built to encourage wildlife habitation, such as protected enclaves for imagined and real bird nests (Barger, 2020).

Like the growth of our own bodies, the growth of a forest, or emerging climate risks, the giants take shape across multiple co-constituting interactions but because of the embodied form the giants take, at first glance we see them not as processes but as solidified bodies. *Co-constituting interactions* are disparate but combine to form a new thing.[1] Of course, the boundaries between industrial transport systems, nature, and the giants serve only as *heuristics*, or mental shortcuts to solving problems. Just as the influence of humans is never removed from the environment—with metal machines strewn across galactic bodies in the near and distant solar systems and the microplastics and leaked chemicals in every nook and cranny of the earth—the giants, too, feel this influence and shape their environment, in return by solidifying walking paths, providing shade to fish in a pond, and providing habitation for small critters that seek refuge from the Kentucky heat in the folds of giant clothing.

Taking Jane Bennett's vital materialism to heart, we follow her lead in conceptualizing ourselves and the giants, in their pseudo-human image, as "an array of bodies" (2010, p. 113). In other words, we see Dambo's message not only pointing toward the potentiality of refuse to be more than trash, but also an introspection to look at the processes and matter that makes up our own bodies. For instance, where do we draw the line between the callouses on a person's hands and their job in an Amazon warehouse? Sitting with the giants and investigating their bodies invites us to question the limits of our own bodies and ask: How have industrial processes and commercial systems shaped the body I now occupy? In this way, we follow Dambo's art-activism not merely as reusing and recycling trash materials but instead as potentially raising the same question as Bennett: If we took our multiplicity seriously, "if we were more attentive to the indispensable foreignness that we are, would we continue to produce and consume in the same violently reckless ways?" (2010, p. 113).

The multiple ontologies of the forest giants reanimate the vitality of environmental rhetoric that technical writing professionals must constantly wrestle with. This is to suggest that the giants themselves stand in as a discursive tool for understanding the complexity and inter/intra-relations of climate change. This second form of reanimation is more nefarious.

In their coming to life, the giants represent a process of multiple imbricated natural, unnatural, intended, and unintended processes that mirror the pattern of global climate change. This process is seen in some climate change

---

1 For instance, think about how gender normativity, social media trends, and drought conditions came together in 2020 when a smoke-bomb at a gender reveal party seemingly started the El Dorado fire in California.

processes that create feedbacks that further increase the speed and intensity of climate change. For instance, desertification increases the aridity and instability of temperature of local conditions; such "feedbacks can alter the carbon cycle, and hence the level of atmospheric CO_2 and its related global climate change, or they can alter the surface energy and water budgets, directly impacting the local climate" (Mirzabaev, 2019, p. 268). Additionally, the feedbacks of climate change have impacted consumption and production systems. For instance, the increased market for lightweight (and often "green") batteries has reanimated new regimes of imperialism in the Global South (Frankel & Whoriskey, 2016). The degree to which this imperialism is naturalized by many is perhaps most saliently declared by Elon Musk, popularizer of an electric car, in his tweet that "we'll coup whoever we want" (Rozsa, 2020).

In short, we view the giants as a product of reanimated relationships between people and materials that ask us to consider our relationship with the environment and the material waste we produce. Additionally, we view the giants as a discursive potential for illustrating the larger risks of climate change as a process of reanimating waste material, spilled chemicals, and new regimes of imperialism (Empson, 2009; Patel & Moore, 2018; Plumber, 2018).

Coupled with an ecological perspective, an ontology of multiplicity only makes more apparent the complexity involved with the potential implications of human action. Yet, so often, human perspective remains just that: human. It is difficult to get people to take multiplicity seriously—especially if we limit our modes of communication. The complexity stemming from multiplicity, however, cannot simply be extended to our ontological perspectives. Our epistemological practices must incorporate such thinking as well if we are to see real difference. In other words, perhaps we must extend our knowledge-making and sharing practices (technical communication) to account for these complexities. Technical communication theories as an analytical lens might be a useful starting place. Yet, we can take this further if we understand "language and texts are not simply the means by which individuals discover and communicate information, but are essentially social activities, dependent on social structures and processes not only in their interpretive but also in their constructive phases" (Cooper, 1986, p. 366). Such a view suggests that discursive and rhetorical practices must be understood to exist within the larger ecology, meaning myriad elements within said ecology can be affected by and can affect our rhetorical and discursive practices.

In our mind, risk communication has a duty to attend to these forms of reanimation. It is not simply that reanimation is a process to be pointed out. Technical documents that demonstrate the far-reaching ends of climate change and the increasingly present self-constituted feedback systems of climate change exist in abundance. Nor is it merely the duty of risk communication to reanimate risks anthropomorphically (or otherwise) as some way to have discourse serve as more salient and useful reminders. Instead, it is to question our role in creating the means of reanimation: what new lives do objects get when they are dumped in the landfill; when they are burned instead of recycled?

Conclusion

Like Bennett, we worry about the self-interests implied in some environmental turns,

> whether environmentalism remains the best way to frame the problems ... or for inducing, more generally, the political will to create more sustainable political economies in or adjacent to global capitalism. Would a discursive shift from environmentalism to vital materialism enhance the prospects for a more sustainability oriented public?
>
> *(Bennet, 2010, pp. 110–111)*

A shift toward vital materialism emphasizes the notion that every part of the whole has some sort of agency, even if the part requires some sort of stimuli— ultimately promoting a shift toward multiplicity. The issue with an environmental rhetoric approach is that it quickly becomes politicized. Killingsworth and Palmer call this limit in our discourse *Ecospeak*, which they define as the discourse that outlines specific positions in the public debate about the environment (2012, p. 8). Ecospeak so often falls into a simple binary like much political discourse. On one side you have the environmentalists who, allegedly, care not for the economy and seek to protect the environment at all costs; on the other side, you have the developmentalists who allegedly care about economic gain over all else (2012, p. 9). Allowing this binary to taint our analytical approach, we risk oversimplifying the arguments. It is imperative to understand that there is nuance in public debates. Such a view complicates research, but "[r]hetorical exchange is a bloody mess, a living thing, or, more accurately, a confluence of many living things: an ecology" (Rivers & Weber, 2011, p. 193).

Additionally, risk communication must look beyond the immediate, obvious elements. The "activities, actors, situations, and phenomena" that exist in a given system or ecology can be understood as "interdependent, diverse, and fused through feedback" (Fleckenstein et al., 2008, p. 389). Such directions ask that we extend our research to include nonhumans and the various alterations of bodies in which processes of climate change, industrial and consumer systems, and art intermingle, while remaining cognizant that our discursive practices exist within a larger rhetorical ecology.

Climate change is a world-altering threat and with this new world will come new risks. In many ways, risks are already embodied in the symbols of people on signs, the infographics showing water levels rising above cities and parks, and in the images of migrants fleeing food insecurity. Of course, not all embodiment is good embodiment. As we continue to address the risks of climate change, we must also recognize that the risks are not merely local environmental risks, nor are they global risks that only exist in some abstracted disruption of microorganisms and far-off food chains. The risks also have a very real effect on people—individually and collectively. As scarcity and food and water insecurity destabilize people across the globe—disrupting the everyday "normal"—will our reaction be based on human self-interest, local self-interest? Or will we rightfully see our collective place in the larger environment? Will we shift away from appeals to anthropocentrism, and the complications that arise there, and understand that these risks have implications for all?

We suggest that we begin to frame risk, especially environmental risk, as risks of reanimation. With or without further intervention, the process of climate change is underway. We are not calling for a radical change to disaster communication, such imminent and definitive risks require a different approach, but instead an attention to the everyday risks of further perpetuating environmental collapse and disaster. The giants have a place in the forest and a responsibility to the forest—not by virtue of being human-adjacent but because they are of the forest. As the giants decay away and become a potential eyesore, will they too eventually reach a final end-of-life spot in a landfill or will we have taken the message? Similarly, we ask that risk communication practitioners recognize the responsibility we have to the world because we are of it. The giants embody their responsibility, and we can witness the very material realities of being of and with this world. Even the matter humans are made of is not new. We, as humans, are experiencing our own decomposing process as every second passes; we, as humans, are made of recycled materials as well. In a vital materialist sense, our whole is made up of parts; yet, we too, are parts to a larger whole.

In real terms, furthering our communicative effectiveness necessitates embracing more artistic (or less readily defined as technical) communication. There is no lack of information about the effects of our actions on the environment—nevertheless, less is communicated about the hurdles to action—and even less about the affective dimensions of climate action (Walsh, 2019; Walsh & Walker, 2016). We frame reanimation and vital materialism as calling for an effective response to climate change that does not settle for simple one-shot technological fixes but instead looks toward systemic change. In other words, we might have learned to see the single-use water bottles with the same disgust as garbage floating down the river. Of course, this has given way to a plethora of reusable water bottles. But, more importantly, might we learn to reflect on this disgust for single-use water bottles alongside our cultural and political willingness to allow conditions, like in Flint, Michigan, that make single-use water bottles necessary? When entire cities have their water contaminated in an attempt to cut-costs and when disabled people are stigmatized for using plastic straws, then it is clear that our current approaches to meaning-making are flawed. Looking beyond singular objects or balance sheets and toward the interactions and emerging lives alongside humans and nonhumans enables technical communicators to make more informed and ethical decisions. Understanding environmental risks as risks that are reanimated and rearranged through the complex systems of contemporary life points toward new ways of thinking about our impact and interactions with the environment. Such new insights, to be sure, are tentative and relational but we accept such flexibility as a welcome feature and not a shortcoming.

Discussion Questions

1   The materials that make up the giants have surreptitious origins. We know that discarded pallets and other industrial wooden materials are the primary materials; however, the complete history of the materials are a mystery. In line with our method of thinking about the lives of objects, what social inequities are hidden when the history of materials remains impossible to trace?

2   The giants are created alongside industrial and commercial systems that leave indelible tracks on their bodies. How are our own bodies and affectivities shaped in processes that mirror the creation of the giants? For instance, how has the process of your life changed your body? As a technical communicator, how might you begin to answer this question for stakeholders to the projects you work on? How might the stakeholders of the project be the nonhuman elements that are, ultimately, impacted the longest?

3   What are the complications scholars and practitioners encounter by extending environmental justice to nonhumans? How can we work to make nonhuman impacts visible alongside the human impact in technical communication?

Assignments

1   Product Tracing

While the wooden pallet may get reused and/or repurposed, it can be difficult to trace the life of these workhorses. However, other products such as the shoes you are wearing, the computer you use for class, or the phone you carry at all times have materials that often can be traced. Have you ever thought about where the materials and labor come from for these everyday items? Try choosing one of these common items you use and trace the materials and/or labor behind it. Some companies will make this information accessible to the public—while others may not. However, the materials used may be so specific that they are extracted from only one place. Think of solar panels for example: where do the materials that are used to making these come from? Many of the chemicals in these panels are extremely toxic and are difficult to dispose of; some of the materials used are also quite finite. So, while solar may seem like a viable alternative, it actually gets complicated when tracing this product. Think of a product you use regularly—or that you would simply like to learn more about—and try tracing the materials back to their origin. What is more, what labor goes into the production of said product? Where is this labor done and by whom?

2   Reanimation Environmental Empathy Map

Create an empathy map for a part of your environment. Think about a local source of water or point of elevation—if it could talk what would it say it thinks and feels; what would it say brings it pleasure and what brings it displeasure; what are its goals?

Consider the longevity of this environmental feature: was your river a source of food and water for indigenous peoples; was your hilltop part of folklore? Consider the difficulty you might have in finding such information: whose stories and relationships with the environment are left out? What purpose do potential occlusions serve? Compare your field report with your peers—what differences do you see?

References

Andersson, A., Winslott Hiselius, L., & Adell, E. (2020). The effect of marketing messages on the motivation to reduce private care use in different segments. *Transport Policy, 90*, 22–30.

Barger, J. (2020). This Copenhagen artists turns trash into trolls. *National Geographic.* https://www.nationalgeographic.com/travel/article/this-copenhagen-artist-turns-trash-into-trolls.

Bennett, J. (2010). *Vibrant matter: A political ecology of things.* Duke University Press.

Bernheim Arboretum and Research Forest (2021). https://bernheim.org.

Bivens, C., & Heilig, L. (2020). The activist syllabus as technical communication and the technical communicator as curator of public intellectualism. *Technical Communication Quarterly, 29*(1), 70–89.

Buchanan, R. (1992). Wicked problems in design thinking. *Design Issues, 8*(2), 5–21.

Cooper, M. M. (1986). The ecology of writing. *College English, 48*(4), 364–375.

Cryer, D. A. (2018). Withdrawal without retreat: Responsible conservation in a doomed age. *Rhetoric Society Quarterly, 48*(5), 459–478.

Dambo, T. (2021). *Troll map.* https://trollmap.com.

Druschke, C. A. (2019, February 20). A trophic future for rhetorical ecologies. *Enculturation: A Journal of Rhetoric, Writing, and Culture.* http://enculturation.net/a-trophic-future.

Edbauer-Rice, J. (2005). Unframing models of public distribution: From rhetorical situation to rhetorical ecologies. *Rhetoric Society Quarterly, 35*(4), 5–24.

Empson, M. (2009). Climate change: Capitalism's inbuilt obsolescence. *Socialist Review, 342.* http://socialistreview.org.uk/342/capitalism-and-climate-change-accumulating-chaos

Frankel, T. C., & Whoriskey, P. (2016). Tossed aside in the 'white gold' rush: Indigenous people are left poor as tech world takes lithium under their feet. *Washington Post.* https://www.washingtonpost.com/graphics/business/batteries/tossed-aside-in-the-lithium-rush.

Harman, G. (2005). *Guerilla metaphysics: Phenomenology and the carpentry of things.* Open Court.

Hawaway, D. J. (1985). Manifesto for cyborgs: Science, technology, and soclaist feminism in the 1980s. *Socialist Review, 80*, 65–108.

Hawaway, D. J. (2016). *Staying with the trouble: Making kin in the chthulucene.* Duke University Press.

Jackson, Z. I. (2020). *Becoming human: Matter and meaning in an antiblack world.* NYU Press.

Killingworth, M. J., & Palmer, J. S. (2012). *Ecospeak: Rhetoric and environmental politics in America.* SIU Press.

Lancaster, A. (2018). Identifying risk communication deficiencies: Merging distributed usability, integrated scope, and ethics of care. *Technical Communication, 65*(3), 247–264.

Mirzabaev, A., et al. (2019). *Climate change and land: An IPCC special report on climate change, desertification, land degradation, sustainable land management, food security, and greenhouse gas fluxes in terrestrial ecosystems.* https://www.ipcc.ch/srccl.

Paliewicz, N., & McHendry, G. (2017). When good arguments do not work: Post-dialectics, argument assemblages, and the networks of climate skepticism. *Argumentation and Advocacy, 53*(4), 287–309.

Patel, R., & Moore, J. W. (2018). *A history of the world in seven cheap things: A guide to capitalism, nature, and the future of the planet.* University of California Press.

Plumber, B. (2018, September 4). You've heard of outsourced jobs, but outsourced pollution? It's real, and tough to tally up. *New York Times.* https://www.nytimes.com/2018/09/04/climate/outsourcing-carbon-emissions.html.

Propen, A. D. (2018). *Visualizing posthuman conservation in the age of the anthropocene.* Ohio State University Press.

Rickert, T. J. (2013). *Ambient rhetoric: The attunements of rhetorical being.* University of Pittsburgh Press.

Rivers, N. A. (2015). Deep ambivalence and wild objects: Toward a strange environmental rhetoric. *Rhetoric Society Quarterly, 45*(5), 420–440.

Rivers, N. A., & Weber, R. P. (2011). Ecological, pedagogical, public rhetoric. *College Composition and Communication, 63*(2), 187–218.

Rozsa, M. (2020, October 20). Elon Musk becomes twitter laughingstock after Bolivian socialst movement returns to power. *Salon.* https://www.salon.com/2020/10/20/elon-musk-becomes-twitter-laughingstock-after-bolivian-socialist-movement-returns-to-power.

Sackey, D. J. (2018). An environmental justice paradigm for technical communication. In A. M. Haas & M. F. Eble (Eds.), *Key theoretical frameworks: Teaching technical communication in the twenty-first century* (pp. 138–159). University Press of Colorado.

Spinuzzi, C. (2003). *Tracing genres through organizations: A sociocultural approach to information design.* MIT Press.

Thackara, J. (2017). *How to thrive in the next economy: Designing tomorrow's world today.* Thames & Hudson.

van Dooren, T. V., Kirksey, E., & Münster, U. (2016). Multispecies studies: Cultivating arts of attentiveness. *Environmental Humanities, 8*(1), 1–23.

Walsh, L. (2019). A zero-sum politics of identification: A topological analysis of wildlife advocacy rhetoric in the Mexican gray wolf reintroduction project. *Written Communication, 36*(3), 437–465.

Walsh, L., & Walker, K. C. (2016). Perspectives on uncertainty for technical communication scholars. *Technical Communication Quarterly, 2*, 71–86.

Warren, K. J. (1987). Feminism and ecology: Making connections. *Environmental Ethics, 9*(1), 3–20.

Wickman, C. (2014). Wicked problems in technical communication. *Journal of Technical Writing and Communication, 44*(1), 23–42.

12

TECHNICAL WRITING AS EMBODIMENT

iFixit

Elizabeth Baddour

Problem: Although technology has offered humankind unprecedented access to information, entertainment, and global connectivity, discarded e-waste poses a significant threat to the environment and to the humans who mine our refuse for precious metals. Carcinogenic toxins from discarded e-waste seep deep into the earth and into the bodies of those who are powerless to resist the aftereffects of mass consumerism.

Solution: Through a collaborative effort, technical writers work with students and instructors to experience the personal empowerment of learning and then teaching others how to repair their broken electronics. Sharing knowledge of repair exposes the danger that e-waste poses while facilitating the reduction of discarded e-waste to impoverished areas of our world.

Within a span of a few decades, technology has transformed from convenience to necessity as evidenced by the sheer number of people publicly tethered to their cell phones in the course of everyday existence. As technology continually evolves—aided by consumerism and abetted by functional obsolescence—consumers inadvertently contribute to the 50 million tons of electronic waste (e-waste) generated every year as devices are discarded and replaced for newer versions (Dais, 2020). Americans, for example, throw away 416,000 cell phones daily for a total of 151 million phones discarded annually (Dais, 2020). It is difficult to underestimate the need to slow the global consequences of consumerism that our behavior inflicts on our planet and on fellow human beings who mine our e-waste here and abroad for tiny bits of precious metals. Although technology has offered humankind unprecedented access to information, entertainment, and global connectivity, discarded e-waste poses a significant threat to the environment and to the humans who mine our refuse for precious metals. Carcinogenic toxins from discarded e-waste seep deep into the earth and into

DOI: 10.4324/9781003266549-15

the bodies of those who are powerless to resist the aftereffects of mass consumerism. Through a collaborative effort with iFixit, technical writers work with students and instructors to experience the personal empowerment of learning and then teaching others how to repair their broken electronics. Sharing knowledge of repair exposes the danger that e-waste poses while facilitating the reduction of discarded e-waste to impoverished areas of our world.

Foreground

The ubiquity of cellular telephones and electronic technologies testifies to the permanence of the devices and our reliance upon them for communication, commerce, education, and entertainment. As of 2019, 96% of adult Americans have a cellphone of some kind, and 75% of adult Americans own a laptop or desktop computer. Data from the UN International Telecommunications Union, the World Bank, and the UN itself indicates that although 1.1 billion people around the globe live without electricity, active cell phone subscriptions outpace world population (Murphy, 2019). Statistics indicate that there are more cell phones in the world than live human bodies to utilize them.

Although the devices we depend on may seem innocuous, they foreshadow personal and planetary peril through their planned or perceived *functional obsolescence*, where an object's outdated design makes it less useful or desirable. Rapidly advancing technological change and the cultural imperative to upgrade often means that discarded electronic devices end up as toxic waste in the United States and beyond. *E-waste* is defined as any electronic device—from refrigerators, circuit boards, TVs, and cellphones to DVD players—that are scrapped after their usefulness ends. The Environmental Protection Agency (EPA) estimates that for every 1 million cell phones recycled, 35 pounds of copper, 772 pounds of silver, 75 pounds of gold, and 35 pounds of palladium are recovered (EPA, n.d.). However, an estimated 80% of electronics are not recycled, instead ending up in landfills at home and abroad where their toxicity leaches into the environment (Baldé et al., 2017). Apart from repair, recycle, or repurpose, the remains of devices once deemed practical permute into 48 million tons of pernicious poison that endangers the environment and those who mine the megatons of e-waste for sustenance (iFixit, 2020). Although technology undeniably contributes to societal advancement, ironically, it quietly contributes to the gradual decline (or *entropy*) of the fragile ecosystem upon which we depend.

One countermeasure against the mounting problem of rapidly accumulating e-waste is through the intervention of technical writing and the embodied knowledge passed through a network of free repair manuals produced via iFixit, a technical writing firm that specializes in combining service-learning with sustainability. California-based iFixit is the hub of a wiki-based website that teaches people "how to fix almost anything" (iFixit, 2019a) through their online distribution of free repair manuals that are created by individuals across the globe under the aegis of iFixit's staff of technical writers. Through service-learning, iFixit fosters a comprehensive network of college students and their professors who are invested in the firm's curriculum and its *ethos,* or moral character, because of the benefit students receive in obtaining "real-world" experience through their contribution to iFixit's repair catalog. In exchange, the students either repair recommissioned e-waste sent by iFixit to instructors or document the repair of students' choosing under the firm's approval. In re-turn, iFixit freely supplies classrooms with their specialized toolkits to facilitate repair, the sale of which funds their mission of policy advocacy and the docu-mentation of clear, easy to understand technical writing that enables anyone, anywhere to fix almost anything—for free.

This chapter argues that technical writing instruction promulgated by iFixit, when viewed through the lens of Actor Network Theory (ANT), contextu-alizes embodiment with the right-to-repair movement to resist conformity to the "take-make-use-dispose" pattern of manufacture and consumption that is common in a *linear economy,* where raw materials are used to make a prod-uct and then, it is thrown away once it is used (Vickers, 2019, para. 1). This chapter explores technical writing as a disruption to the deleterious effects of a consumer-driven response to manufacturers who transform raw materials into products that consumers ultimately toss out as waste—as contrasted with a *circular economic model* that values "take, make, use, re-use and re-use again and again" (Vickers, 2019, para. 1)—a pattern that coheres with sustainability ef-forts that fight global warming. After describing the term *embodiment* in relation to technical writing, this chapter briefly revisits the historical connection be-tween rhetoric, writing instruction and the body, and the ways in which social justice is manifested through technical writing. Next, the essay explores the way in which ANT informs the collaborative partnership between academics and industry in the resistance against the entropy created by e-waste. Finally, this chapter reasserts the necessity of considering the objects that we surround ourselves as co-agents in creating a more sustainable world for our planet and those with whom we share it through risk communication and iFixit.

iFixit and Social Justice

Using open-access service-learning, iFixit bridges educational contexts with environmental activism to demonstrate global citizenry, ethics, and ecolog-ical activism to college students at no cost. Environmental activism through the medium of repair embodies a rhetorical response to the agency of com-monplaces that threaten the planet and its inhabitants through environmental degradation generated from abandoned technological devices. Recognizing what Jane Bennett (2010) describes as the "thing power" of discarded objects, iFixit acts as a mediator in resisting e-waste through service-learning centered upon creating what is arguably the world's largest free and easily accessible repair manuals. Partnering with college instructors and their students, iFixit's technical writing professionals work with undergraduates, graduate students, and their professors each semester in a comprehensive network that fosters social justice through knowledge-making founded on environmental activism to produce change through the agency of embodiment. In other words, students learn how to create change by making repairs, then teaching others how they can also make repairs. It is through teaching others how to repair objects that technical writing is recognized as a form of risk communication as well as an act of social justice.

Risk communication is broadly defined as conveying information to the public about health or environmental hazards that exist to control such risks. Framed in context of this definition, it is easy to understand how the collaborative efforts of iFixit and their matrix of students and instructors are at work to diminish the flow of e-waste into our world through the simple act of repair.

In understanding technical writing as a manifestation of social justice, it is helpful to contemplate philosopher Iris Marion Young's conception of the relationship of domination and oppression in a social context. It is also important to recognize theory in naming and understanding concepts—such as risk communication and social justice—to help identify a problem and then work toward its solution. At the beginning of this chapter, we identified the problem of e-waste. We also identified a solution to the problem of e-waste, which is found in teaching others the value of repair to combat the growing problem of e-waste and its toxic effects, particularly on marginalized populations. *Marginalization* is the treatment of certain groups as insignificant within a society. Typically, the voices within marginalized groups are not accorded recognition, and consequently, they and their problems are relegated to the fringes or margins of society. Marginalized groups are often powerless, subjugated to the will of those in power, and they may experience exploitation and violence.

In considering the problem of e-waste disposal, think about where our garbage most often ends up: near the most expensive homes in town, or in places or countries that lack (or choose to ignore) regulations designed to protect their citizens? According to United Nations University, millions of tons of our discarded electronics—from motorized toothbrushes to computer screens—end up in poorer countries whose political systems have fewer health and environmental laws. Highly toxic forms of lead and mercury are prevalent in e-waste. As villagers burn circuit boards over fires to extract valuable copper wire for example, they unknowingly expose their bodies and their environment to cancer-causing agents such as dioxins (Leung, 2007). This is a form of exploitation and hidden violence against their bodies and their environment. People in developing nations often have few economic opportunities and recovering precious metals from e-waste is a means to feed their families despite the health risks that primitive recycling poses. Absent laws and regulations prohibiting such injustice, this form of cheap labor benefits the powerful at the expense of the oppressed and marginalized.

In her book *Justice and the Politics of Difference*, Young theorizes that oppression is a key term that exists in a framework of unjust circumstances. She describes "five faces" of oppression as a means to understand the ways in which oppression operates and thrives. Young's five faces of oppression are marginalization, powerlessness, violence, exploitation, and *cultural imperialism*, which is the imposition of a dominant culture's values upon another culture. Young's "five faces" serves as a foundation for analysis about the reverberations of irresponsible consumerism and the dangers inherent in rendering fellow humans "invisible" so that their problems are subsumed to the pursuit of economic prosperity of the status-quo (or dominant group).

Repair through technical writing helps us to "see" marginalized others and begin the process of staunching the harmful flow of e-waste into the lives of those whose existence is likely markedly different from our own. iFixit produces manuals through two primary means—through ordinary individuals who edit existing entries in iFixit's catalog of more than 13,500 devices, or through any individual's addition of new repair information to iFixit's ever-expanding repair guide compendium (McCrigler, 2019).[1] Partnering with more than 87 colleges and universities in the United States, Canada, and Europe, iFixit's Technical Writing Project (iTWP) works with student teams to affect and realize repair documentation for an unlimited range of devices. In exchange for iFixit's editing and oversight of its content, students complete the iTWP with published evidence of their contribution to iFixit's free repair manual and an empirical appreciation for the inherent value of repair.

In addition to supporting technical education and the sale of specialized tools, iFixit is a strong advocate of right-to-repair legislation both domestically and in Europe (iFixit, 2019b). The medium of repair is a window through which students engage with broken objects to embrace a first-hand perspective that ethical ownership involves action; it is a form of resistance to the disruption imposed by manufacturers' restricted access to repair documentation and is a challenge to the barriers intentionally created in device design to prevent restoration. The right-to-repair movement challenges manufacturers who restrict the repair of items people own—from game consoles, cell phones, and farm equipment to much-needed hospital equipment like ventilators—through limiting parts availability or making device repair available only through an authorized dealer. One glaring example is the tractor company John Deere, whose licensing agreement threatens the risk of breach of contract should a farm

---

1 iFixit's in-house technical writing team writes repair guides and tear-downs (taking a device apart to learn "how stuff works") for devices such as smartphones and tablets that are new to the marketplace.

equipment owner tamper with the software that runs their investment (John Deere, 2016). Tech giant Apple, like many other manufacturers, requires authorized parts or specific tools to implement repair; Apple's pentalobe screw—unlike its ubiquitous cousins (the flat-head and Phillips head screwdriver)—is designed to keep users out rather than allowing users access to initiate repair. The right-to-repair movement seeks to foster an end to such restrictions by making repair a right and redefining ownership to include the entitlement to repair any item that is purchased. In addition to assisting in the repair of everyday items, iFixit offers the world's largest free medical repair database. As hospitals faced an overwhelming need for ventilators during the nascent stages of the 2020 pandemic, iFixit gave hospitals information necessary to quickly repair broken ventilators and other essential hospital equipment. As an already overwhelmed medical system struggled to cope with a rapidly increasing patient case load during the 2020 pandemic, some hospitals found themselves constrained by "lucrative licenses and maintenance contracts" that impeded repair of essential, lifesaving equipment (Purdy, 2020). This scenario involving ventilator repair illustrates the amorphous nature that the term "own" represents even in the most dire of conditions—a worldwide pandemic: hospitals possess the equipment, but not the right to fix the very devices necessary to assist sick and struggling COVID-19 patients with breathing.

In *The End of Ownership*, Perzonowski and Schultz insist that the modern definition of ownership has changed to a form of possession that is limited to "licensing, a term that restricts consumers' ability to resell, lend, transfer, and even retain the digital media they acquire" (2018, para. 1). The modified definition of ownership is particularly evident in digital products that delimit the behavior of owners via restrictive codes, software locks, and restrictive licensing terms. In addition to digital music, games, movies, and software, Perzonowski and Schultz cite the ebook as a prime example of an object and the evolving nature of the term *ownership*. Contrasted by a hardcover or paperback book counterpart that a consumer freely owns and may redistribute at will, the ebook is restricted by licensing terms that limit a consumer's lending and resale—and may even confine reading to a certain platform. This restricted concept of ownership is a contested definition that the right-to-repair movement seeks to ameliorate in part through the mediation of shared knowledge via the network of technical writing and iFixit.

Theoretical Foundation

Understanding theories behind a subject is a way to come to an understanding of what we observe. Much like the network of a spider's web, theory helps to establish connections between concepts in the identification of a problem. Naming and explaining a problem is an early step toward initiating change. In your reading below, notice the networks that are active in the relationships between iFixit and the global network of technical writers like you who work to create change through educating others about the personal and planetary value of repair.

As we consider theories found in extant technical communication scholarship, this chapter is framed through Clay Spinuzzi's definition of distributed work, which he describes as "sociotechnical networks [that] hold together and form dense interconnections among and across work activities that have traditionally been separated by temporal, spatial, or disciplinary boundaries" (2007, p. 268). Spinuzzi's observations are foundational in understanding the mutual connections formed through the iFixit triad of professional technical writers working with a nationwide matrix of co-ed collaborators under the facilitation of college instructors, all actors in an *assemblage* (a collection of things or people) whose common focus centers on environmental ethics. The reciprocal network of college students producing knowledge under the demanding standards of iFixit technical writers is a means to manifest Bruno Latour's conception of actors who are able to "translate, transform, distort and modify other actors" to reduce the residual impact of nonhuman actants upon the global web of humankind (1993, p. 39). In other words, mediating the ready disposal of e-waste through repair has a ripple effect on people and communities far away. Your action—in choosing to repair—makes you an "actor" in reducing the stream of e-waste to developing nations through your active participation in resisting entropy, or decay.

Spinuzzi's observation that distributed work transcends "temporal, spatial and disciplinary borders in the activity of a sociotechnical network" accurately characterizes the iFixit technical writing project as a conduit for shared knowledge, education, and environmental action (2007, p. 268). Typical classroom constraints such as time, location, or even national borders are limited only by availability of internet connectivity and the desire to teach others how to restore that which is broken.

At the intersection of technical writing, Actor Network Theory (ANT), and the right-to-repair movement runs the undercurrent of embodiment. Broadly defined, *embodiment* is the manifestation of an abstract idea that flows concomitant with the "physical and mental experience of existence" (Cregan, 2006, p. 5). In essence, embodiment is the architecture linking the *telos* (ultimate object or aim) of repair to the ethos of sustainability. Students' work in documenting and returning broken objects to usefulness illustrates Abby Knoblauch's conception of embodiment as "physical motion and the knowledge that might stem from such motion, sensory or bodily response, and a metaphorical and physical connection between the body and writing" (2012, p. 51). The tactile facility of teaching others to repair through technical communication is a clear example of the intersection of language and *soma*, or body. Students (or individuals) in the process of documenting and photographing steps in a repair process for inclusion in a professionally produced repair manual integrate elements of their minds, their bodies, and their language to effectuate and perpetuate the sharing of knowledge. Repair reflects the essence of embodied knowledge both in its direction and in its result. Rhetoric flows through the repair process as current through an electrical wire; the act of repair illustrates the reinforcement of cultural values championed by iFixit via its iTWP. iFixit's advocacy of the right-to-repair movement makes repair easier and less costly, and in the case of ventilators during a time of pandemic, iFixit's repair guides are also lifesaving (Rosa-Aquino, 2020).

Historical Context of Embodiment and Language: Rhetoric and Progymnasmata

Historians of rhetoric have long conflated writing instruction with the embodiment of ethical action. For example, Patricia Bizzell and Bruce Herzberg observe, "Rhetoric, in Isocrates' opinion, was a powerful tool for investigating problems... and for moving people to action for the common good" (2001, p. 67). Similarly, Richard Leo Enos argues that writing, for Isocrates, "is a way of coming to understand... with the intent of resolving social issues" (2012, p. 31). And much more recently than Isocrates, David Fleming argues that an ethical approach to the teaching of writing is less about "process, product and writing-across-the curriculum paradigms" but rather the cultivation of "deep-seated, intellectually powerful, and socially valuable habits of discourse that they [students] acquire" (2003, p. 106). From a rhetorical perspective, the aim of writing is effecting change-inducing persuasion in its various forms; for the iTWP, persuasion takes the form of knowledge-sharing through instruction and guidance that promotes socially valuable discourse. Language and the body have a relationship extending to antiquity—for body/mind synergy lies at the heart of agency-producing action. For context, Isocrates, a 4th-century BCE rhetorician and writing teacher, is well noted for his assertion of the value of incorporating ethics and writing instruction. The word *rhetoric* still describes the ability to write well and speak convincingly.

Progymnasmata is a rhetorical exercise used by rhetoric teachers such as Isocrates and St. Augustine in antiquity. As the etymology of the word *progymnasmata* implies, it shares the same root as the words *gymnasium* and *gymnastics*, and it means "preliminary exercises" (Perrin, 2016, para. 3). In *Antidosis,* Isocrates links rhetoric with athletics in his treatise on the art of discourse in which he espouses his theory of rhetorical education as a type of gymnastics for the mind. *Progymnasmata* is defined as a primary tool for teaching the basic rhetorical techniques of invention, style, and arrangement that were particularly useful in the Greco-Roman sociopolitical world (Lanham, 2001). As James J. Murphy notes, *progymnasmata* exercises offered Roman teachers a "systematic yet flexible tool for incremental development of student abilities" while offering students the opportunity to "absorb ideas of morality and virtuous public service from the subjects discussed and from their recommended amplifications on themes of justice, expediency and the like" (Murphy, 2001, p. 69). Similarly, the iFixit TWP provides students with a humanitarian basis for their rhetorical training by incorporating concern for the earth and its inhabitants along with writing instruction.

Embodiment at the intersection of writing instruction and social consciousness informs my conflation of *progymnasmata* in antiquity with technical writing in the present. The iFixit TWP immerses students in the elements of invention, style, and arrangement while providing them with a rhetorical foundation anchored in the embodiment of service-learning. Invention begins as students focus on selecting their choice of "fix" and its approval by iFixit. The next several weeks are spent on the style and arrangement of the developing repair guide, a process that on the surface may appear simple. For example: photographs must be shadow-free, focused only on a specific step and shot at a particular aspect ratio. In the process of documenting and photographing steps in their repair process, students use an embodied approach to teach others how best to affect repair. Hands or fingers must not occlude the focus of the photograph—actions that are far from easy when holding a micro-screw easily dwarfed by a fingernail. Similar to progymnasmata exercises, as students advance through the invention, style, and arrangement phases of the iFixit TWP, they develop resiliency in maneuvering around inevitable challenges, which in turn enables them to document and share their experiences with others engaged in similar device repair. The body and mind work together in the iFixit TWP to offset and hopefully reduce the ill effects of toxic e-waste upon the planet and other human bodies half a world away.

Understanding iFixit's advocacy of ethical action combined with performance via the repair advocacy that rhetorically reflects its ethos, we can acknowledge that iFixit's TWP is informed by a social constructivist perspective. Environmental entropy and the interruption of entropy share a commonality in origin and in its moderation: social constructivism. Matthew Lynch writes that social constructivism holds that

> social worlds develop out of individuals' interactions with their culture and society. Social constructivism teaches that all knowledge develops as a result of social interaction and language use and is shared. Constructivist learning attaches as much meaning to the process of learning as it does to the acquisition of new knowledge.
>
> *(2016, para. 3)*

A social constructivist perspective of learning involves more than the acquisition of a new skill. Learning is facilitated through challenging students to interrogate their existing beliefs—in this case, by creating awareness of the effects of our throw-away culture upon bodies—to consider new ways of being good stewards of our planet through the writing of repair guides that resist the entropy to which our cultural attitudes contribute. Concomitant with a social constructivist perspective of technical writing is awareness of the necessity of networks within the web of actors and actants.

Our cultural proclivity to replace rather than repair is a socially constructed problem whose antidote is also socially constructed through the intervention of technical writing. Technical writing is the intermediary between Isocrates' call to activism and that of 21st-century rhetoricians such as Enos and Fleming through the modern adaptation of the ancient rhetorical exercise, progymnasmata. From this viewpoint, I affirm the value of technical writing that focuses upon teaching others to repair as a modern version of *progymnasmata*. Repeated exercises in composing and studying entries in the iFixit catalog of easy fixes during a semester continues the relevance of *progymnasmata* in bridging ethics with composition to achieve an embodied, impactful result upon the community of beings to which we all belong.

With every repair iFixit and its network of technical writing students produce, the opportunity benefit for the planet increases as repair facilitates the possibility of a proportionate reduction in landfill waste. The more instruction created for the iFixit technical writing project, the greater the likelihood that the linear cycle of "take-make-use-dispose" is interrupted. With each semester, teaching others how to repair items becomes a 21st-century iteration of progymnasmata as the rhetorical exercise in service-learning yields planetary civility through repair.

Similar to students of rhetoric in antiquity, the iFixit TWP, as the modern counterpart of progymnasmata, is also subject to exercises that are extraordinarily detailed. By working with two- or three-person student teams across the globe from their base in California, iFixit posts milestones for students to aspire to as they work their way through the iTWP in a given semester. Each milestone must adhere to iFixit's exacting standards for style and arrangement to accurately reflect the company's ethos of consistency and quality in its repair guide. Students must create a draft of the steps in their repair, photograph each major step in a repair process using their own hands, then add colored markups in sequenced order if clarity or safety dictates. Before moving on to the next milestone, each step is evaluated and approved by iFixit, beginning with the choice of repair. While this description is an oversimplification of the iFixit TWP, each student team within the iFixit's technical writing network of 87 colleges and universities must actively communicate with members of the California iFixit team on a consistent, weekly basis as their repair guide develops. This student/industry partnership is illustrative of the complex relationship linking human individuals with the nonhuman actors under repair to facilitate change.

The iFixit TWP serves as a heuristic not only for writing tech repair manuals, but for advocating the role of a circular economy as a form of resistance to entropy. Technical writing that advances environmental and social justice is performative through its rhetorical impact in a network of people, technologies, and other components that "reciprocally influence the actions of one another" (Getto et al., 2014, p. 187). The next section examines the nexus of interrelated bodies both human and nonhuman to explore the entanglement of embodied knowledge through the lens of ANT.

Learning as Embodied Response: A Case Study

Embodied knowledge, as defined by A. Abby Knoblauch, is "knowledge that is very clearly connected to the body" and often originates with a bodily response, or "gut reaction" (2012, p. 52). Knoblauch writes that embodied knowledge is "that sense of knowing something through the body" (62). Emotions, ranging from fear, excitement, or apprehension are not uncommon bodily responses to a project outside the realm of familiarity to a student, and the meticulous details implicit in the iFixit TWP are no exception.

In the first days of a new semester, my description of the iFixit TWP project with its collaborative work, the requirement that each two or three-person student team must find an item to repair is not currently addressed in the iFixit catalog; the student-centered focus of the work that places the instructor in the role of facilitator (for most of the semester), and the prospect of a meticulous photographing process involving a light box is sometimes perceived as intimidating by some students, particularly so during the nascent days of the project on our university's campus. In other words, before students gained confidence with the assignment and its goals, my introduction of the iFixit project elicited a "gut reaction" not unlike that described in Knoblauch's (2016) definition of embodiment. At semester's end, however, students uniformly expressed the collective response of excitement mixed with affirmation in completing a major assignment while contributing to the betterment of their world. For most students, the iFixit TWP was their first experience in service-learning and their satisfaction in successfully completing their project and its many phases—from selection of their fix to field testing—was palpable.

Emotional responses such as that registered by my students at the conclusion of the iFixit TWP are not unique within academia. Poetry can speak to the soul and bring tears to the eyes. Essays, photographs, or lectures can spark anger or rage. Frustration can flush the skin or raise blood pressure. These examples of visceral bodily responses are not confined to the realm beyond the classroom, for bodily reactions ranging the gamut from anger to empathy have the propensity to inspire and foment change within and without institutions of higher learning. However, as Knoblauch observes, "embodied response is rarely legitimated in academia" (2016, p. 54). Eliding of the role of bodily responses such as pre-project panic (an emotion that may inspire or motivate) or elation over accomplishment (a reward in itself) is to reinforce the binary between body and mind, blurring the role of motivation and the important ways humans learn from sensory responses even in academia.

To the contrary, visceral responses, "gut feelings," and strong emotional responses are instructive for a quickening of the mind that produces awareness, reaction that foments action, curiosity that yields satisfaction, academic learning that spawns change, or even a sixth sense that ensures survival itself. Sara Ahmed expounds upon the connection between bodies and knowledge creation when she observes that

> knowledge cannot be separated from the bodily world of feeling and sensation; knowledge is bound up with what makes us sweat, shudder, tremble, all those feelings that are crucially felt on the bodily surface, the skin surface where we touch and are touched by the world.
>
> *(2014, p. 171)*

Ahmed's assertion merges with academia during the first days of teaching the iFixit TWP. The course introduction includes a short video clip of iFixit co-founder Kyle Wiens's research trip to Africa in which he explains that 100 million tons of electronics are illegally imported into Africa every year. The clip features children as young as six years old mining mounds of e-waste over burning fires that, unbeknownst to them, produce toxic fumes that they inadvertently inhale. Weins insists that "electronics recycling is rarely done the way you and I would expect," because as the children work, they breathe in arsenic, mercury, lead, and other pernicious fumes that are commonly found in e-waste such as burning keyboards and computer terminals (iFixit, 2019c). Wiens's brief clip highlighting the devastating effects of our throw-away culture concludes with the argument, "They are burning our waste and it's our responsibility [to find a solution]" (iFixit, 2019c).

Weins's video clip succinctly encapsulates iFixit's ethos through the elicitation of an emotional response connecting students' bodies to the bodies of others through the agency of the iFixit TWP. Technical writing that intervenes even in small ways to disrupt human and planetary suffering is effective because of the network of people and things that make it possible—the iFixit technical writing staff, college instructors across the United States, Europe, and Canada who facilitate the iTWP, technology that makes boundaries of time and location irrelevant, and of course, the students who invest themselves in the work of learning while offering solutions to countervail the devastating effects of e-waste induced entropy.

iFixit as Entropy's ANT(idote)

ANT illustrates iFixit's "symmetry between humans and nonhumans" in its goal of advancing repair as a first response to the entropy that a broken object presents to a human owner. From this perspective, ANT is useful in understanding how iFixit's TWP works within a systematized matrix of students, instructors, and professional technical writers to interfere with the entropy caused by discarded e-waste. The iFixit TWP exemplifies elements of Actor Network Theory through the work of one of ANT's early influencers, Alfred North Whitehead, an early 20th-century thinker credited with developing process philosophy, a movement that greatly influenced Bruno Latour and eventually led to the conceptual framework that is a hallmark of ANT. While ANT is a complex, evolving theory of the social world that is difficult to succinctly quantify, I offer a serviceable definition. ANT arose in the 1980s and has its roots in the social studies arm of the sciences. It is recognized by several key characteristics—namely, that "science is as a process of heterogeneous engineering in which the social, technical, conceptual and textual are puzzled together and transformed" (Crawford, 2005, p. 1). Actors/actants are defined as any human or nonhuman entity existing within an ontological network. ANT does not differentiate between nature and society and is "preoccupied with the process of ordering or the ways in which societal order is achieved and the role material elements and other nonhumans play in the process" (Bærenholdt et al., 2009, p. 15). Rather, ANT advances the assumption that "all entities achieve significance in relation to others" as illustrated by the relationship of our toxic discarded electronic devices to our fellow beings and the world that we all inhabit (Crawford, 2005, p. 1). Scientific knowledge in ANT depends on aggregate communities of human and nonhuman elements that form networks in the production of society (Michael, 2017, p. 18). ANT is important in this chapter because within its structure, the power that entropy wields over our fellow humans, ourselves, and the world we share is interrupted. Risk communication, as discussed earlier in this chapter, operates within the network of *being* (or ontology) with the goal of disrupting power structures that permit dangerous conditions to exist unchecked.

Risk communication is the exchange of information to the public about health or environmental hazards that exist to control or mitigate such risks. Risk communication strives to educate people about public hazards ranging from environmental dangers to those that jeopardize personal health. Let's take a look at the operation of risk communication within the context of iFixit.

iFixit as Risk Communication

On its website, iFixit's clearly articulated mission is to "make repair accessible and easy for as many people around the world as possible. We want to show the world how to fix everything. The easier it is to fix something, the more people will do it" (iFixit, 2020b). In this succinct statement, iFixit conveys the role of risk communication as an agent in creating a more sustainable world. By sharing knowledge of repair—from the tools needed to each repair's degree of difficulty—iFixit's global audience is equipped with skills that keep e-waste, one object at a time, out of the world's garbage heap. For free. Because e-waste is a global problem, iFixit's guides can be translated into many languages to increase the scope of its lifesaving mission of knowledge creation and sharing. The partnership of iFixit with university students and instructors illustrates the essence of ANT and the importance of identifying our actions as either active or passive consumers. Will we adopt iFixit's ethos of sustainability and move away from a linear "throw-away" economy, or will we move toward the ideology of responsible consumerism and global ethics through repair? Perhaps, this chapter has quickened you to cultivate a deeper relationship with your stuff as you consider that "fixing a device is actually helping to fix the world" (M. Rippens personal communication, July 2, 2021).

Conclusion

Theorizing the mutuality between human beings and their nonhuman counterparts is a small step toward acknowledging the role of technical writing as a form of intervention in the growing global and personal crisis that perpetuates environmental degradation. Since humans have been thriving upon the earth, we have been entangled with material things that influence, enhance, and advance our lives. From stone-age cookware to cell phones, bricks and basketballs, humans and things are intertwined in a relationship that is dynamic and continually evolving—impelled by our innate and unquenchable human desire to consume that which adds value to our existence. At the crux of our present society's penchant for the linear economy of "take-make-use-dispose" is the human yearning for the "good life," or what the Greeks called *eudaimonia*. Aristotle famously claimed that virtue is necessary for living well, but more importantly, *eudaimonia* (or to "live well") according to Aristotle is the use of rational faculties in the application of virtues to moral dilemmas (Taylor, 1998).

In the context of this chapter, ANT theorist Scot Barnett's definition of the good life complements this discussion of ethics and the role of technical writing in advancing care of our planet and each other. Barnett asserts that

> the good life is not the life lived alone in contemplation... nor is it the life lived apart from the things of the world. The good life is where humans and things come together to make a home—where they are at home with each other.
>
> *(Barnett, 2017, p. 189)*

Understanding the definition of *eudaimonia* in the contemporary sense of ethics merges technical writing bodies with the bodies of those who wish to repair rather than discard—contributing in some small way to the health of our home, and that of children who mine e-waste while living in other nations.

Discussion Questions

1 Political philosopher Iris Marion Young theorizes that *oppression* is a key term that exists in an unjust framework. Young asserts that domination occurs in the presence of unjust circumstances. In considering the five faces of oppression, list how you see each of those five faces at work in the scenario involving the exportation of e-waste to developing nations.

2 (a) After rereading the section describing Actor Network Theory (ANT) entitled "iFixit as entropy's (ANT)idote," discuss how ANT operates within the framework of iFixit and its collaborators. (b) Describe what the author does with the title of this subsection.

3 In your own words, describe your understanding of the ancient rhetorical practice of *progymnasmata*. Can you link other subjects you have studied with this form of skill development that builds upon repetition and gradual mastery? Describe in detail.

Assignments

1 Kyle Wiens's video (https://edu.ifixit.com/what-we-do) provides an overview of the ways in which our disposable e-waste continues to negatively impact our planet and its inhabitants long after it is discarded. Using this video and the "Learning as Embodied Response: A Case Study" section as your point of reference, write a description of any "gut feelings" that you have after watching this video. List and elaborate upon five key impressions the video leaves you with. Will your "gut reaction" to what you've learned lead to an embodied response? Why or why not?

2 Following this link to iFixit's "Fix Your Stuff" Repair Guide (https://www.ifixit.com/Guide).

Using your imagination, locate an item that you own that needs repair. Here is an example: https://www.ifixit.com/Guide/How+to+Fix+a+Hole+in+Mesh+Shoes/136828. Write a summary of your experience affecting your own repair and your thoughts on the ease or difficulty of the repair. Through this action, are you a passive or active consumer?

References

Ahmed, S. (2014). *The cultural politics of emotion*. Routledge.

Baldé, C. P., Forti, V., Gray, V., Kuehr, R., & Stegmann, P. (2017). *The global e-waste monitor 2017: Quantities, flows, and resources*. International Telecommunication Union.

Bærenholdt, G. T., & Jóhannesson, J. O. (2009). Actor-network theory/network geographies. In R. Kitchen (Ed.), *International encyclopedia of human geography* (pp. 15–19). Elsevier.

Barnett, S. (2017). *Rhetorical realism*. Taylor & Francis.

Bennett, J. (2010). *Vibrant matter: A political ecology of things*. Duke University Press.

Clarke, L. D. (1957). *Rhetoric in Greco-Roman education*. Columbia University Press.

Crawford, C. (2005). Actor network theory. In G. Ritzer (Ed.), *Encyclopedia of social theory* (Vol. 1, pp. 1–3). Sage. https://www.doi.org/10.4135/9781412952552.n1.

Cregan, K. (2006). *The sociology of the body: Mapping the abstraction of embodiment*. Sage.

Dais, D. (2020, December 17). *The ugly truth about toxic e-waste*. Freethink.com. https://www.freethink.com/videos/e-waste.

Enos, R. L. (2012). The art of literate rhetoric. In J. J. Murphy, R. L. Enos, R. A. Lanham, & C. D. Lanham (Eds.), *Greek rhetoric before Aristotle* (pp. 143–175). Parlor Press.

Environmental Protection Agency (2020, December 9). *Electronics donation and recycling.* www.epa.gov/recycle/electronics-donation-and-recycling.

Fleming, J. D. (2003). The very idea of a progymnasmata. *Rhetoric Review, 22*(2), 105–120. https://doi.org/10.1207/s15327981rr2202_1.

Getto, G., Franklin, N., & Ruszkiewicz, C. (2014). Networked rhetoric: iFixit and the social impact of knowledge work. *Technical Communication, 61*(3), 185–201.

iFixit (2019a). *About iFixit.* https://www.ifixit.com/Info/index.

iFixit (2019b). *Right to repair.* https://www.ifixit.com/News/category/right-to-repair.

iFixit (2019c). *Why repair matters.* https://edu.ifixit.com/what-we-do.

iFixit (2020a). *E-waste is the toxic legacy of our digital age.* http://www.iFixit.com/Right-to-repair/E-waste.

iFixit (2020b). *Why repair? What would change if.* https://www.ifixit.com/Info/why.

John Deere (2016). *License agreement for John Deere embedded software.* https://www.deere.com/assets/pdfs/common/privacy-and-data/docs/agreement_pdfs/english/2016-10-28-Embedded-Software-EULA.pdf.

Knoblauch, A. A. (2012). Bodies of knowledge: Definitions, delineations, and implications of embodied writing in the academy. *Composition Studies, 40*(2), 50–65.

Lanham, R. A. (2013). Progymnasmata. In *A handlist of rhetorical terms* (p. 120). University of California Press.

Lanham, C. D., & Murphy, J. J. (2001). Writing instruction: Late antiquity to twelfth century. In *A short history of writing instruction: From ancient Greece to modern America* (pp. 79–122). Erlbaum.

Latour, B. (1993). *We have never been modern.* Harvard University Press.

Leung, A. O., Luksemburg, W. J. Wong, A. S., & Wong, M. H. (2007). Spatial distribution of polybrominated diphenyl ethers and polychlorinated dibenzo-p-dioxins and dibenzofurans in soil and combusted residue at Guiyu, an electronic waste e-cycling site in southeast China. *Environmental Science & Technology, 41*(8), 2730–2737. doi: 10.1021/es0625935.

Lynch, M. (2016, November 19). Social constructivism in education. *Edvocate.* https://www.theedadvocate.org/social-constructivism-in-education.

McCrigler, B. (2019). *Achievement unlocked: 50,000 repair guides now on iFixit.* https://www.ifixit.com/News/14090/50000-guides.

Murphy, J. J. (2001). Habit in roman writing instruction. In *A short history of writing instruction: From ancient Greece to modern America* (pp. 35–78). Erlbaum.

Murphy, M. (2019, April 29). *Cellphones now outnumber the world's population.* Quartz. https://qz.com/1608103/there-are-now-more-cellphones-than-people-in-the-world.

Perrin, C. (2016, April 14). *Writing & rhetoric: The method, the philosophy & the progymnasmata.* Classical Academic Press. https://classicalacademicpress.com/blogs/classical-insights/writing-rhetoric-the-method-the-philosophy-the-progymnasmata.

Perzanowski, A., & Schultz, J. (2018). *The end of ownership: Personal property in the digital economy.* http://www.theendofownership.com/related-research.

Pew Research Center (2019, June 12). *Demographics of mobile device ownership and adoption in the United States.* http://www.pewinternet.org/fact-sheet/mobile.

Purdy, K. (2020, December 23). *Top 5 right to repair wins of 2020.* iFixit. https://www.ifixit.com/News/47829/top-5-right-to-repair-wins-of-2020.

Quintilian (1968). *Institutio Oratoria* (H. E. Butler, Trans.). Harvard University Press.

Rosa-aquino, P. (2020, October 23). Fix, or toss? The "right to repair" movement gains ground. *New York Times.* https://www.nytimes.com/2020/10/23/climate/right-to-repair.html.

Spinuzzi, C. (2007). Guest editor's introduction: Technical communication in the age of distributed work. *Technical Communication Quarterly, 16*(3), 265–277. https://doiorg.ezproxy.memphis.edu/10.1080/10572250701290998.

Taylor, C. (1998). Eudaimonia. In *The Routledge encyclopedia of philosophy.* Taylor & Francis. Retrieved 7 February 2021 from https://www.rep.routledge.com/articles/thematic/eudaimonia/v-1. doi: 10.4324/9780415249126-A125-1.

Vickers, E. (2019, August 08). In the circular economy, it's take, make, use, reuse. Retrieved 20 January 2021 from https://www.bloomberg.com/professional/blog/circular-economy- take-make-use-reuse-2.

13

CHANGING PLACES

Understanding Climate Change Risk Communication and Comprehension through Socially Constructed Features of Place

Zachary Garrett

Some political challenges in the United States enjoy broad support for collective action, such as public education, policing, national defense, and Social Security. In these cases, the problems are systemic and the necessary resources are vast. Accordingly, a stable majority accepts the need for collective action. However, for other similar problems, like climate change and the provision of health care, splintered public opinion discourages action.

Problem: Climate change communication's failure to launch in the face of clear risk is one of the trickiest riddles in scientific and political communication. As evidence for human-induced climate change has mounted, public opinion has barely moved; in fact, there have been significant setbacks (for instance, the temporary 2020 withdrawal of the United States from the Paris climate agreement). The inability to connect is particularly challenging for technical communicators because they serve on the frontlines of climate communication. Climate data is analogous to other scientific data, but it must be handled with greater care since its reception is more politicized. Public trust in environmental scientists differs along partisan lines in the United States, with many Republicans denying climate change and many Democrats supporting the science. However, scientists in fields other than environmental science are respected on both sides of the aisle (Funk et al., 2019, p. 21).

Solution: One promising communication framework involves engagement with users through the concept of place to reduce the psychological distance of climate change impacts and make climate risks concrete and present. An event or experience is psychologically distant when it is not instantly present and must be experienced indirectly. The experience of climate change is subject to significant psychological and physical distance, and this has had an impact on

DOI: 10.4324/9781003266549-16

the effective communication of risk. Psychological distance also provides a convenient foil for motivated skeptics: those who have economic or cultural reasons to deny the existence of or human contributions to climate change. Place is fundamental when describing the interaction among human bodies, minds, and surroundings, helping us differentiate place from space (Tuan, 1977).

Place is unavoidable; Allan Pred calls places "the essence of human geographic inquiry" (1984). Observers often assume that illustrating the threats to place from climate change would positively influence climate action. However, this ignores the degree to which place can also reinforce the cognitive and behavioral barriers that frustrate climate change information uptake. In short, place as a complex factor in climate change risk communication is essential because people are not disembodied minds but are persons situated in space and relationship to other people. I seek to understand the nature of that complexity and establish place in the conversation. In this chapter, I will describe and model socially constructed, physically enacted, and embodied notions of place consisting of the interaction of human behaviors and geographical/cultural features. The purpose of these models is to improve the quality of climate risk communication and give voice to cultural expressions that may not be accounted for in the status quo of climate risk communication. Further, since place and climate change is an active area of inquiry, I will review current understanding and raise ongoing questions.

Assumptions of Climate Change Risk Communication

Scientists share a near-consensus about the reality of human-induced climate change, and it is incorporated into high-stakes business decisions such as insurance forecasts and drainage design. Given these realities, the relative stability of public opinion (Krosnick & MacInnis, 2020a, 2020b), exhibiting a minimal, slow increase in consensus acceptance over time with a tenacious core of denialism, may be the most remarkable aspect of climate change communication. Stanford University's *National Survey of Public Opinion on Global Warming* indicates that a substantial majority of Americans believe that earth's temperature has "probably been increasing" over the past century (Krosnick & MacInnis, 2020b). That "probably" allows for significant departures from the scientific consensus and reflects tepid public support for a serious national or multinational response.[1] However, this survey does indicate widespread belief in the existence and future unfolding of climate change and a conviction that

---

1   These compelling numbers hide a strong partisan factor in climate opinions in the United States (Dunlap et al., 2016); these partisan differences also exist in other countries, but the gap is considerably wider in the United States according to a 2018 Global Attitudes Survey by Pew Research (Poushter & Huang, 2019).

the government and the private sector ought to respond. Given such extensive support, why are leaders unable to spend this apparent political capital?

While a majority of Americans accept climate change and support mitigating actions on paper, the "issue public" (Krosnick & MacInnis, 2020a), or those who are most "alarmed" or "concerned" (Leiserowitz et al., 2009) are a much smaller proportion. In the American political system, motivated minorities wield immense power while majorities with halfhearted convictions fail to turn their weak adherence into decisive action. While climate change denialists are scarcer than those who are motivated to support mitigation (Leiserowitz et al., 2009), denialists' influence is amplified by those with economic interests in the status quo (Oreskes & Conway, 2010), by a media and cultural environment that seeks to describe simplistic contrasts in expert opinion, and by powerful psychological and somatic (or bodily) forces built into human nature that promote inaction. Motivated skepticism is the most prominent barrier to broader acceptance of the consensus.

If the goal is action, the climate change risk communicator's task is not to leap into the fray and compel denialists to repent. Instead, the greatest gains can be achieved by increasing the existing adherence of those already weakly convinced that climate change is a problem. In so doing, the denialist position will not disappear, any more than flat earth beliefs (for those who believe that the earth is flat instead of round) or dangerous Holocaust denialism has fully disappeared. However, increasing adherence will defuse the power of the denialist extreme while leaving room for the healthy skepticism that energizes scientific inquiry. This audience is that sensibly moderate middle[2] who have not yet made climate change knowledge a high priority, who have no conscious motivations to deny climate change, who are turned off by the heat of the "debate" as portrayed in the media, and who have trouble refereeing conflicting claims to credibility by participants in the public discussion. One problem is that public discussion is not limited to scientists, academic authorities, and vetted media sources, but includes anyone who can publish online. In that environment, the pyramid of credibility is not apparent to the casual observer, making it difficult to know who to trust. What looks like easily distinguishable credentials to a highly interested party (i.e., one who is alarmed about climate change consequences or whose job requires them to engage with questions of science or argumentative credibility) looks like a legitimate disagreement to the outsider, especially if other parties (media outlets, convinced denialists) have an interest in portraying the discussion accordingly.

Finally, communication problems are usually characterized as worsened by cognitive barriers (lack of understanding) or motivated reasoning, and these are useful, if imperfect, models for understanding the problem because the barriers

---

2 Philip Eubanks provides a compelling profile of this audience in Chapter 2 of *The Troubled Rhetoric and Communication of Climate Change*.

are not merely cognitive. Generally, humans more readily modify behavior in response to immediate crises and rationalize away events in the distant future; if prevention seems impossible, humans tend to adapt instead. For example, we do not try to prevent death by extending our lifespans beyond the apparent maximum, but we adapt to that lifespan since avoiding death is impossible. Climate change communication is subject to both of these obstacles to change. If these barriers are construed as cognitive barriers alone, the problem is internal to each person. However, people are not brains or minds. Instead, our mind–body[3] interacts with the spaces and people around it as we decide what to do and believe, thereby drawing in other people and things. Actions and beliefs arise from this interaction. The interaction between mind–body and environment can likewise be conceived as a unity: the "person-place" (Seamon, 1979). For these reasons, "place" as a concept is underappreciated in developing climate change engagement strategies. It helps us incorporate embodiment in addition to cognition in our communication efforts.

Efforts to Connect Place and Climate Change Communication

Unfortunately, the scale of climate change as typically presented (a global phenomenon) does not match with an individual's perception of place. A global view introduces abstraction and ambiguity. Theoretically, then, to make climate change real—to "proximize" it, in the words of Brügger et al. (2016)—could motivate action. The concept is sound, but efforts to study the phenomenon have provided inconsistent results. The fact that such efforts have failed to find sure footing invites deeper analysis. Given the centrality of place in the human perception of environment, it is curious that the place-based approach has not been studied or used more deliberately in climate risk communication. It is perhaps less surprising when you consider the reality of climate risk communication. Climate change is either formulated in scientific terms, with caution and a disembodied air of objectivity, or in political terms, lost in inflexible ideologies and motivated reasoning. Something personable and identifiable is often excluded in these lines of approach. In this section, I explore a sampling of efforts to study or understand place and engagement with climate change.

Reviews of Existing Scholarship/Theoretical Frameworks

Since the study of the relationship between place, climate change, and engagement is in a nascent stage, there is still room for development of theoretical frameworks to describe and understand the underlying phenomena. In a review of 66

---

3 The "mind-body" is a more significant phenomenon than the mind or body alone; each, alone, is impossible. A mind without a body is undefined, like dividing by zero, and a body without a mind is an automaton or corpse.

studies, Emily Nicolosi and Julia Corbett (2018) found that leveraging place attachment is an effective strategy for climate risk communication, but many variabilities are hidden in the "is." The number of studies (25) conducted specifically on the intersection of place relations and climate change is relatively small, and the methodologies were highly variable and mainly focused on non-minority "Global North" populations. This exposes the need for more resources directed toward the understanding of "Global South" and majority–minority communities, and highlights the disadvantages of living on the margins of academic interest. The studies that have been conducted are narrow in their scope (as would be expected given resource limitations and the exigencies of empirical research), and as such, place-based communication strategies reflect minimal impact. Another serious problem is the attempt to study affective phenomena with quantitative methods; as they note, qualitative or mixed methods studies would be more likely to "tap the depth and intricacy" of place meanings (Nicolosi & Corbett, 2018, p. 94).

Those who have developed other theoretical frameworks include Hess and colleagues (2008), Groulx (2017), and Adger (2016). Hess's work was developed for the preventive medicine community; understanding the local impacts of climate change has significant importance in public health planning, since public health issues can differ from county to county, or even block to block. The core of their insight is to emphasize the extent to which place is one of the primary variables in individual and community climate change risks; these risks are "place-specific and path-dependent." They illustrate a definition of place rooted in human relationships and recognize its importance in physical and mental health. Further, they draw out the added value of place in a public health response: it provides a basis upon which to prioritize certain responses based on where the impacts will be most significant, and it emphasizes the importance of adaptation buy-in from community institutions (municipalities, planning organizations, regional health agencies, etc.). Groulx, a professor at the University of Northern British Columbia, extends the concept through his research in the small town of Churchill in northern Manitoba. The town's location on Hudson Bay makes it a center for ecotourism—viewing polar bears and the northern lights—and thus is impacted seriously by climate change. Among other things, Groulx's framework envisions place as a way to assign value to non-economic assets in a community: place is one collective value that most people in a given locality share apart from the need for subsistence and economic flourishing. Finally, Adger, a professor of Human Geography at the University of Exeter (UK), shares an editorial explaining how place prioritizes adaptations to climate change by influencing values and perception of the fair allocation of impacts.

Studies of Place Relationships or Impacts on Specific Communities

While it is challenging to test the interaction of place and climate change engagement experimentally, it is possible to understand the existing perception of place-based climate impacts through various qualitative approaches. Of these studies, some are designed to inform communication efforts, test models of understanding or communication, or understand community response through theoretical frameworks.

Many of these studies focus on unique communities that are particularly vulnerable to climate change. These communities often have multiple vulnerabilities: economic, cultural, etc. These studies include the initial presentation of the previously mentioned work from Groulx et al. (2014) in Churchill, Manitoba, impressions gathered from residents in Thunder Bay, Ontario, another vulnerable northern city (Galway, 2019), an ecolinguistic analysis conducted in North Frisia, a German region on the border with Denmark (Döring & Ratter, 2018), a study of place/climate connections among the Inuit in Nunatsiavut, Canada (Cunsolo Willox et al., 2012), a study of the impact of natural disasters on the perception of climate change risk in rural Australia (Boon, 2016), a study of migration motivations among residents of highland villages in Peru (Adams, 2016), and a qualitative study among shrimp farming villages in coastal Bangladesh (Kais & Islam, 2019). Other studies have investigated larger or less vulnerable communities, such as farmers in the American Midwest (Arbuckle et al., 2013) and private forest owners in Germany and Sweden (Blennow et al., 2016).

Findings in these studies of place/climate relationships in particular communities are as diverse as the communities in which they were held. For instance, in Groulx's (2014) study in Churchill, Manitoba, in a survey of 25 residents (8.2% of the total population over 18), one conclusion was that socially salient messages more effectively motivate action; the role of place here is to determine what matters to a community. In Thunder Bay, Ontario, Galway's (2019) 18 walking interviews—a strategy designed to enhance the understanding of place by allowing the environment to inform the discussion (Evans & Jones, 2011)—indicated conceptualizations of climate change shaped by environmental factors, which support the connection but do not suggest a communication strategy. Döring and Ratter's (2018) ecolinguistic analysis in North Frisia, where climate change is contested in a tourism and agriculture-driven economy, uncovered the interaction between place and climate change perception and the tensions produced between insiders and outsiders in a tight-knit rural community. Döring and Ratter identified what they call "conceptual metaphors": climate change is an enemy, preventing climate change is war, climate change is punishment for human sins, climate change is overheating, climate change is hot air (or a hoax), and climate change facilitates eco-dictatorship. This framework provides one window into climate change perception as a socially constructed phenomenon formed by features of place and culture. Adams (2015) considered why residents of certain villages in highland Peru would choose not to migrate in the face of serious, climate-related environmental degradation—place attachment was an important factor—suggesting directions for future adaptations within particular populations. Adams surveyed 433 people in four villages; some of these villages were very small, and thus she was able to survey virtually all households.

In rural Australia, Boon (2016) found that prior disaster experience had no impact upon climate change risk perceptions, and that risk perception was more connected with trust in climate change communication and geographical location, which suggests cultural influences rather than event-based influences. Kais and Islam (2019) identified a possible connection between climate extremes and perceptions of local climate change. The authors describe a perception process rooted in embodiment, a "psychophysical process" that involves the uptake of light, sounds, and other signals from the environment and, filtered through the body's parts capable of perceiving them, they are converted by the brain into perception. The person then uses this information to inform action; this process is heavily mediated by culture. Thus, "people from different cultures may perceive climate risks differently" (3). While this study does not replicate Boon's work in Australia, considering them together raises questions about what kinds of weather or climate-related experiences would impact climate change perception and raises questions as to the mechanism: Is it proximization? Reinforcement of trust in climate communicators? Or something else altogether?

These studies' diverse purposes and designs make it challenging to draw unified conclusions about the connection between place and climate change at this stage; however, they provide some nascent, location-specific insights and provide methods for studying similar phenomena in other locations. Additionally, they offer several examples of how a researcher can gather hard-to-access information beyond surveys or structured sit-down interviews. Finally, they indicate research gaps; these gaps include the emphasis divide observed by Nicolosi and Corbett (2018) in their review, where the prevalence of studies is heavily weighted toward the "Global North" even though developing countries are most vulnerable to the impacts of climate change. Overall, the insights from these studies, beyond their considerable value for the communities in which they were conducted, cast a halting and flickering light on the broader connection between place, engagement, and climate change.

Experiments

Only a few experiments have been conducted on the place, climate change, and engagement or communication connection. Such experiments are challenging to design because of the complexity of influences on one's prior beliefs and risk perception. The mechanisms involved in changing one's mind are not well understood, and the results have been mixed. However, these studies are helpful for understanding certain specific phenomena, testing cognitive models, and raising more questions for future research.

In a UK-based study of 161 psychology students, Spence and Pidgeon (2010) sought to study the transfer of existing findings from behavioral science and health research to the particular case of climate change; their study is relevant here because one of their two frames was a "local" versus "distant" frame (the other examined "gain" versus "loss" outcomes). In their study, the perceived severity of climate change was greater in the "distant frame" group, but attitudes toward climate change mitigation were stronger in the local frame group. The distant frame appeared to negatively impact climate change mitigation compared with perceptions of severity. In British Columbia, Scannell and Gifford (2013) found a connection between place attachment and the effectiveness of a locally framed message. Climate change engagement was greater among those who had received a locally framed message. However, engagement did not differ among those who had read a globally framed message and those in the control group (no message). A study (Schoenefeld & McCauley, 2016) of similar phenomena took place among 99 Vermonters randomly assigned to three conditions: receipt of a local narrative, a global narrative, and no narrative. Their goal, in part, was to study the impact of communication on individuals with a "self-transcendent" versus a "self-enhancing" outlook and found (as mentioned previously) that locally framed information had a negative impact on those with a self-transcendent perspective—contradicting the effect seen in Scannell and Gifford's study, at least for those with a self-transcendent view. Finally, de Boer et al. (2016) hypothesized that proximal, place-interactive information—in this case, flood risk, significant in the Netherlands, where 26% of the land lies below sea level (Schiermeier, 2010)—would tip the balance in favor of a more accurate risk assessment, especially if the information provided did not directly threaten one's existing worldview. They found, with some limitations, that it was possible to increase the participants' local climate risk perception in combination with increased motivation for flood damage prevention, despite a certain level of climate change skepticism.

Of course, the major limitation of all of these studies is a lack of ecological validity; the real world is not a laboratory. Communicators rarely seek to change minds with one impression of a message—as advertising professionals can attest—and messages are rarely received in a standardized context. Communication and uptake of information are complex phenomena that interact with the mind, body, culture, environment, and multiple other factors. On the experimental front, more extensive, naturalistic, or qualitative studies are needed to understand the phenomenon of belief/behavior change and the experienced impacts of climate change. However, studies of the language that people use concerning climate change are valuable; language must be used as a mediator of climate change information because the direct bodily impacts of the change are rarely felt strongly enough to promote action. Climate change is real, but climate change must also be experienced rhetorically, which is the case with most phenomena that extend beyond immediate perceptual boundaries (e.g., patriotism, faith, love for friends who have moved away).

Place and Embodiment: Understanding the
genius loci (Spirit of Place)

The cognitive model is not wrong, but it is incomplete. Human persons are inherently embodied. We are shaped, defined, and limited by the physical space we take up and the technology we use to extend our presence artificially. Place directs our actions; we can only be in one space at a time, and by acting in specific ways in time and space, we close off other possibilities (Pred, 1984). Accordingly, consideration of the entire human physical environment is a necessary component of an effective climate communication effort since it defines the space occupied by the whole human person: "the self cannot be an agent ... without place" (Sack, 1997, p. 127). This interaction between the human body, its environment, and the human meaning-making apparatus is a phenomenon we call *place*. Definitions of *place* include both physical boundedness and human meaning; place is space experienced (Sack, 1997, p. 16). Environment and humanness come together to generate place. As anthropologist Marc Augé observes,

> ...the layout of the house, the rules of residence, the zoning of the village, placement of altars, configuration of public open spaces, land distribution, correspond for every individual to a system of possibilities, prescriptions, and interdicts whose content is both spatial and social.
>
> *(Augé, 1995, p. 52)*

The increasing volume of human activity conducted digitally (accelerated by the COVID pandemic) complicates the concept of physical boundedness: How can physical boundedness be understood in an environment where many critical human relationships are sustained with disregard to their physical location? However, this merely extends a problem that is already well understood: actualizing meaning requires human perception. It applies to the technology of videoconferencing software as much as the technology of eyeglasses on a near-sighted person: bringing elements of the human environment into awareness that could not be apprehended before. Technology can distort as well as extend our perception; as Robert Sack notes, "[o]ur actions make things happen so extensively, quickly, and powerfully that we seem to be on the verge of rending the very fabric of nature, social relations, and meaning" (1997, p. 1). The difference is that these technologies encourage a more selective treatment of one's private environment and a more significant disregard for the public environment around a person's physical location; this has important implications for the interaction of embodiment and place.

The more remarkable element of place, however, is meaning. *Meaning*, as a human phenomenon, is influenced by the needs and interests of the body and mind and relates to human significance, proximity, identification of patterns (whether natural or imposed), reason, emotional associations, links to bodily movements, and uptake and interpretation of external stimuli. Place cannot be easily abstracted but is best described in terms of "places" (Geertz, 1996). As Neil Adger explains, "places give meaning to lives and hence place attachment is a core element of individual well-being" (Adger, 2016, p. A1). The meanings invested in place accumulate and grow over time, and they become more persistent as one spends more time in a place (Hess et al., 2008).

In many cases, meaning can be substituted for the term *significance*, and parts of the environment can gain and lose significance based on argument and reprioritizing objects within one's perception. Significance requires a subject and an object. The polar bear's plight as a meaningful symbol in the public climate change conversation provides an example. Most people do not live near polar bears, and they have little significance within the immediate human sphere apart from their importance to small populations of Arctic peoples and their role as apex predators in fragile ecosystems. However, through technologies that extend the limits of human perception (photography, video, scientific instruments) and through *argument*, or an appeal to increase the prioritization of the subject in one's inventory of objects that deserve attention, many people have assigned more human meaning to polar bears than they would otherwise possess.

Meaning, then, requires significance or prioritization and the nature of the significance assigned; the nature of the significance can be described through the choices and behaviors of people. Meaning is created; perhaps, this is close to what the urbanist Edward Soja (1996) presents in his concept of Thirdspace, the interplay of the real (Firstspace) and the imagined (Secondspace) that elevates the spatial ways of knowing to equal significance with historical and social ways of knowing. In the context of place, the nature of significance has to do mainly with proximity, practicality, possession, and particularity, as shown in Figure 13.1.

Proximity, at its core, is simple: How accessible is a particular locale to the human senses or mind–body? Human scale is essential to proximity; as anthropologist Clifford Geertz observes, "no one lives in the world in general" (Geertz, 1996, p. 262). Attempts to create a sense of responsibility attaching to the entire world community are rarely successful because this is not how our minds and bodies work. We are impacted most by what we can sense; this is how our environment becomes present to us.

Practicality is a measure of how humans sort the significance of objects and events based on their function. We define some places in broad, inelegant strokes because of their impracticality: consider "outer space" in American culture or "wilderness" in the Jewish and Christian scriptures. They are poorly defined, stark, and austere, avoided by humans. They induce a sense of foreboding, provoking an uneasy rush of adrenaline. These are places, but they are not finely grained. They are defined mainly by what they are not, representing

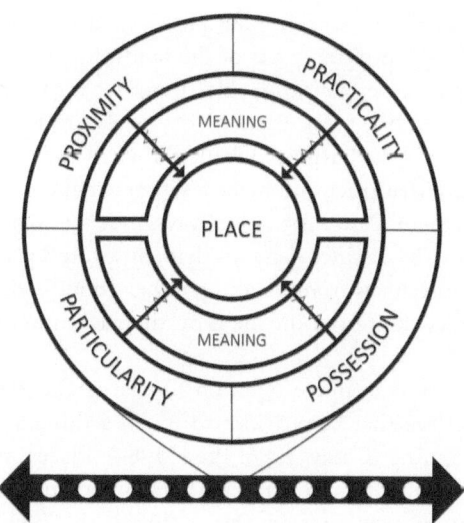

FIGURE 13.1 Place as a combination of person/environment interactions processed through the human meaning-making apparatus; the timeline indicates that place is experienced as a series of events.

something dangerous, impractical, or useless. The meaning assigned to them, like a desert, which for most is understood in scientific terms rather than through lived familiarity, reflects their impracticality (Sack, 1997, p. 66).

Possession operates through power and social relations. Residents of a city or country have an awareness of possession that exceeds any legal claim they have to possession; this sense can motivate material, moral, or spiritual progress ("what can we do to improve *our* city") or make residents protective and suspicious ("look what *they* have done to *our* country"). Control – or possession – is the sense in which one may be able to distinguish "home" from other places: home is where you are "in a position to arrange and control a wide sector of what occurs within, for a significant portion of time" (Sack, 1997, p. 13). Fragmentation in our culture – often revealed through divisions between people based on socioeconomic, political, religious, or even marketing identities – particularly impacts the reach of possession (Seamon, 1980), which leads to defensiveness about what remains. Finally, possession is not just about ownership but apprehensibility; to understand something is, in part, to "have" it, and thus concrete places are scaled to the human sensory and conceptual apparatus.

Particularity can be defined as much by what it is as by what it is not. Descriptions of people, places, and things have limits; the question "What is Cleveland?" (The city proper? The whole metropolitan area? The collective people who live in it?) may have a fuzzy answer, but whatever the answer is, Cleveland is undoubtedly not Cincinnati. Defining particularity seems to be a universal human impulse: "Carving out places ... occurs in the simplest pre-literate societies [...i]dentifying parts of the landscape, clearing sites, erecting shelters, bounding areas, establishing rules about what should or should not be in this place" (Sack, 1997, p. 7). Particularity is also related to specialization; places are designated for a purpose, which allows us to leave that activity behind when we leave that place and make a statement about how much we care about what we do there. Particularity is always understood in terms of relationships among things; the distinct elements that constitute a place do not, as Marc Augé explains, "prevent us from thinking either about their interrelations, or their shared identity conferred on them by their common occupancy of the place" (1995, p. 54).

Finally, the previous example question "What is Cleveland?" has answers that lead us to believe that place is not so much a thing as it is an event, an unfolding, or becoming (Casey, 1996; Pred, 1984). Places are not "persistent" (Groulx, 2017) in a static sense. A place, then, is perhaps more like an organism, an ecosystem, or even an event than a single "that" which can be distinguished fully from other places or the conditions which created it. However, the event of place is characterized by its stability (Augé, 1995). The amount of physical persistence necessary to sustain a sense of place is not easy to describe. It appears to have no upper bound, though successive generations have constantly renewed places that have been sustained for millennia (Jerusalem, Athens, Kathmandu) to the extent that the characteristics of the place as originally constituted have little resemblance to the places that exist today.

On the surface, this model has certain limitations; it does not incorporate the influence of power relationships, struggle, and resistance (Feld & Basso, 1996) and provides no mechanism for adjudicating when the meaning of a place is contested (and it almost always is). It does not include any reference to the types of place meanings that people generate, which span the range of human emotions. However, this model, when used in practice, can and must include all of these features. In fact, the model is well suited to explaining and communicating the human impacts of climate change, especially since climate change will impact impoverished and disadvantaged communities around the world with greater destructive force than the wealthy communities that enjoy the resources to adapt. Furthermore, an inventory of climate impacts cannot be limited to the physical assets of a community, which are often designed to be ephemeral in any case. Rather, the cultural endowments of a community will be at risk, and these cultural endowments are expressed in part through place. Often, disadvantaged or minority cultures have places that are given little consideration in the important conversations in a society and are therefore plundered, and this often occurs because the language for understanding these places is underdeveloped or framed in negative terms. Place as a conceptualization strategy can contribute to social justice efforts by providing the language to define what it is that needs to be protected and why.

Place and Climate Change

Place occupies a smaller space, so to speak, than it should in the communication of climate risks. In some ways, this is expected: Climate change is studied and communicated scientifically, privileging the tangible (Groulx et al., 2014) and measurable, while place combines tangible objects, persons, and affects or experiences. Climate change communication occurs in a mode that only overlaps partially with the phenomena that comprise place. However, place has enormous value for describing the proximal impacts of climate change and developing a climate change conceptualization strategy that is meaningful to non-expert users.

A projected increase in global temperatures of (for instance) 1°C conceals significant local variation in specific places where the impacts would be more significant than a 1°C temperature hike. Jeremy Hess and colleagues (2008), elaborating on the importance of place in the public health response to climate change, explain that in the United States, coasts, riverine environments, deserts, and Arctic areas will be disproportionately affected by climate change. These are the environments that are the most ecologically diverse and (save for the Arctic) the densest human populations. In sum, physical location is perhaps the most critical factor in the level of exposure to climate risks. These risks do not involve higher temperatures alone, but more intense storms, flooding, agricultural disruption, impacts on water availability, and—in the future—human migration. In addition to the physical impacts, disruption of one's place connection can have serious mental health consequences. Responses and adaptations to climatic changes impact our sense of place more than a stretch of summer over time for air conditioners. As Mark Groulx explains, "many sociocultural values and traditions are rooted in ... relationships with the landscape that will be impacted by climate change" (2017, p. 1378). However, it is essential to note that these climate-impacted changes result from our placemaking behaviors, as will be our responses to them. The climate change that disrupts our place relationships also originates from our placemaking behaviors.

A *placemaking behavior* is an action that results in the phenomenon of place, including interaction with proximity, practicality, possession, and particularity in the environment. For instance, economic factors tie a person to a location ("proximity"), creating a point where increasing distance from the center represents greater or lesser accessibility ("practicality"). Historically, this connection was linked to the agricultural productivity of the land, but in industrialized nations, a person's immediate relationship to land resources is more distant. However, economic production is not the only factor that ties a person to a particular location; family, social links, financial investments (house, mortgage), and cultural inertia also support proximity and practicality, creating several paths of minimal resistance in the moves of everyday life. All of these ties involve a set of decisions, behaviors, and habitual movements that contribute to placemaking for each individual and group over time, forced by life's needs.

Placemaking also involves possession, and this possession is distinct from legal title to a piece of property. In fact, legal title is only a small part of the possession aspect of placemaking. Possession also involves what (or who) belongs in a particular place, reflecting the cultural, ideological, and hierarchical values (Cresswell, 1996). Levels of possession vary in different places depending on function and role. For example, a child at a public school during the school day is in the "right" place; legally and socially, they should not be anywhere else.

Particularity is the element of placemaking that is often overlooked in American places because of technological change and mass production. One example is the reduced attention to climate in building practices. Today, local American building practices are influenced more by mass production of materials (the same spruce or fir lumber; the same platform framing), air conditioning and central heating, building codes, and teams of commercial contractors building from stock plans reproduced hundreds of times. Thus, they are more uniform; local climatic or cultural features are only included as an evocation of something otherwise absent, often as a marketing motif. These new influences are also an expression of particularity, but on a larger scale and for new reasons.

Anthropogenic climate change is created by human activity, and placemaking is a core social and cultural subset of human activity. As a result, climate change as we know it, is a product of placemaking and thus requires great sensitivity to address. Communicating climate risk is not as simple as projecting the impacts and waiting for people to change. Because of the connection between ourselves and our sense of place—something we have possession of and contribute to—climate change mitigations can be perceived as a threat to home in a broader sense. Following is a model of the placemaking interaction; from there, place and climate change can be connected through placemaking behaviors.

The model of the interaction between placemaking and climate change, as portrayed in Figure 13.2, is as follows:

1　Humans have what could be described as a placemaking impulse. Physical acts of existing in the world inherently create a sense of place.
2　Some placemaking behaviors are intentional; others are unintentional. Sometimes we splash, and sometimes we ride the waves.
3　Placemaking behaviors change the environment, leading to a modified environment (a different place) that interacts with further placemaking behaviors.
4　Placemaking behaviors are directed and regulated by the inherent limitations of the human body, the reasoning faculties, risk assessment values influenced by culture, and the placemaking behaviors of others.

There are intentional and unintentional behaviors that contribute to placemaking and climate change. Since the start of the industrial revolution, the common

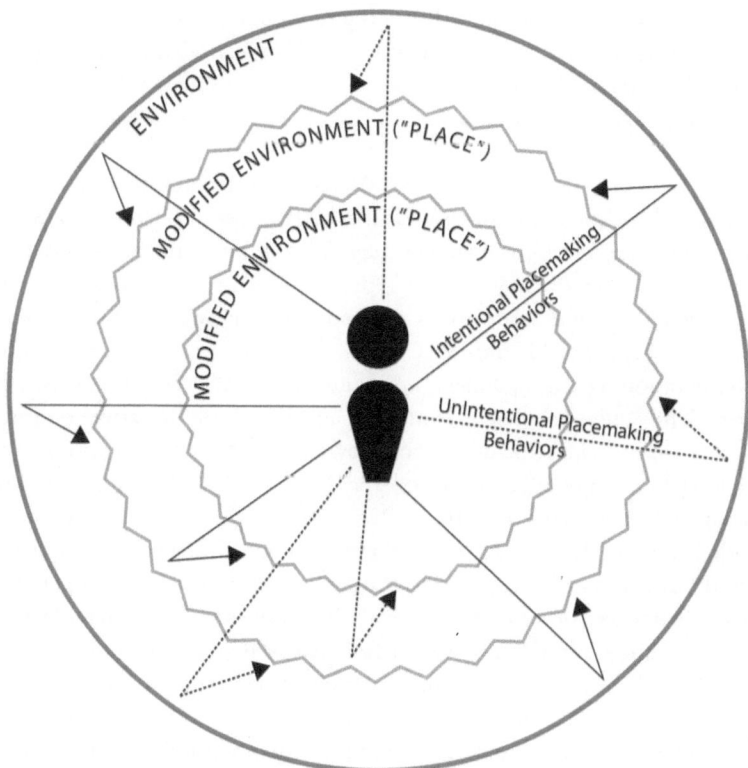

FIGURE 13.2   Place as an ongoing process of intentional and unintentional place-making behaviors in interaction with one's environment.

factor in all of these interactions, with few exceptions, is the use of fossil fuels in everyday life. In the United States, transportation and electricity generation account for over half of our immense greenhouse gas emissions (U.S. EPA, n.d.). This proportion reflects the energy intensiveness of American lives: priorities, preferred products, and activities, all framed by the automobile-centric built environment. Air conditioning means that transom windows, ten-foot ceilings, and deep porches are no longer necessary to stay cool indoors, even in the middle of the sweltering sunbelt, but cooling that space is incredibly energy-intensive. However, these changes have an impact on our cultural values and the manifestation of place. The landscape of the American South, which was once physically and culturally represented by heavy tree shade and deep porches to promote cooling, is now marked by artificially cooled interior spaces and sunbaked subdivisions, bypasses, and highways that facilitate the movement and storage of cars and trucks. These are the new places, and they have obtained that air of inevitability that makes them feel like home, even to those who understand their profound climate impact. This is what the communicator is up against: envisaging a new place displacing what one already possesses and has created can be taken as an attack upon the self. Social identity often outplays measurable physical risk in the complex scrum of personal and collective risk assessment (Adger, 2016); it does not necessarily matter that climate change will have an impact.

Leveraging place attachment in climate communication, then, is a double-edged sword. Schoenefeld and McCauley (2015), based on research with 99 Vermonters, found that local frames may actually be counterproductive with those whose value orientations are "self-enhancing" (as opposed to "self-transcendent"). This framework is derived from Schwartz (1992, 1994); according to Schoenefeld and McCauley, Schwartz provides two continua of conservation versus openness to change and self-enhancing versus self-transcendent values. Schoenefeld and McCauley focus on the latter continuum. *Self-enhancement* values include social power, status, and recognition, while *self-transcendent* values include justice and concern for a broader community (Schoenefeld & McCauley, 2016, p. 726). Devine-Wright and Howes (2010) provide a case study, for example, where place attachment fueled opposition to wind farms in north Wales; similar projects will inevitably be a part of the climate change response. Unfortunately, there are situations where "place" as a communication and risk assessment framework may result in less change in behavior and belief, not more. The accretions of placemaking layers are much deeper than those illustrated in Figure 13.2. The depth of the placemaking actions of others that have come before and the individual and collective response to events and trauma can lead to a passive sense of inevitability; Adger explains that weather hazards, for instance, can feed "[p]erceptions of powerlessness and low self-efficacy" that can arrest action (2016, p. A2). However, the concept of place can reinforce a sense of commitment to others, reifying cultural bonds expressed through places that are more durable and concrete than the monetized impacts (measured in money, an abstraction) used in traditional climate risk communication. Economic terms are the language of cost–benefit analyses and lowest-common-denominator arguments for action that unites disparate stakeholders with little else to connect them (Garrett, 2021; Groulx, 2017).

Reflection

What is the role of place? Considering the diversity of perspectives and mixed results from limited studies, at this point, little more can be said except that place has explanatory power. It serves as a scaffold for exploring the interaction between climate change, engagement, cognition, and embodiment. We need more than an engagement with a sense of place to determine how to address climate change and upon what basis. As Robert Sack explains, "[s]ituatedness does not show us where to go, or even what might be a good destination" (1997, p. 5).

He argues for a holistic, comprehensive vision of space and social relations. I hold that this insight must be supplemented by an understanding of ethical foundations and risk prioritization, both individual and collective. However, since our embodiment and situatedness is so much a part of who we are and how we act, an approach to climate change communication that emphasizes scientific observations and probable forecasts is only likely to influence those who make that way of knowing meaningful to them. In articles, white papers, and reports with a specialized audience expecting an esoteric treatment of a topic, the positivistic approach has identifiable strengths, but it is not enough to communicate risks to a broader audience, especially one with a built-in partisan skepticism.

Place is not a panacea. However, using place to conceptualize the interaction between minds–bodies and the environment provides the basis for a more particularized analysis of where the climate communication project has failed. It also promotes empathy toward those who reject the genuine threat of climate change; if climate change results from what we call "placemaking behaviors" in which all people engage, it is not surprising that prevention and mitigation actions could be perceived as a threat to one's place. Empathy does not ameliorate the problem, since misinformation, cognitive barriers, and lifestyle lock-in (Boucher, 2016) persist, but the empathetic stance can motivate a stronger pursuit of genuinely effective communication strategies, rather than those which merely increase polarization. The difference is this: with rare exceptions, there are few qualitative lifestyle differences between those who accept the scientific consensus and those who do not in terms of responsibility for climate change and mitigating actions. All of our placemaking behaviors have contributed to the problem, and all of our places will alter as a result. Disruption to place has happened and will happen whether we try to prevent and mitigate climate change or not. We can choose our adventure, but ethically, our communicative choices should be directed at changing those beliefs and behaviors that result in loss of life, reduced quality of life, and place detachments. However, we must understand that the impact of those choices is complicated and affects people unevenly at the point of the deepest connections between ourselves and our environments. Place is one meaningful lens for understanding and expressing this impact.

Discussion Questions

1 How does place unite people in an area who are divided in other ways (politically, culturally, etc.)?
2 Provide three examples of possession of a place.
3 Using examples, describe the connection between placemaking behaviors and climate change? In your response, consider the four elements of proximity, practicality, possession, and particularity.

Assignments

1 Find an example of climate change risk communication intended for public consumption. Recreate the artifact to account for elements of place or placemaking.
2 A city manager has hired you to produce a 2- to 3-page report on the projected impact of climate change over the next 50–100 years in key areas of the city. Using a specific place near you as an example, identify the possible impacts of a changing climate (e.g., shifts in plant and animal life, drainage concerns, population increase or decrease, access to water). The report will be used to build a proposal for action at an upcoming city council meeting. Consider your audience: is it culturally or ideologically diverse? Is it skeptical or welcoming?

References

Adams, H. (2016). Why populations persist: Mobility, place attachment, and climate change. *Population and Environment, 37*(4), 429–448. https://doi.org/10.1007/s.

Adger, W. N. (2016). Place, well-being, and fairness shape priorities for adaptation to climate change. *Global Environmental Change, 38*, A1–A3. https://doi.org/10.1016/j.gloenvcha.2016.03.009.

Arbuckle, J. G., Prokopy, L. S., Haigh, T., Hobbs, J., Knoot, T., Knutson, C., Loy, A., Mase, A. S., McGuire, J., Morton, L. W., Tyndall, J., & Widhalm, M. (2013). Climate change beliefs, concerns, and attitudes toward adaptation and mitigation among farmers in the Midwestern United States. *Climatic Change, 117*(4), 943–950. https://doi.org/10.1007/s10584-013-0707-6.

Augé, M. (1995). *Non-places: Introduction to an anthropology of supermodernity.* Verso.

Blennow, K., Persson, J., Persson, E., & Hanewinkel, M. (2016). Forest owners' response to climate change: University education trumps value profile. *PLoS ONE, 11*(5), 1–13. https://doi.org/10.1371/journal.pone.0155137.

Boon, H. J. (2016). Perceptions of climate change risk in four disaster-impacted rural Australian towns. *Regional Environmental Change, 16*(1), 137–149. https://doi.org/10.1007/s10113-014-0744-3.

Boucher, J. L. (2016). Culture, carbon, and climate change: A class analysis of climate change belief, lifestyle lock-in, and personal carbon footprint. *Socijalna Ekologija, 25*(1–2), 53–80. https://doi.org/10.17234/SocEkol.25.1.3

Brügger, A., Morton, T. A., & Dessai, S. (2016). "Proximising" climate change reconsidered: A construal level theory perspective. *Journal of Environmental Psychology, 46*, 125–142. https://doi.org/10.1016/j.jenvp.2016.04.004.

Bryson, B. (1989). *The lost continent.* Perennial.

Casey, E. S. (1996). How to get from space to place in a fairly short stretch of time: Phenomenological prolegomena. In S. Feld & K. Basso (Eds.), *Senses of place* (pp. 13–52). School of American Research P.

Cresswell, T. (1996). *In place/out of place: Geography, ideology, and transgression.* University of Minnesota Press.

Cunsolo Willox, A., Harper, S. L., Ford, J. D., Landman, K., Houle, K., & Edge, V. L. (2012). "From this place and of this place": Climate change, sense of place, and health in Nunatsiavut, Canada. *Social Science and Medicine, 75*(3), 538–547. https://doi.org/10.1016/j.socscimed.2012.03.043.

de Boer, J., Botzen, W. J. W., & Terpstra, T. (2016). Flood risk and climate change in the Rotterdam area, The Netherlands: Enhancing citizen's climate risk perceptions and prevention responses despite skepticism. *Regional Environmental Change, 16*(6), 1613–1622. https://doi.org/10.1007/s10113-015-0900-4.

Devine-Wright, P., & Howes, Y. (2010). Disruption to place attachment and the protection of restorative environments: A wind energy case study. *Journal of Environmental Psychology, 30*(3), 271–280. https://doi.org/10.1016/j.jenvp.2010.01.008.

Döring, M., & Ratter, B. (2018). The regional framing of climate change: Towards a place-based perspective on regional climate change perception in north Frisia. *Journal of Coastal Conservation, 22*(1), 131–143. https://doi.org/10.1007/s11852-016-0478-0.

Dunlap, R. E., McCright, A. M., & Yarosh, J. H. (2016). The political divide on climate change: Partisan Polarization widens in the U.S. *Environment, 58*(5–22), 4–23.

Eubanks, P. (2015). *The troubled rhetoric and communication of climate change: The argumentative situation.* Routledge.

Evans, J., & Jones, P. (2011). The walking interview: Methodology, mobility and place. *Applied Geography, 31*(2), 849–858.

Feld, S., & Basso, K. H. (Eds.). (1996). *Senses of Place.* School of American Research P.

Funk, C., Hefferon, M., Kennedy, B., & Johnson, C. (2019). *Trust and mistrust in Americans' views of scientific experts.* Pew Research Center.

Galway, L. P. (2019). Perceptions of climate change in Thunder Bay, Ontario: Towards a place-based understanding. *Local Environment, 24*(1), 68–88. https://doi.org/10.1080/13549839.2018.1550743.

Garrett, Z. (Forthcoming). A gateway drug for science literacy and moral action: Climate change in the composition classroom. In Zachary Garrett (Ed.), *Exigence in the anthropocene: Teaching ecocomposition in the age of climate change.* Parlor Press.

Geertz, C. (1996). Afterword. In S. Feld & K. H. Basso (Eds.), *Senses of place* (pp. 259–262). School of American Research P.

Groulx, M. (2017). "Other people's initiatives": Exploring mediation and appropriation of place as barriers to community-based climate change adaptation. *Local Environment, 22*(11), 1378–1393. https://doi.org/10.1080/13549839.2017.1348343.

Groulx, M., Lewis, J., Lemieux, C., & Dawson, J. (2014). Place-based climate change adaptation: A critical case study of climate change messaging and collective action in Churchill, Manitoba. *Landscape and Urban Planning, 132*, 136–147. https://doi.org/10.1016/j.landurbplan.2014.09.002.

Hess, J. J., Malilay, J. N., & Parkinson, A. J. (2008). Climate change: The importance of place. *American Journal of Preventive Medicine, 35*(5), 468–478. https://doi.org/10.1016/j.amepre.2008.08.024.

Kais, S. M., & Islam, M. S. (2019). Perception of climate change in shrimp-farming communities in Bangladesh: A critical assessment. *International Journal of Environmental Research and Public Health, 16*(4). https://doi.org/10.3390/ijerph16040672.

Krosnick, J. A., & MacInnis, B. (2020a). *Climate insights 2020: Surveying American public opinion on climate change and the environment.* Resources for the Future. https://doi.org/10.1007/978-3-030-29966-8_11.

Krosnick, J. A., & MacInnis, B. (2020b). *National survey of public opinion on global warming.* https://www.resources.org/archives/climate-insights-2020-survey.

Leiserowitz, A., Maibach, E., & Roser-Renouf, C. (2009). *Global warming's six Americas: An audience segmentation analysis.* Yale Program on Climate Change Communication.

Máté, F. (1998). *The hills of Tuscany: A new life in an old land.* Albatross.

National Association of Realtors (2013). *2013 community preference survey.* https://www.nar.realtor/articles/2013-community-preference-survey.

Nicolosi, E., & Corbett, J. B. (2018). Engagement with climate change and the environment: A review of the role of relationships to place. *Local Environment, 23*(1), 77–99. https://doi.org/10.1080/13549839.2017.1385002.

Oreskes, N., & Conway, E. M. (2010). *Merchants of doubt: How a handful of scientists obscured the truth on issues from tobacco smoke to global warming.* Bloomsbury Press.

Poushter, J., & Huang, C. (2019). Climate change still seen as top global threat, but cyberattacks rising concern. *Pew Research Center,* 1–36. https://www.pewresearch.org/global/2019/02/10/climate-change-still-seen-as-the-top-global-threat-but-cyberattacks-a-rising-concern.

Pred, A. (1984). Place as historically contingent process: Structuration and the time-geography of becoming places. *Annals of the Association of American Geographers, 74*(2), 279–297.

Sack, R. D. (1997). *Homo geographicus.* Johns Hopkins University Press.

Scannell, L., & Gifford, R. (2013). Personally relevant climate change: The role of place attachment and local versus global message framing in engagement. *Environment and Behavior, 45*(1), 60–85. https://doi.org/10.1177/0013916511421196.

Schiermeier, Q. (2010). Few fishy facts found in climate report. *Nature, 466.* https://doi.org/10.1038/466170a.

Schoenefeld, J. J., & McCauley, M. R. (2016). Local is not always better: The impact of climate information on values, behavior and policy support. *Journal of Environmental Studies and Sciences, 6*(4), 724–732. https://doi.org/10.1007/s13412-015-0288-y.

Schwartz, S. H. (1992). Universals in the content and structure of values: Theory and empirical tests in 20 countries. In M. Zanna (Ed.), *Advances in experimental social psychology,* 25, 1–65. Academic Press. http://dx.doi.org/10.1016/S0065-2601(08)60281-6.

Schwartz, S. H. (1994). Are there universal aspects in the content and structure of values? *Journal of Social Issues, 50,* 19–45. http://dx.doi.org/10.1111/j.1540-4560.1994.tb01196.x.

Seamon, D. (1979). *A geography of the lifeworld.* St. Martin's Press.

Seamon, D. (1980). Body-subject, time-space routines, and place-ballets. In A. Buttimer & D. Seamon (Eds.), *The human experience of space and place* (pp. 148–165). St. Martin's Press.

Soja, E. W. (1996). *Thirdspace: Journeys to Los Angeles and other real-and-imagined places.* Blackwell Publishing.

Spence, A., & Pidgeon, N. (2010). Framing and communicating climate change: The effects of distance and outcome frame manipulations. *Global Environmental Change, 20*(4), 656–667. https://doi.org/10.1016/j.gloenvcha.2010.07.002.

Theroux, P. (1996). *The pillars of Hercules.* Ballantine.

Tuan, Y. (1977). *Space and place: The perspective of experience.* University of Minnesota Press.

U.S. Environmental Protection Agency (n.d.). *Sources of greenhouse gas emissions.* Retrieved 4 March 2021 from https://www.epa.gov/ghgemissions/sources-greenhouse-gas-emissions.

14

AN ANTIRACIST RHETORIC OF EMBODIED RISK

Samuel Stinson

Problem: Disasters such as climate change and the COVID-19 pandemic are difficult if not fully impossible to present to a general audience in their fullest extent. Given changes of scientific knowledge over time, audiences can become oversaturated with messaging that can fail to persuade. This leads to an impasse in which audiences are placed at higher embodied risk.

Solution: I propose the antiracist concept of the *post-disaster imperative* as a means of rhetorical listening to audiences of such messaging as one means of focusing more on embodied risk and listening to individual narratives and not only on disaster to overcome this impasse.

In this edited collection are chapters focusing on risk and embodiment in technical and professional communication contexts. Each chapter has viewed risk through the lens of decisions made in human systems and leading to consequences experienced in lived, embodied experiences. These contexts range from messaging and risk in university contexts, environmental and climate concerns, to risks from political settings related to COVID-19, and to risks related to land and water resource management.

This chapter reconsiders the claim made in Bernard-Donals's 2001 article "The Rhetoric of Disaster and the Imperative of Writing" that disaster rhetoric frequently leads to a rhetorical impasse where it becomes highly difficult to represent disaster through writing. The additional purpose here is to consider how contemporary disaster rhetoric surrounding both the COVID-19 pandemic and climate change creates an impasse from this perspective. This chapter will conclude by theorizing how practitioners of technical and professional writing may work constructively to pass through this barrier using an explicitly antiracist rhetorical approach that focuses on embodied risk as a means of overcoming the impasse.

DOI: 10.4324/9781003266549-17

Due to its potential for great risk of environmental destruction, climate change and its effects remain a serious issue deserving careful discussion and widespread promotion. Communicating the reality of disaster through disseminating scientific facts is a problematic and complex task, frequently leading to an impasse, a place in which reality cannot be accurately represented. In this chapter, I will assume three complicating factors with respect to disaster rhetoric: (1) the use of general overstatements through media representing science; (2) a public distrust of scientific and journalistic institutions; and (3) the potential anonymity of sources that deprives them temporarily of credibility. These three factors generally work to negate the effect of scientific information through public disbelief.

Moreover, instead of media and political interests, the medium of writing and language is also to blame for this failure of communication of risk. Language and its technologies of communication and representation may not provide a suitable means of representing such risk as climate change deserves. Buell (1995) reminds us of Charles Darwin's coining of the term *natural selection*. He mentions Darwin's later wish that he could go back to rename the concept of natural selection as *natural perseveration* instead, so that it would have carried more the connotation of being self-organizing and less purposeful (p. 283). A similar concern exists regarding the disaster of climate change, its rhetorical framing to the public, and the embodied risk that exists for those who will be impacted by its effects.

Yet disaster, in the exact value of the term, is an unknowable thing and impossible to represent in its fullness, a claim made in Bernard-Donals's 2001 article "The Rhetoric of Disaster and the Imperative of Writing." Building on Bernard-Donals (2001) and revisiting this claim, the purpose here is to consider how disaster rhetoric problematizes rhetorical agency and composition. This is necessary because public communication frequently reaches a rhetorical and compositional impasse, the impossibility of representing disaster through writing. For this reason, I will be considering the following questions: How are individual choices made more complicated as a result of the way public experts discuss and communicate about disasters? And how does this lead to a situation in which communication breaks down? What can be done to improve this situation so that embodied risk from disasters is decreased? These are important questions to consider since the public messaging of disasters and crises affects the work of technical and professional writers in a direct and systemic way.

Accurate information distributed through public sources impacts day-to-day decisions of workers who are technical and professional writing practitioners. Problematically, failure to ask these questions may lead to ignoring lived experience, narrative, and personal accounts of individuals at risk and affected by such disasters. At the conclusion of this chapter, I will propose the concept of the *post-disaster imperative* as a means of rhetorical listening. This concept provides one means of focusing more on embodied risk instead of disaster itself. By doing this, it may be possible to overcome the impasse that prevents the much-needed communication about disaster.

Bernard-Donals (2001) raises the specter of disaster as being an unknow-able, unrepresentable construct that cannot be fully contained or represented through the act of writing (p. 73). Disaster, with this definition in view, is something that extends past a human ability to understand and know, since disaster exceeds the capacity for attention and comprehension. And because of this rhetorical incapacity to fully grasp disaster, it is impossible to truly uti-lize writing to inform the public about the fullness of disaster. This is what is termed "a rhetoric of disaster" (p. 74). Complications within a rhetoric of di-saster come from the challenge inherent in language that promises to represent the real circumstances in which embodied risk occurs, yet this promise may not be kept as disaster escapes the capacity for language to explain and make plain (p. 79).

For Bernard-Donals, writing becomes a form capable only of representing a part for the whole and never reality itself. What is most challenging and in need of reconsideration, now 20 years after Bernard-Donals, is the claim that the more problematic issue in representing disaster is the perspectives of those who experience disaster, not language itself:

> The traditional view of historical narratives and testimony is that their veracity was linked to their transparency: the language of history is meant to provide a window through which we see clearly the events them-selves. But if language doesn't yield the events of history this simply— particularly events in whose effect upon the witness or historical actor is brutally traumatic—there must be some way to convey not just the events but also to register or indicate the traumatic kernel of their effect.
>
> *(Bernard-Donals, 2001, p. 79)*

From a less charitable perspective, it may appear that Bernard-Donals is blam-ing the victims of disaster for not producing narratives or accounts for historians that are comprehensible to the public for the purposes of representation. How-ever, as Bernard-Donals further states, the purpose of making this statement is to direct the reader into possibilities in which verisimilitude, the appearance of something represented, is not the main goal of communication but metonymy, the use of one term in part as with a structure to point to a greater wholeness (p. 80). In this way, language may potentially come to the rescue with respect to the impasse preventing writing from representing disaster. For Bernard-Donals, metonymy may not be ideal, but it is a pragmatic solution to this problem.

During the past several decades, numerous ecological, environmental, and biological disasters have appeared on the world stage via mass media. In the context of this chapter, it is important to consider how a rhetoric of disaster has extended to the general media landscape and attempt in which disaster is represented to inform audiences about impending or imminent disasters. Key metaphors become common in media representation of disasters, as happened before with H1N1 and swine flu (Angeli, 2012; Ding, 2014). Certain approaches to representing disasters have become less effective over time. In the media, invoking the trope or commonplace of environmental apocalypse has become a common tool of disaster rhetoric. For instance, the CDC made use of apocalyptic rhetoric in an apparent mock zombie apocalypse scenario for preparedness training (Cheek, 2019). This tool may in some cases be viewed less directly as metonymy and more as a strategic overstatement for purposes of persuasion. Yet similar to the use of apocalyptic framing of disaster, a public that has been saturated with overstatement may not be receptive to this approach.

As Buell (1995) observes, "the rhetoric of apocalypticism implies that the fate of the world hinges on the arousal of the imagination to a sense of crisis" (p. 285), noting further that this social imagination requires a focus not only on the reality of impending environmental disaster but the scaffolding of widespread commonplace awareness to prevent impending disaster and to promote change. Thus, during recent decades, through the Internet, television, and radio, a general messaging campaign has promoted awareness of environmental disaster in the face of human actions, pollution, and consumer choices. Science has noted that there are trackable changes in climate over time, and these changes have led scientists to observe that carbon emissions have directly caused and contributed to these changes (Watts, 2018).

As scientific knowledge grows, the gap between experts and audiences widens. The environmental apocalypse trope has frequently appeared in an attempt to work against the negative effects of this worldwide impending disaster. The implied hope in doing this is inspiring or otherwise terrifying individuals into choosing to protect the environment, supporting policies that would reduce carbon emissions, and selecting leadership in government that would support this important work. However, a valid concern exists that messaging invoking environmental apocalypse may have overstimulated the public and may have lessened its impact with the public. As Buell has observed, too much arousal of the environmental imagination may lead to despair and inaction in avoiding eventual environmental disaster. In the next section of this chapter, I will expand this discussion to include this discussion of representation and disaster rhetoric to more recent disasters, specifically to more recent developments in climate change and the COVID-19 pandemic.

Contemporary Disaster Rhetoric and the Rhetorical Impasse

In this section, I will discuss how disasters and contemporary disaster rhetoric creates a rhetorical impasse in messaging concerning climate change and

COVID-19. Put simply, disasters themselves are culpable for generating a rhetorical impasse. But disaster rhetoric is also culpable. Primarily relying on an apocalyptic frame to promote awareness of climate change or COVID-19 may actually build resistance against a desensitized public hearing this messaging. In addition to resistance, repeated statements about disasters may frequently lead to negative emotional outcomes and perception about the ability to lead to desired goals, and further public apathy (Andre, 2015). In recent years, a movement has begun that questions whether or not it would be possible to slow the growth of world economies for the sake of reducing the impact of climate change (Jackson, 2017; Rathi, 2020). The messaging of this movement has been a hard sell, since asking a weary public to combat climate change by inviting economic changes has not fully found a clear appeal. What is needed is an approach that considers not just the desired outcome of messaging but the emotional capacity of the public to push through the impasse in representing the fullness of disasters.

Discourse around disasters demonstrates that the mass media public is generally oversaturated with sometimes conflicting information, disinformation, or in the case of the recent pandemic, rapidly changing representations of data within public and social media spheres. Rhetoric provides a means of risk communication and instilling belief among audiences so that the public is better informed about recent developments in scientific research related to embodied health. However, some public communication along these lines may incorrectly represent risk or oversaturate messaging of risk in ways that inspire resistance among audiences. Such messaging can lead the public to apathy or even a fatalistic view instead of dispositions that may more constructively lead to sustained mitigation or resolution measures. The public at large may believe in response to such disaster rhetoric that nothing can be done, or that they lack agency due to institutional control and power.

Careful disaster rhetoric notes that resolving the current climate crisis should be informed by scientific data and expert data that can be disseminated widely among the public so that the risks of such imminent disaster may be more widely understood. For the purposes of this chapter, the risk of climate change involves the imminent disaster that is predicated by the broad consensus of scientists observing the direct role of human behavior in leading to and also in potentially preventing and reducing the ill effects of this environmental change. In essence, there is something that human beings can do in very practical, determined ways to do this by following credible science. But it is important that this message is not oversaturated or overstated.

It is likely that today's experts have reached a rhetorical quandary. In the context of climate change and as observed by noted climate scientists in 2018, the upcoming year 2030 represents a significant time of predicted disaster. Failure to act in the intervening years to reduce human acceleration of climate change may lead to periods of environmental upheaval that will ultimately be harmful for the planet and its inhabitants:

> The world's leading climate scientists have warned that there is only a dozen years for global warming to be kept to a maximum of 1.5C, beyond which even a half a degree will significantly worsen the risk of drought, floods, extreme heat and poverty for hundreds of millions of people.
>
> *(Watts, 2018)*

However, spreading these scientific facts requires the use of rhetoric that both correctly represents and does not overstate this unfavorable fact to a public that may become easily confused by professional scientific messaging or that must already sift through media misinformation. According to the 2021 Working Group I contribution to the UN Intergovernmental Panel on Climate Change, "Each of the past four decades has been successively warmer than any decade that preceded it since 1850" (SPM-5). The report maintains the scientific consensus that human actions have caused and largely contributed to changes in climate and have led to the overall warming effect. From a traditional messaging standpoint, representing the data and conclusions of this report requires the use of rhetorical framing for presenting this information to a general public that may lack expertise in the science presented. Yet the ability to create belief among the public through rhetoric is challenging here due to the ways in which information can be framed. Should the public, keeping them away from the word *doom* with respect to climate change, be provided alternatives such as communicating "in plain language," craft careful slogans, use multiple rhetorical frames, get rid of jargon, or avoid apocalypticism and employ only more positive messaging that does not lead to resistance? (Montenegro, 2009). Not just accurate but rhetorically careful representation of scientific data is called for here. This report created a quick series of media stories representing its results that claim that it may be already too late to mitigate the warming of the climate (Fritz & Ramirez, 2021; Sharma, 2021). Again, the risk here in saying the situation is bleak is that inaction may follow.

While the science of climate change is settled, the emergence of COVID-19 in 2020 led to immediate system-wide lockdowns as a preventative measure for reducing risk of infection as scientists attempted to keep up with advancing knowledge about the virus. Taking these measures had allowed scientific experts time for continued research and development of vaccines while simultaneously cooperating with governments to message cautiously in the face of an unfolding understanding of the unfolding disaster. Yet, messaging, policy, and communication of mitigation strategies during the pandemic have had problems. According to the Pew Research Center, "During the early days of the coronavirus outbreak in the United States, about half of Americans (48%) said they had encountered at least some completely made-up news about it" (Jurkowitz & Mitchell, 2020). Also, noting a rise in misinformation in 2018–2020, 48% of Americans would prefer government intervention against the spread of misinformation (Mitchell & Walker, 2021). In addition, unpredictable changes in messaging have generally led to frustration at large in the U.S. public ("The corona vaccines," 2021). The potential link between vaccine hesitancy and misinformation represents an important call to action for science information experts, as the unclear messaging from the U.S. federal government has led to confusion in the past year (Stolberg & Shear, 2021).

While one-direction messaging from institutions of power have been important to track, scholars in professional and technical writing have also pointed to the systemic view of communication that also includes tacit knowledge, and not just empirical data and its distribution. These scholars state that information in technical documents is impacted by social circumstances that affect the writing in a circular and cyclical manner. Beverly A. Sauer (2003) in *The Rhetoric of Risk* notes that workplace writing, in the general way that this is accomplished, uses a multitude of rhetorical strategies along with traditional alphabetic writing and the creation of documents (p. 7). With this in mind, I will point out here that circumstances in which technical writing is generated are impacted by context-knowledge of environmental and ecological settings, as well as local knowledge and lived experiences of those producing that work. These circumstances are also inclusive of the risks posed by impending disasters.

With respect to the rhetorical cycle in which technical documents are created and disseminated, Sauer (2003, pp. 75–76) maintains that there are six critical moments of transformation in workplace contexts in which embodied experience of people is collectivized, encoded, and inscribed in those documents. This cycle as Sauer describes it moves from documentation, to reports, to policies, to procedures, to training, and back to documentation, noting that the cycle is enacted, when:

1   Oral testimony and embodied experience are captured in writing;
2   Information in accident reports is re-represented in statistical records;
3   Statistical accounts are re-represented as arguments for particular policies;
4   Policies and standards are transformed into procedures;
5   Procedures are re-represented in training; and
6   Training is re-represented to workers.

COVID-19 has impacted technical writing directly due to both the global nature of the disaster and the role that globalism and neoliberalism have played in contributing to the spread of the virus. In keeping with the rhetorical transformation cycle Sauer describes, it is to be understood that changes in the workplace context as a result of disasters affect the embodied experience represented in the first stage of the cycle. In other words, the embodied experience of people is collectivized, encoded, and inscribed in documents in the new distributed workplace setting on account of both COVID-19 and the mitigation strategies of a more decentralized workplace setting.

Before the pandemic, there were already trends toward writing taking place in distributed locations beyond the traditional offices. Technical writers and professional writers have long utilized the Internet to work from a variety of locations. However, with the pandemic, these trends have become accelerated. During 2020–2021, workplace writing in the states had frequently begun to take place in private homes while writers and other workers socially distanced and sheltered in place, and collaborative group meetings and work were accomplished across the Internet using a variety of apps for both synchronous and asynchronous meetings using such live-discussion technologies as Zoom and Slack. Added to this change are the blended contexts in which technical writing takes place in one's living space among other home-based lived experiences, the change of systems due to the pandemic have resulted in changes in the embodied experiences of workers and in many settings, their families. Expectations for how work is done, including how writing is done, have changed to become more decentralized and more dependent upon individual worker judgment and less on centralized managerial control of work.

Sauer (2003) notes that tacit knowledge, a type of information and context that appears and circulates in a workplace setting, informs how work and writing take place and are accomplished there (p. 205). The change of work to living spaces through technological means redistributes tacit knowledge from a site in which workers assemble in a single physical location to one in which work takes place distributed among many sites, business settings, and private homes. Simply put, what workers know about how to do their jobs and the expectations of writing have changed due to the change in setting. Thus, the distribution of knowledge that was once considered tacit or easily at hand in workplaces may be more complexly located in the new workplaces, or perhaps not at all. More research is needed with respect to how disasters such as climate change and pandemics affect the availability of tacit knowledge for workplace writing.

It is important to deal further with problems associated with writing about and disseminating information in the media about disaster, since messaging about disaster affects the lived experiences of writers. This is what is meant by referring to embodied risk, the way that problematic messaging by experts leads to actual consequences for writers. There are necessary differences to consider between an emerging disaster such as COVID-19 and an imminent, impending disaster such as climate change. Yet communicating accurate and factual data surrounding such disasters as these and the embodied risks created by these has been fraught with a dissemination of scientific language through media sources and by means of elected officials and career bureaucrats in governmental organizations.

Elected representatives have a duty to represent scientific facts carefully and meticulously in clear and accurate language to the public. This is done so that the messaging elicits both belief and a support for proposed policies that lead to action and change. Yet these same officials must also carefully represent scientific data in a way that moves emotions for these same reasons and includes an awareness of what the public believes. The balance between evoking in the public a direct emotional response to natural disasters through such messaging on the one hand and carefully representing factuality of data on the other hand can be troubled, having unforeseen side effects. Members of the opposing party can take inaccurate representation within messaging to task, as frequently happens when popular representations of science may appear in political discussions. Inaccurate data and messaging can lead the public at large to lose belief in scientific institutions or even the process of science if messaging skews data too frequently and too obviously.

Overstatement of risk in public messaging may water down scientific messaging. An article headline in *USA Today* had presented this climate knowledge using the following headline: "'The world is going to end in 12 years if we don't address climate change,' Ocasio-Cortez says" (Cummings, 2019). In 2019, it came about that Alexandria Ocasio-Cortez, the U.S. Representative for the 14th Congressional District of New York, was lampooned across the Internet, television, and cable media for her contention that climate change would lead to a sudden, immediate, and calamitous end of the world. The chuckles and guffaws of commentators erupted in ridicule of the inartful notion that the world was coming to an end in 12 years, touching upon the apocalyptic theme in climate change discussions that has aimed to stress the significance of the human impact upon the environment.

With such belief at hand, through an overstatement of the case in which scientifically proffered *risk* is represented as cataclysmic disaster, a journalistically informed population may not appear motivated to do much to prevent something that is predicted in such dark, stark language. After all, what motivation exists for a public to mitigate or prevent climate change if total environmental disaster appears to them entirely menacing and completely destructive? This headline also reduces the story to the apparent claim being made about the world coming to an end, which misses the point being made by the representative that public fear of the effects of climate change is largely being ignored.

Though the representation of scientific facts in question in the elected official's wording can be challenged, a more charitable hearing of the statement may prove to be more inclusive and empathetic toward those listening who experience and will experience the negative impact and risks of climate change in their lives and bodies. Carefully put, the UN report from 2018 states that failure to address global warming within the range of 1.5°C may lead to increased environmental and economic risks across the world. Included in the list are "drought, floods, extreme heat and poverty for hundreds of millions of people" (Watts, 2018). Environmental change and economic consequences may be preventable if acted upon in the time remaining, which the report specifies as 12 years.

Represented with artful words or not with artful words, these scientific facts of impending disaster will place millions of human beings at greater risk. What is worthwhile here is going beyond the base factuality of the words to also consider the public emotional impact: there is much fear related to the climate crisis, and some of this is driven in mass media by means of political narratives. But consideration of individual and public experiences and a felt sense of the impact of climate change is valuable for what it says. With consideration of metonymy, even inaccurate scientific knowledge can prove worthwhile if it represents the public's authentic sentiment of its position of embodied risk. This is not to accept the circulation of misinformation or disinformation in the media but to call for a more inclusive awareness of public reaction to the dissemination of scientific information. By doing this, the use of overstatement in that piece, while initially appearing not to be useful, may be for beneficially including awareness of the public at large and their perceived fear of the impending disaster of climate change.

Theorizing an Antiracist Rhetoric of Embodied Risk

I will end this chapter by theorizing an approach in which practitioners of technical and professional writing may work constructively to pass through this barrier using an explicitly antiracist approach that focuses on embodied risk. I am proposing that a renewed focus on embodied risk in the midst of disaster can be used to further the work of social justice and antiracism. Climate change and the COVID-19 pandemic impact how writing occurs in professional communities due to the influence these have had on increasing embodied risk along with associated racial disparities. There is a link between racism and disasters, as BIPOC individuals may be living in areas already contaminated by toxic facilities as well as may live in locations that may not adequately protect them from the natural consequences of climate change (Patnaik et al., 2020; Rysavy & Floyd, 2021). Additionally, activists have long noted the link between environmental justice and the fight for racial justice (Gardiner, 2020; Schuldt & Pearson, 2016; Thomas & Haynes, 2020). As systemic racism creates conditions for embodied risk by marginalizing perspectives of diverse lived experiences, systemic change and reform that highlights marginalized perspectives can work to eliminate racial disparities.

According to the CDC, there is also a worldwide linkage between race and health equity with respect to a greater embodied health risk during the COVID-19 pandemic, given that "a pathogen can travel from a remote village to major cities on all continents in 36 hours" (2021c). Data collected between March 2020 and June 2021 show that Black Americans have had a higher hospitalization and death ratio than white Americans (2021b). The CDC also reports heightened risk factors for Black Americans with respect to COVID-19 including "poverty," "healthcare access," and being "disproportionately represented among essential workers and industries" (2021a).

As mentioned earlier in this chapter, disaster rhetoric frequently creates a problematic situation in which it becomes difficult to explain or communicate the reality of disasters through language. Drawing from critical race theory, antiracist approaches to achieving social justice aim to identify racial disparities that exist within institutions of power to identify and eliminate systemic racism (Delgado & Stefancic, 2017; Gillborn, 2009; Kendi, 2019). One targeted approach for this work is to locate and address unequal outcomes, places where groups identified by race appear to have an imbalance due to privileged groups possessing a greater chance of success.

Institutions of power create a complex rhetorical impasse that impedes communication. In addition, structures of writing within technical and professional writing discourse communities organize how writing works and may also further inequities and unequal outcomes through those structures. Disidentification, the alignment of identity within those institutions, is also a powerful tool for individuals to pass through this quagmire by careful identification with systems of power (Caughie, 1999). However, scholars of critical race theory have discussed how raising personal stories can be highly effective in persuading audiences to consider or reconsider testimony that impacts the lived and embodied experiences of individuals. In this way, systemic change makes it possible to focus on perspectives that have been previously ignored so that individuals may pass through more equitably.

The proposed post-disaster imperative reflects the need to critique institutions while also to rhetorically listen to individual experiences representing embodied risk through empathy. This approach acknowledges that given that disaster is impossible to represent, it is of greater value to instead consider embodied risk in its messy, complicated forms. This is possible to do through acknowledging the complications that occur as these experiences appear within lived experience and narrative that are shared. Thus, one tool for furthering the work of antiracism includes the use of narratives drawing from and communicating personal experiences of racially marginalized individuals to raise awareness of systemic racism (Delgado & Stefancic, 2017). In using narratives, it is possible to raise awareness of the embodied risk that these individuals find themselves in. These narratives also are framed as pedagogical, teaching tools used to communicate these experiences in creative ways. I will include an example of this approach:

> *I want you to imagine with me for a moment that time travel exists. Not just the one-way time travel we are all accustomed to, but that time travel in the opposite direction is possible. Yes, this is the realm of science fiction, but hear me out. Time*

travel to the past exists. Now, also imagine that you've just received a message from a time traveler who lives in the future, your immediate future. This person however has forgotten to sign this particular message, but it is clear from the writing in their message that they are writing from at least two weeks ahead of you in time. And they have a very special message, and they are asking for you to do something very important.

Now, the time traveler claims to be from an immediate future in which a deadly virus has been unleashed upon the world, a pandemic, a word constructed from Greek meaning all people. Now, this message could very well have been faked. And it is lacking a clear signature. Added to that, the time traveler decided to go through a few back-door channels to get the message to you indirectly. The question for you is: What do you believe? Is the message legitimate? And more importantly, do you act on the message and take the precautions the message suggests?

The rationale for using this example is in making space for personal stories and narratives using the antiracist assumption that individual perspectives and not just one-directional data dispensed from governmental, educational, and business institutions is necessary for implementing systemic change. I will here give the background for this narrative based on my own personal experience. In March 2020, when news of the pandemic began to become more widely known in the United States, two of my friends, colleagues from and living in China, had already been living with the effects of the pandemic for several months' time. I had been in communication with my colleagues frequently for several months as they had mentioned measures such as social distancing and stay-at-home orders that were designed to reduce the spread of COVID-19.

In early March, a friend from China, on WeChat, posted a message that was alleged to be from an American in Bergamo, Italy where social circumstances due to the pandemic were already beginning to become more troubling, with hospitals beginning to be overwhelmed. At the time I had received and shared this message on Facebook, March 13, 2020, I did not know the name of the original poster or where it had originated. I would soon learn that the post was written by an American named Cristina Higgins (Chuck, 2020).

Within the post, Higgins asks Americans who read the message to limit social contact with others in light of the understatement of the U.S. media in covering the severity of COVID-19. Before sharing on my Facebook page, I had briefly evaluated whether to post this since at the time I had viewed it as anonymous. I am not ordinarily in the habit of sharing unattributed sources on social media. But I took the risk of doing this, believing it would be better to share an unverified source that turned out to be true than not to do this. Very soon, the question of authorship and anonymity for the post would arise. I soon saw that the post was shared 43 more times by my friends and friends of friends. One former colleague from my former work in public radio said,

"I hope EVERYONE will take a moment to read this post." He then additionally qualified me as the sharer, noting that I am "a grad of WKU and worked with many of us at public radio. He is very level headed and not one to overstate the facts."

A friend-of-a-friend had this evaluation:

> I have to say this seems awfully suspicious. No name attached, on a chat group where people can post anonymously for a person in Italy who seems to be an expert on the us [sic] media and accusing them of being lax. Plain and simple, if I do not know the source, I cannot place value on it.

Another friend, in direct reply to my post, asked me this: "Is this from you? You there?" When I mentioned that it was not me (and I was not located in Bergamo, Italy), he breathed a sigh of relief. It was then that I realized that I should go back and edit the post to clarify that I was not the post's author and that it was taken from WeChat.

The purpose of sharing this personal story is to illustrate that rhetorical listening can be problematic when the lack of knowing the identity of the speaker makes identification with that person uncertain. In cases such as this, there is an implied embodied risk to reputation in sharing unattributed information that may turn out to be false, but there may also be an embodied risk in putting others at risk in not warning others. Within composition studies, rhetorical listening has been proposed as one approach to support antiracism (Ratcliffe, 2007), because rhetorical listening allows greater awareness of emotional and lived experience contexts, a vital tool in dismantling structural racism. Ratcliffe, like Bernard-Donals, maintains the usefulness of metonymy in this vein of listening to others.

I call attention here to the use of this personal story as a shared space for the efforts of antiracism with respect to disaster rhetoric. What is most fascinating about this exchange, beyond my own involvement, is that it represents an occasion in which there is an absence of available knowledge about imminent disaster, the COVID-19 pandemic that is about to spread further in the United States. But this absence does not prevent the transmission of testimony of warnings from those facing the effects of disaster to those who would soon be at greater risk of health and safety. Utilizing Sauer's (2003) framework, the Higgins post had circulated from Facebook to other social media platforms (WeChat) and recirculated back to Facebook, though missing identifying information. Whether the author information was simply lost due to being an unsigned post on the author's page is a matter of conjecture. However, the circulation of this message reflects a new type of tacit knowledge that invites more engagement and accountability than institutional messaging alone.

Conclusion

Now at the end of this chapter, and this edited collection, I am suggesting the post-disaster imperative as an antiracist approach that acknowledges both the impossibility of fully representing disaster as well as the popular resistance to disaster rhetoric. Moving beyond disaster itself, which is not fully comprehensible yet always changing, the post-disaster imperative emphasizes not just the act of representing and communicating in a speaker-to-audience direction but in listening to the lived experiences of those affected by disaster.

Bernard-Donals's claim that metonymy is a workable solution to the problem of representing disaster allows for the antiracist post-disaster imperative to represent marginalized perspectives beyond just conventional disaster rhetoric as a part-for-the-whole approach that may pass through the impasse. Both the promotion of scientific knowledge and the lived experience perspectives of marginalized voices each metonymically curate experiences drawn from reality that can better inform knowledge concerning embodied risk and provide better preparation of multiple perspectives for disaster preparedness.

The call to action here is to listen to individuals when they represent their embodied risk, since this is many times more certain than the full representation of disaster and is more readily available information in many cases. Thus, a renewed focus on embodied risk in the midst of disaster can be used for the efforts of social justice and antiracism. Representing disaster rhetoric is a never-ending task, but re-presenting alternative perspectives that focus on embodied risk and not just the dissemination of scientific facts in a one-directional institutional manner can also better align with what technical and professional writers know about how documents embed values in a recurring cycle.

Discussion Questions

1 This chapter makes a distinction between disasters and embodied risk being two separate things. Explain what this difference is in your own words. How are these two concepts related? How are these different? Even if it is impossible to fully write about disaster, why is it important to write about embodied risk?

2 This chapter includes sources from both scientific and individual perspectives. Which did you find more persuasive? And why? Does your personal experience with either disaster mentioned in this chapter impact your reading?

3 The word overstate appeared multiple times in this chapter. Review where this word appears and explain how it is used in each instance. What is significant about this word in the context of this chapter?

Assignments

1 Write a memo to your employer (or prospective employer) in which you identify the embodied risks you recognize that you are faced with in your employment. Then, write a reflection in which you answer this question: How challenging is it to talk with your employer about your embodied risk in the workplace? What strategies

would you suggest that someone in your place could use to raise awareness of these concerns?

2 Brainstorm a list of at least ten places in the reading that you felt strongly that you had something to say to the writer or the sources. Then, take this list and write an email to the author in which you frame your response to the chapter. Finally, share this list with your peers in the class and ask for their response.

3 Write your own narrative creatively to tell a story about something that needs to change with respect to embodied risk. Then, share your story with a classmate and ask them to write you an email in which they respond to the story.

References

Andre, E. K. (2015). The need to talk about despair. In A. Brei (Ed.), *Ecology, ethics, and hope* (pp. 1–12). Rowman & Littlefield.

Angeli, E. L. (2012). Metaphors in the rhetoric of pandemic flu: Electronic media coverage of H1N1 and swine flu. *Journal of Technical Writing and Communication, 42*(3), 203–222.

Bernard-Donals, M. (2001). The rhetoric of disaster and the imperative of writing. *Rhetoric Society Quarterly, 31*(3), 73–94.

Buell, L. (1995). *The environmental imagination: Thoreau, nature writing, and the formation of American culture.* The Belknap Press of Harvard University Press.

Caughie, P. (1999). *Passing and pedagogy: The dynamics of responsibility.* University of Illinois Press.

Centers for Disease Control and Prevention (2021a, April 19). Health equity considerations and racial and ethnic minority groups. U.S. Department of Health and Human Services.

Centers for Disease Control and Prevention (2021b, July 16). Risks for COVID-19 infection, hospitalization, and death by race/ethnicity. U.S. Department of Health and Human Services.

Centers for Disease Control and Prevention (2021c). Why it matters: The pandemic threat. U.S. Department of Health and Human Services.

Cheek, R. (2020). Zombie ent(r)ailments in risk communication: A rhetorical analysis of the CDC's zombie apocalypse preparedness campaign. *Journal of Technical Writing and Communication, 50*(4), 401–422.

Chuck, E. (2020, March 11). American on coronavirus lockdown in Italy: "It's surreal. It's dystopian." NBC News. https://www.nbcnews.com/health/health-news/american- coronavirus-lockdown-italy-it-s-surreal-it-s-dystopian-n1155576.

The coronavirus vaccines and misinformation: A board-certified infectious disease physician discusses the vaccines for COVID-19 and some people's reluctance to get vaccinated. (2021, June 28). Pew Research Center.

Cummings, W. (2019, January 22). "The world is going to end in 12 years if we don't address climate change," Ocasio-Cortez says. *USA Today.*

Delgado, R., & Stefancic, J. (2017). *Critical race theory: An introduction* (3rd ed.). NYU Press.

Ding, H. (2014). *Rhetoric of a global epidemic: Transcultural communication about SARS.* Southern Illinois University Press.

Fritz, A., & Ramirez, R. (2021, August 9). Earth is warming faster than previously thought, scientists say, and the window is closing to avoid catastrophic outcomes. CNN.

Gardiner, B. (2020, June 9). Unequal impact: The deep links between racism and climate change. *YaleEnvironment360*.

Gillborn, D. (2009). Who's afraid of critical race theory in education? A reply to Mike Cole's The color line and class struggle. *Power and Education, 1*(1), 125–131.

The Intergovernmental Panel on Climate Change (2021). Working group I contribution to the sixth assessment report. United Nations.

Jackson, T. (2017). *Prosperity without growth*. Routledge.

Jurkowitz, M., & Mitchell, A. (2020, April 15). Early in outbreak, Americans cited claims about risk level and details of coronavirus as made-up news. Pew Research Center.

Kendi, I. X. (2019). *How to be an antiracist*. One World.

Mitchell, A., & Walker, M., (2021, August 18). More Americans now say government should take steps to restrict false information online than in 2018. Pew Research Center.

Montenegro, M. (2009, May 21). Is there a better word for doom? Six experts discuss the merits of framing climate change. *Seed Magazine*.

Patnaik, A., Son, J., Feing, A., & Ade, C. (2020, August 15). Racial disparities and climate change. *PSCI*.

Ratcliffe, K. (2007). *Rhetorical listening: Identification, gender, whiteness*. Southern Illinois University Press.

Rathi, A. (2020, August 18). *These folks think eternal economic growth will lead to unstoppable climate change*. Bloomberg.

Rysavy, T. F., & 00, A. (2021). People of color are on the front lines of the climate crisis. Green America.

Sauer, B. A. (2003). *The rhetoric of risk: Technical documentation in hazardous environments*. Routledge.

Schuldt, P., & Pearson, A. (2016). The role of race and ethnicity in climate change polarization: evidence from a U.S. national survey experiment. *Climatic Change, 136*(3–4), 495–505.

Sharma, A. (2021, August 7). World is on the brink of catastrophe, warns government climate chief. *The Telegraph*.

Stolberg, S. G., & Shear, M. D. (2021, August 2). Americans suffer pandemic whiplash as leaders struggle with changing virus. *New York Times*.

Thomas, A., & Haynes, R. (2020, June 22). Black lives matter: The link between climate change and racial justice.

Watts, J. (2018, October 8). We have 12 years to limit climate change catastrophe, warns UN. *The Guardian*.

INDEX

Note: **Bold** page numbers refer to tables; *italic* page numbers refer to figures and page numbers followed by "n" denote endnotes.